Sludge Reduction Technologies in Wastewater Treatment Plants

A perfect system does not stimulate new ideas

Sludge Reduction Technologies in Wastewater Treatment Plants

Paola Foladori, Gianni Andreottola and Giuliano Ziglio
University of Trento
Italy

Published by   IWA Publishing
Republic - Export Building
1 Clove Crescent
London E14 2BA, UK
Telephone: +44 (0)20 7654 5500
Fax: +44 (0)20 7654 5555
Email: publications@iwap.co.uk
Web: www.iwaponline.com

First published 2010
© 2010 IWA Publishing

Cover image: Studies of flowing water, with notes (c. 1510-13) by Leonardo da Vinci. By permission of The Royal Collection © 2009, Her Majesty Queen Elizabeth II.

Apart from any fair dealing for the purposes of research or private study, or criticism or review, as permitted under the UK Copyright, Designs and Patents Act (1998), no part of this publication may be reproduced, stored or transmitted in any form or by any means, without the prior permission in writing of the publisher, or, in the case of photographic reproduction, in accordance with the terms of licences issued by the Copyright Licensing Agency in the UK, or in accordance with the terms of licenses issued by the appropriate reproduction rights organization outside the UK. Enquiries concerning reproduction outside the terms stated here should be sent to IWA Publishing at the address printed above.

The publisher makes no representation, express or implied, with regard to the accuracy of the information contained in this book and cannot accept any legal responsibility or liability for errors or omissions that may be made.

Disclaimer

The information provided and the opinions given in this publication are not necessarily those of IWA Publishing and should not be acted upon without independent consideration and professional advice. IWA Publishing and the Author will not accept responsibility for any loss or damage suffered by any person acting or refraining from acting upon any material contained in this publication.

British Library Cataloguing in Publication Data
A CIP catalogue record for this book is available from the British Library

Library of Congress Cataloging-in-Publication Data
A catalog record for this book is available from the Library of Congress

ISBN: 9781789065305

Contents

Preface xiii
1   INTRODUCTION 1
2   SLUDGE COMPOSITION AND PRODUCTION IN
    FULL-PLANTS 7
    2.1   INTRODUCTION 7
    2.2   SLUDGE COMPOSITION 9
          2.2.1   Sludge fractionation as COD 10
    2.3   SLUDGE PRODUCTION 12
          2.3.1   Primary sludge production 12
          2.3.2   Biological excess sludge production 13
          2.3.3   Calculation of biological excess sludge production 13
    2.4   TYPICAL SLUDGE PRODUCTION DATA 18
3   CURRENT SLUDGE DISPOSAL ALTERNATIVES AND
    COSTS IN CRITICAL AREAS 21
    3.1   INTRODUCTION 21

© 2010 IWA Publishing. *Sludge Reduction Technologies in Wastewater Treatment Plants.* By Paola Foladori, Gianni Andreottola and Giuliano Ziglio. ISBN: 9781789065305. Published by IWA Publishing, London, UK.

3.2 TOTAL COSTS FOR SLUDGE TREATMENT AND DISPOSAL 25
3.2.1 Sludge treatment costs in Europe 25
3.2.2 Sludge disposal costs in Europe 26

4 PRINCIPLES OF SLUDGE REDUCTION TECHNIQUES INTEGRATED IN WASTEWATER TREATMENT PLANTS 29
4.1 INTRODUCTION 29
4.2 CELL LYSIS AND CRYPTIC GROWTH 32
4.3 UNCOUPLED METABOLISM 35
4.4 ENDOGENOUS METABOLISM 37
4.5 MICROBIAL PREDATION 39
4.6 BIODEGRADABILITY INCREASE IN INERT SOLIDS 40
4.7 HYDROTHERMAL OXIDATION 41

5 OVERVIEW OF THE SLUDGE REDUCTION TECHNIQUES INTEGRATED IN THE *WASTEWATER HANDLING UNITS* 43
5.1 ENZYMATIC HYDROLYSIS WITH ADDED ENZYMES 45
5.2 ENZYMATIC HYDROLYSIS BY THERMOPHILIC BACTERIA (THERMOPHILIC AEROBIC REACTOR) 46
5.3 MECHANICAL DISINTEGRATION 47
5.4 ULTRASONIC DISINTEGRATION 48
5.5 THERMAL TREATMENT 50
5.6 CHEMICAL AND THERMO-CHEMICAL HYDROLYSIS 50
5.7 OXIDATION WITH OZONE (OZONATION) 51
5.8 OXIDATION WITH STRONG OXIDANTS (DIFFERENT FROM OZONE) 53
5.9 ELECTRICAL TREATMENT 55
5.10 ADDITION OF CHEMICAL METABOLIC UNCOUPLERS 57
5.11 SIDE-STREAM ANAEROBIC REACTOR (AT AMBIENT TEMPERATURE) 57
5.12 EXTENDED AERATION PROCESS 59
5.13 MEMBRANE BIOLOGICAL REACTORS 60
5.14 GRANULAR SLUDGE 61
5.15 MICROBIAL PREDATION 62

6 OVERVIEW OF THE SLUDGE REDUCTION TECHNIQUES INTEGRATED IN THE *SLUDGE HANDLING UNITS* 65

- 6.1 ENZYMATIC HYDROLYSIS WITH ADDED ENZYMES 67
- 6.2 MECHANICAL DISINTEGRATION 68
- 6.3 ULTRASONIC DISINTEGRATION 69
- 6.4 THERMAL TREATMENT 70
- 6.5 MICROWAVE TREATMENT 71
- 6.6 CHEMICAL AND THERMO-CHEMICAL HYDROLYSIS 73
- 6.7 OXIDATION WITH OZONE (OZONATION) 74
- 6.8 OXIDATION WITH STRONG OXIDANTS (DIFFERENT FROM OZONE) 74
- 6.9 ELECTRICAL TREATMENT 76
- 6.10 AEROBIC DIGESTION 77
- 6.11 DIGESTION WITH ALTERNATING AEROBIC/ANOXIC/ANAEROBIC CONDITIONS 78
- 6.12 DUAL DIGESTION 79
- 6.13 AUTOTHERMAL THERMOPHILIC AEROBIC DIGESTION 81
- 6.14 ANAEROBIC DIGESTION 82
- 6.15 THERMOPHILIC ANAEROBIC DIGESTION 83
- 6.16 MICROBIAL PREDATION 83
- 6.17 WET AIR OXIDATION 84
- 6.18 SUPERCRITICAL WATER OXIDATION 86

7 PROCEDURES FOR ESTIMATING THE EFFICIENCY OF SLUDGE REDUCTION TECHNOLOGIES 89

- 7.1 INTRODUCTION 89
- 7.2 COD AND TSS SOLUBILISATION 91
- 7.3 DEGREE OF DISINTEGRATION 92
  - 7.3.1 Degree of disintegration based on COD solubilisation (DD_{COD}) 93
  - 7.3.2 Degree of disintegration based on oxygen consumption 94
- 7.4 BIODEGRADABILITY EVALUATED BY RESPIROMETRY 95
- 7.5 DENITRIFICATION RATE EVALUATED BY NUR TEST 97

	7.6	ANAEROBIC BIODEGRADABILITY EVALUATED BY BIOGAS PRODUCTION	99
	7.7	BACTERIA INACTIVATION	100
	7.8	EFFECT ON SLUDGE RETENTION TIME (SRT)	102
	7.9	MAXIMUM GROWTH YIELD, OBSERVED BIOMASS YIELD, OBSERVED SLUDGE YIELD	103
	7.10	EVALUATION OF SLUDGE REDUCTION	105
	7.11	TREATMENT FREQUENCY	105
	7.12	PHYSICAL PROPERTIES OF SLUDGE	106
8	**BIOLOGICAL TREATMENTS**	109
	8.1	INTRODUCTION	109
	8.2	RECENT INSIGHTS ON DEGRADABILITY OF SLUDGE UNDER AEROBIC AND ANAEROBIC CONDITIONS	112
		8.2.1 Aerobic conditions	112
		8.2.2 Anaerobic conditions	113
		8.2.3 Disintegration of sludge flocs under anaerobic conditions and in the presence of sulphides	114
	8.3	THE INFLUENCE OF AEROBIC/ANOXIC/ANAEROBIC CONDITIONS ON HETEROTROPHIC MAXIMUM GROWTH YIELD	115
		8.3.1 Process of denitrification + nitrification	117
		8.3.2 Digestion with alternating aerobic/anoxic/anaerobic conditions	118
	8.4	SIDE-STREAM ANAEROBIC REACTOR (AT AMBIENT TEMPERATURE)	121
		8.4.1 Oxic-Settling-Anaerobic process	121
		8.4.2 Cannibal® system	125
	8.5	THERMOPHILIC ANAEROBIC DIGESTION	129
	8.6	THERMOPHILIC AEROBIC REACTOR	131
		8.6.1 Integration in the wastewater handling units (S-TE process®)	134
		8.6.2 Integration in the sludge handling units. Dual digestion	137
		8.6.3 Integration in the sludge handling units. Autothermal thermophilic aerobic digestion	138
	8.7	ENZYMATIC HYDROLYSIS WITH ADDED ENZYMES	140
	8.8	ADDITION OF CHEMICAL METABOLIC UNCOUPLERS	144

	8.9	PREDATION BY PROTOZOA AND METAZOA	148
		8.9.1 Types of predators	150
		8.9.2 Process configuration and sludge reduction	152
		8.9.3 Pros and cons of microbial predation	159
	8.10	EXTENDED AERATION PROCESSES	161
	8.11	MEMBRANE BIOLOGICAL REACTORS (MBR)	162
		8.11.1 MBR + physical, chemical treatments	163
	8.12	GRANULAR SLUDGE	164
9	**MECHANICAL DISINTEGRATION**		167
	9.1	INTRODUCTION	167
	9.2	TYPES OF EQUIPMENT FOR MECHANICAL DISINTEGRATION	168
	9.3	ENERGY LEVELS REQUIRED FOR SLUDGE DISINTEGRATION	170
	9.4	LYSIS-THICKENING CENTRIFUGE	171
	9.5	STIRRED BALL MILLS	174
	9.6	HIGH PRESSURE HOMOGENISER	178
	9.7	HIGH PRESSURE JET AND COLLISION SYSTEM	182
	9.8	ROTOR-STATOR DISINTEGRATION SYSTEMS	183
	9.9	COMPARISON OF MECHANICAL DISINTEGRATION TECHNIQUES	185
10	**ULTRASONIC DISINTEGRATION**		189
	10.1	INTRODUCTION	189
	10.2	CONFIGURATIONS AND EQUIPMENT FOR ULTRASONIC DISINTEGRATION	191
	10.3	EVALUATION OF ENERGY APPLIED IN ULTRASONIC TREATMENT	193
	10.4	THE INFLUENCE OF ULTRASOUND FREQUENCY ...	195
	10.5	COD SOLUBILISATION	197
		10.5.1 The influence of sludge concentration	199
	10.6	INFLUENCE ON MICROORGANISMS	200
	10.7	INFLUENCE ON SLUDGE SETTLEABILITY AND DEWATERABILITY	203
	10.8	INTEGRATION OF ULTRASONIC DISINTEGRATION IN THE WASTEWATER HANDLING UNITS	203
	10.9	INTEGRATION OF ULTRASONIC DISINTEGRATION IN THE SLUDGE HANDLING UNITS	205

11 THERMAL TREATMENT 209
11.1 INTRODUCTION 209
11.2 COD SOLUBILISATION 211
11.2.1 COD solubilisation at moderate temperatures (<100°C) 212
11.2.2 COD solubilisation at high temperatures (>150°C) 214
11.3 INCREASE OF BIODEGRADABILITY 215
11.4 NITROGEN AND PHOSPHORUS SOLUBILISATION 216
11.5 INFLUENCE ON MICROORGANISMS 218
11.6 INFLUENCE ON SLUDGE SETTLEABILITY AND DEWATERABILITY 219
11.7 INTEGRATION OF THERMAL TREATMENT IN THE BIOLOGICAL PROCESSES 220
11.7.1 Integration of thermal treatment in the wastewater handling units 220
11.7.2 Integration of thermal treatment in the sludge handling units 221
11.7.3 Full-scale applications 223
11.8 MICROWAVE TREATMENT 229

12 CHEMICAL AND THERMO-CHEMICAL TREATMENT 233
12.1 INTRODUCTION 233
12.2 TYPES OF ACIDIC OR ALKALINE REAGENTS 234
12.3 COD SOLUBILISATION 235
12.3.1 Effect of temperature 236
12.3.2 Effect of pH 237
12.3.3 Effect of contact time 238
12.3.4 Comparison of solubilisation levels under different conditions 240
12.4 NITROGEN AND PHOSPHORUS SOLUBILISATION 240
12.5 INFLUENCE ON SLUDGE DEWATERABILITY 243
12.6 INTEGRATION OF THERMO-CHEMICAL TREATMENT IN THE WASTEWATER HANDLING UNITS 243
12.7 INTEGRATION OF THERMO-CHEMICAL TREATMENT IN THE SLUDGE HANDLING UNITS 244

13 OZONATION 249
- 13.1 INTRODUCTION 249
- 13.2 PARAMETERS INVOLVED IN OZONATION 251
- 13.3 CONFIGURATION OF OZONATION REACTORS 252
  - 13.3.1 Ozone transfer in sludge 255
- 13.4 DEFINITION OF OZONE DOSAGE 257
- 13.5 EFFECT OF SOLIDS MINERALISATION 260
- 13.6 COD SOLUBILISATION AND TSS DISINTEGRATION 261
- 13.7 NITROGEN AND PHOSPHORUS SOLUBILISATION 267
- 13.8 INTEGRATION OF OZONATION IN THE BIOLOGICAL PROCESSES 268
- 13.9 INTEGRATION OF OZONATION IN THE WASTEWATER HANDLING UNITS 269
  - 13.9.1 Initial studies and ozone dosage calculation 271
  - 13.9.2 Results on sludge reduction 273
  - 13.9.3 Influence on WWTP effluent quality 278
  - 13.9.4 Influence on sludge pH 280
  - 13.9.5 Influence on sludge flocs and microorganisms 280
  - 13.9.6 Influence on nitrification 283
  - 13.9.7 Influence on denitrification 285
  - 13.9.8 Influence on sludge settleability 286
- 13.10 INTEGRATION OF OZONATION IN THE SLUDGE HANDLING UNITS 287
  - 13.10.1 Ozonation + anaerobic digestion 288
  - 13.10.2 Ozonation + aerobic digestion 292
  - 13.10.3 Influence on sludge dewaterability 293

14 COMPARISON OF PERFORMANCE OF SLUDGE REDUCTION TECHNIQUES 295
- 14.1 INTRODUCTION 295
- 14.2 COMPARISON OF COD SOLUBILISATION 300
- 14.3 COMPARISON OF DEGREE OF DISINTEGRATION 302
- 14.4 COMPARISON OF REDUCTION OF SLUDGE PRODUCTION 303
- 14.5 COMPARISON OF IMPACTS 309
  - 14.5.1 Impacts of techniques integrated in the wastewater handling units 309
  - 14.5.2 Impacts of techniques integrated in the sludge handling units 313

14.6 COMPARISON OF INSTALLATION/OPERATIONAL
ASPECTS 316
15 NOMENCLATURE 323
16 REFERENCES 329
INDEX 351

Preface

Sludge produced in wastewater treatment plants amounts to a small percentage (around 1%) of the volume of treated wastewater, while the processes for sludge treatment and disposal represent from 20% to 60% of operating costs, considering manpower, energy and sludge disposal. Since the mid '90s, it is well recognised that sludge disposal is one of the most critical issues. The present tendency (for example in EU countries) is towards a common strategy for biowaste prevention: the principles of prevention (reduction of production), maximum reuse, material recycling or energy recovery, instead of mere disposal. Thus, according to this strategy, the need to prevent or reduce the amount of sludge produced in wastewater treatment plants, and which requires disposal, is a priority. The strategies to reduce sludge production are implemented by using specific technologies integrated in the wastewater handling units and in the sludge handling units. At present there are many technical solutions proposed in the scientific literature and commercially available and they are extremely diversified.

The creation of a review – the aim of this book – to take into consideration all the current or proposed sludge reduction techniques is not easy, because of the heterogenous nature of the techniques, and may not be exhaustive.

© 2010 IWA Publishing. *Sludge Reduction Technologies in Wastewater Treatment Plants.* By Paola Foladori, Gianni Andreottola and Giuliano Ziglio. ISBN: 9781789065305. Published by IWA Publishing, London, UK.

Indeed, many techniques remain at experimental pilot or laboratory stage while others have achieved a maturity, which allows commercial use as demonstrated by fully operational full-scale plants.

This book includes certain introductory chapters relating to the quantification of sludge production (Chapter 2) and the current disposal costs for certain international situations (Chapter 3). Chapter 4 illustrates the principles on which the various sludge reduction techniques are based and the principal mechanisms involved.

Chapters 5 and 6 are intended to give two overview sections where the various technologies for sludge reduction are briefly described and divided into techniques integrated in the wastewater handling units (Chapter 5) and techniques integrated in the sludge handling units (Chapter 6). In these chapters most of the existing technologies are schematically introduced by: main characteristics, flow sheet, advantages and drawbacks and macro-categories to which they are belonging (biological, chemical, thermal, physical, mechanical treatments).

In Chapters 8–13 the most widely applied, most promising and innovative techniques are described in depth on the basis of experimental findings referred to in the literature and current lines of research are addressed.

Finally, a performance comparison among the various techniques is presented in Chapter 14 and some references are made to side-effect or impact comparisons among the principal technologies in use.

Our research group within the Department of Civil and Environmental Engineering at the University of Trento (Italy) has acquired an extensive knowledge of this field through experimental work on sludge reduction techniques over the last 5 years. This research field was developed together with the staff of the Wastewater Treatment Agency of the Province of Trento and the authors would like to express their gratitude to the director Eng. Paolo Nardelli for his continuous co-operation and precious and creative contributions to our research.

It would not have been possible to write this book without the contribution of our collaborators at the University of Trento. Ilaria Nobile and Veronica Menapace have made a special contribution, producing the excellent drawings and flow sheets included in the book. Sabrina Tamburini contributed with her stylish hand-drawn pictures included in the book. Special thanks go to our collegue Roberta Villa for her efforts in assisting our sometimes difficult research. We further express our thanks to prof. Marco Ragazzi, for sharing so much of his experience and knowledge of sludge treatment and disposal.

The authors

1
Introduction

In municipal and industrial wastewater treatment plants (WWTPs) the removal of biodegradable compounds and organic or inorganic particulate matter by means of settling and filtration, produces large amounts of waste sludge for disposal. Most WWTPs are based on activated sludge systems, in which sludge is produced in two ways:

- primary sludge: originates from the initial physical separation of settleable solids in the primary settlers;
- secondary sludge: produced in the final settlers where activated sludge is separated from the treated effluent. Secondary sludge is the result of net biological growth and the accumulation of inert and organic refractory compounds.

© 2010 IWA Publishing. *Sludge Reduction Technologies in Wastewater Treatment Plants.* By Paola Foladori, Gianni Andreottola and Giuliano Ziglio. ISBN: 9781789065305. Published by IWA Publishing, London, UK.

Sludge is characterised by a high percentage of volatile solids and a high water content even after dewatering (>70–80% by weight), which results in extremely large volumes of sludge, even though the volume of sludge produced in a WWTP only represents 1% of the volume of influent wastewater to be treated.

As WWTPs become more widespread the volume and mass of sludge generated is expected to increase continuously in the next decade, due to the increase of the population connected to the sewage network, the building of new WWTPs and the upgrading of existing plants because of more stringent local effluent regulations (Ødegaard, 2004; Paul and Debellefontaine, 2007).

In Europe 7.7 and 8.4 million tonnes per year (dry solids) were produced respectively in 2001 and 2003. An increase in sludge production of about 40% from 1998 to 2005 has been observed, reaching a current sludge production of 9.4 million tonnes per year (dry solids). In 2010, production is expected to exceed 10 million tonnes. In Europe, particularly among the new EU members or candidate countries, the increasing choice of biological treatment is leading to higher sludge production. Sludge production increased by about 65% in a decade in the Czech Republic, while in Poland the current half a million tonnes per year of sewage sludge (dry solids) is expected to double by 2015 due to the extension of services (Jenicek, 2007).

The trend is also similar in North America: in the USA sludge production is estimated at 6.9–7.6 million tonnes (dry solids) per year in the period 2005–2010 (Turovskiy and Mathai, 2006; Dentel, 2007), while in Canada sludge generation is approximately 0.4 million tonnes and in Japan it has reached more than 2 million tonnes per year (Okuno, 2007). In China, the current amount of sewage sludge production is expected to increase dramatically to around 11.2 million tonnes of dry solids per year by 2010 (State Environmental Protection Administration of China, Chu et al., 2009).

So far the main options for sludge disposal are agricultural use, landfill, incineration or composting. One of the major destinations for sludge in many areas, especially in the past, was the nearby ocean or sea, although recently many countries have introduced laws for marine pollution control, which do not permit sea dumping.

The disposal of sludge in landfill site will be limited in Europe as a consequence of stricter regulation, aimed to ban the disposal of biodegradable waste in landfills. Among the many drawbacks in sludge disposal in landfill are: gaseous emissions which contribute to global warming, hazardous compounds in leachate to be treated, nutrients and organic matter lost to recycling. For example, in Switzerland the disposal in landfill is already banned and sludge has to be dewatered, dried, incinerated and only the ash disposed of in landfills

(Böhler and Siegrist, 2004). In Germany only material with a loss on ignition <5% may be disposed of in landfill. Sewage sludge, even after anaerobic treatment, still shows a loss on ignition of about 50% (Scheminski et al., 2000). Furthermore the selection of suitable sites for landfill is becoming more and more problematic due to the ever more scarce land availability and consequent increasing costs.

In Central and Eastern Europe the amount of sludge disposed of in landfill is also expected to decrease, while a slow increase in the adoption of more expensive technologies, such as incineration, can be expected (Jenicek, 2007).

Recently, there is a broad consensus of opinion that sludge must be managed in a more sustainable way to recover some energy or to make use of organic and nutrient content on land, but both these options present problems due to public concern.

The amount of sludge used on land or farmland may vary widely between countries, depending on the need to provide an important source of fertilisers or on the need to prevent the overloading of the soil and the environment. For example the percentage of sludge used in agriculture varies between 0 and 66% in European countries: ~0% in the Netherlands, 9% in Sweden, 28% in Austria, 32% in Italy, 33% in Germany, 58% in France, 59% in Denmark, 61% in the UK, 63% in Ireland, 66% in Spain (Müller, 2007). The future use of sludge in agriculture in Western Europe will be strongly influenced by EU directives on sewage sludge use in agriculture, which, notwithstanding the general awareness that agricultural applications could be a better way of recycling valuable compounds from sludge, will always include restrictive maximum values for many hazardous compounds. In fact, the use of stabilised sludge in agriculture has several advantages to reintegrate the progressive loss of organic matter and nutrients in soils, but its use is often strictly regulated due to the potential health risks associated with the presence of pathogens and contaminants such as heavy metals, micro-pollutants, antimicrobial agents or contaminants of pharmaceutical origin. In many areas, the application of sludge to land is regulated by specific guidelines developed at national level, which represent a conservative approach to the management of the health risk. For example, agricultural use of sludge has been limited in Sweden since 1999 and a total ban has been proposed in Switzerland, due to eco-toxicological considerations (Böhler and Siegrist, 2004; Lundin et al., 2004).

Another limiting factor for the agricultural use of sewage sludge in developed countries is the declining public acceptance of crops produced using sludge, a factor which seems to be becoming more important than the development of legal regulatory standards (Müller, 2007). The pros and cons of agricultural use are currently under discussion, considering the benefit of nutrient recycling in relation to the potential risk of contamination with hazardous compounds.

Incineration is another final option for sludge disposal but this process remains cost intensive and leads to the loss of organic matter and nutrients such as the valuable phosphorus resource – finite resource on earth – which is difficult to be recovered, since it is incorporated in the sludge ash.

The prospect for the near future – associated with the continuous increase in sludge production and the regulation limits for disposal alternatives in some countries – is an increase in costs for sludge disposal, especially in some critical areas. Nowadays, in some WWTPs the cost of treatment + disposal of sludge reaches 25–65% of total operating costs of WWTPs, including energy, personnel and management (Pérez-Elvira *et al.*, 2006). The cost for treatment + disposal in European countries has been estimated to reach, on average, approximately 500 € per tonne of dry mass, according to the type of treatment and disposal (Ginestet, 2007a), but a further increase is expected in the near future.

As a consequence, sludge production is increasing, whereas disposal routes are narrowing and costs are increasing (Bougrier *et al.*, 2007a).

From the perspective of an approach compatible with sustainable development, there are two aims with regard to sludge:

- the recovery of materials or energy from sludge, if sludge is considered a resource;
- the reduction of the amount of sludge produced, if sludge is considered waste.

For many producers, there are not enough economic advantages to make beneficial reuse of sludge an attractive investment (Egemen Richardson *et al.*, 2009).

Even though a sludge-zero process remains the utopia in sludge management, a more realistic and feasible practice is to continue to reduce the volume and mass of sludge produced. The current approach to sludge reduction addresses the two following aspects:

(1) reduction of volume of wet sludge;
(2) reduction of dry mass of sludge.

The increase of the solid content in sludge by dewatering significantly reduces the volume of wet sludge for disposal.

The reduction of dry mass of sludge leads to the reduction of solid content and volume and this strategy should be favoured, because it allows the immediate reduction of sludge dry mass during its production in the biological treatment stage. The fundamental aim of this book is the reduction of sludge dry mass (as opposed to the mere reduction of the volume of wet sludge).

The emphasis on the development of processes to reduce sludge production is currently motivated by the various problems/restrictions of the disposal routes mentioned above.

Many of the techniques for sludge reduction described here are suitable for implementation in the wastewater handling units or in the sludge handling units of an existing WWTP, often by simply retrofitting the specific additional equipment.

The extent to which the most appropriate technique for sludge reduction is chosen will depend upon many factors including local conditions, existing configurations of the treatment facility, the number and skill of the operators, the attitude of the water companies, the involvement of suppliers of the technology and equipment in question.

In many situations, a case-by-case solution has to be evaluated in order to optimise the choice and to avoid unexpected costs. In fact the techniques proposed for the on-site sludge reduction in a WWTP lead to additional costs for investment and operations: this aspect requires a detailed evaluation of the economical advantages by comparing the performance of the selected technique (and the effective reduction in dry mass of sludge) with the cost for the final disposal of sludge.

This book aims to help in this field, describing the outcomes from several hundred research projects reported in the literature to help potential users in the applications of these findings in practice.

2
Sludge composition and production in full-plants

2.1 INTRODUCTION

The volume of sludge produced in a WWTP is only about 1% (dewatered sludge is 0.5‰) of the volume of influent wastewater to be treated. To manage WWTPs effectively and efficiently, it is absolutely necessary to extract waste sludge, including inert solids and excess biomass, in order to prevent their accumulation within the system.

In a WWTP, wastewater is treated using various units based on mechanical, physical, chemical or biological methods, producing:

- *primary sludge* – produced by settleable solids removed from raw wastewater in primary settling; characterised by high putrescibility and

good dewaterability when compared to biological sludge; TS content in primary sludge is in the range 2–7% (Turovskiy and Mathai, 2006);
- *secondary sludge* (also called *biological sludge*) – produced by biological processes such as activated sludge or biofilm systems; contains microorganisms grown on biodegradable matter (either soluble or particulate), endogenous residue and inert solids not removed in the primary settling (where a primary settler is present) or entering with the raw wastewater (where no primary settler is present); TS content in secondary sludge is in the range 0.5–1.5% (Turovskiy and Mathai, 2006);
- *chemical sludge* – produced by precipitation of specific substances (i.e. phosphorus) or suspended solids. With regard to P removal by chemical precipitation, this process is based on the addition of salts which leads to additional sludge production (typically 5–7 kgTSS/kgP) corresponding to an overproduction of about 15% (observations from 16 plants worldwide; Woods *et al.*, 1999; Battistoni *et al.*, 2002).

In the sludge handling units, a combination of any two or three of the above types is introduced.

The term "raw sludge" is used to identify sludge not treated biologically or chemically in a specific sludge handling unit. After a specific handling unit, the sludge is classified by the treatment: e.g. anaerobically/aerobically digested sludge, dewatered sludge, thickened sludge or thermally dried sludge.

Other residuals produced in a WWTP, such as scums or material separated by sieving, screening or degritting are not considered as sludge, and not considered in this text.

During the biological treatment of municipal wastewater, in addition to cell biomass, a large amount of non-biodegradable (inert) solids in particulate form, incoming with the influent raw wastewater, contributes significantly to sludge production. This inert particulate material can be of organic or mineral origin. The solids are non-biodegradable in a process sense, because they are not metabolised under the conditions present in activated sludge or biofilm systems (van Loosdrecht and Henze, 1999).

An example is cited by Paul *et al.* (2006b): around 60% of VSS in sludge is composed of inert organic material at sludge retention time (SRT) of 10 days and for an inert COD particulate fraction of 20% in the influent wastewater.

In more detail, this presence of inert organic solids in sludge is due either to their presence in the influent wastewater (e.g. hairs, fibers, etc...), to the endogenous residue produced in microbial decay or to protozoan activity, which may not fully degrade the bacterial cell walls leaving behind inert material (van Loosdrecht and Henze, 1999).

Sludge composition and production in full-plants

Furthermore, when a very high sludge reduction is desired, the conversion of a significant part of refractory particulate organic material into a biodegradable fraction is needed.

The above reasons mean that it is very important to know the composition of sludge in order to understand and approximate the potential efficiency of a sludge reduction technique.

2.2 SLUDGE COMPOSITION

The sludge is commonly quantified with reference to analyses of TS, VS, TSS, VSS, total COD or particulate COD. These measurements are different because they take into account the different constituents of sludge.

Total solids (TS) comprise: (1) soluble and suspended (particulate) fractions, (2) organic (volatile) and inorganic (inert) fractions, as indicated in Figure 2.1.

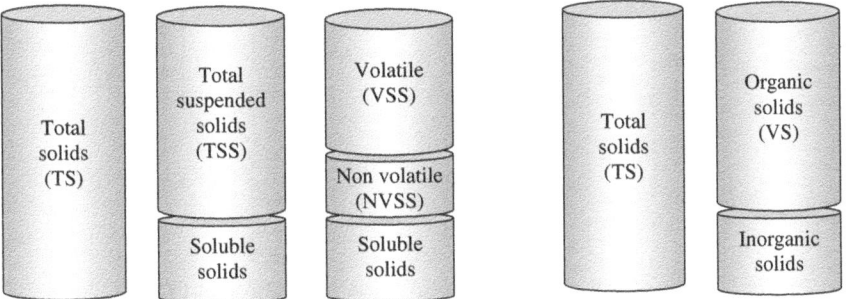

Figure 2.1. Physical fractionation of total solids in sludge.

Total COD does not consider inorganic compounds, only organic ones and it is made up of a soluble fraction and a particulate one, as indicated in Figure 2.2. Particulate COD is a measure of organic matter in suspended form and therefore a relationship exists with the VSS value, throughout the conversion factor, f_{cv}, which typically assumes the value 1.48 mgCOD/mgVSS.

These measurements can be summarised as follows:

(1) *TS (Total Solids)* = quantification of solids both in soluble and in particulate form, and both organic and inorganic;
(2) *VS (Volatile Solids)* = quantification of organic solids, both in soluble and particulate form;

(3) *TSS (Total Suspended Solids)* = quantification of particulate solids, excluding soluble solids both organic and inorganic;
(4) *VSS (Volatile Suspended Solids)* = quantification of particulate organic solids, excluding soluble solids and inorganic solids;
(5) *Total COD* = chemical oxygen demand including both particulate or soluble COD;
(6) *Soluble COD* = chemical oxygen demand of soluble compounds.
(7) *Particulate COD* = chemical oxygen demand of particulate compounds: estimated as the difference between total COD and soluble COD.

Particulate COD can be related to VSS because both measure analogous fractions (particulate organic solids) (see Figure 2.2).

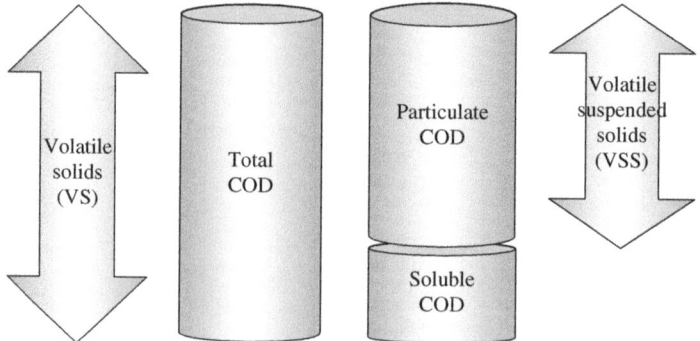

Figure 2.2. Simplified physico-chemical fractionation of total COD in sludge.

The measurements TSS and TS are the most common to quantify the dry mass of sludge. The composition of sludge can be further detailed as described in the following section.

2.2.1 Sludge fractionation as COD

With regard to organic content of sludge expressed as total COD, sludge is composed of a soluble part (S) and a particulate part (X):

$$\text{total COD} = S + X$$

Particulate COD can be determined as the difference between the results of two chemical analyses (X = total COD−S).

Particulate COD of sludge is subdivided in fractions as per Figure 2.3. All the fractions are indicated with the symbol X_y, where X indicates the particulate form.

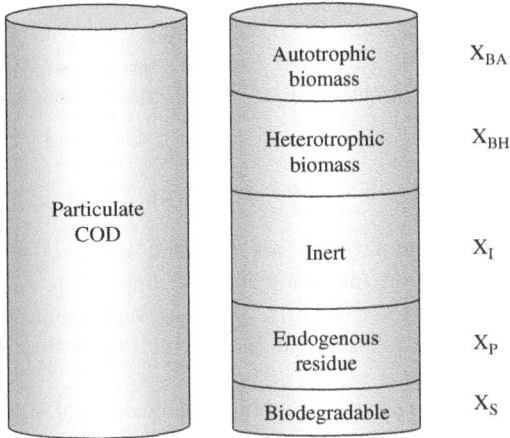

Figure 2.3. Fractionation of particulate COD of sludge.

Particulate COD of sludge is made up of the terms involved in the following sum:

$$\text{Particulate COD} = X_I + X_P + X_S + X_{BH} + X_{BA}$$

Proportions of the various COD fractions in sludge depend on the the operating conditions and the characteristics of the influent wastewater. Details of each fraction of sludge, expressed as COD, are given as follows:

(1) *heterotrophic biomass* (X_{BH}): made up of heterotrophic bacteria involved in the biodegradation of organic matter;
(2) *autotrophic biomass* (X_{BA}): made up of nitrifying bacteria;
(3) *inert particulate COD* (X_I): derives from the inert particulate COD present in the influent wastewater and entering the plant. When it reaches the activated sludge stage, it is not affected by the biological treatment and accumulates in sludge;
(4) *endogenous residue* (X_P): residue of the decay process of bacterial biomass which accumulates in sludge;

(5) *biodegradable particulate COD* (X_S): the slowly biodegradable COD; in activated sludge with sufficiently long SRT this fraction is often small.

In activated sludge, the autotrophic microorganisms grow very slowly and therefore their biomass (X_{BA}) is easily overlooked in the mass balances. It is considered that active biomass mainly consists of heterotrophic organisms, X_{BH}. Omitting the two smaller fractions, X_S and X_{BA}, the composition of activated sludge can be approximated taking into account the following terms:

$$\text{Particulate COD} = X_I + X_P + X_{BH}$$

In particular, the amount of X_I accumulated depends on the content of influent wastewater entering the plant and the amount of X_P depends on the length of the operation period (SRT). In some cases, especially when the SRT in a WWTP is long, the fractions X_I and X_P can be greater than the fraction X_{BH} itself.

With regard to sludge reduction techniques, an optimal strategy should lead to the following desirable objectives:

- to attack, solubilise or reduce the inert fractions, X_I and X_P, converting them into solubile or, better still, biodegradable compounds;
- to reduce the net grown biomass, but to keep the active biomass X_{BH} high in order to ensure that the biological process remains efficient.

2.3 SLUDGE PRODUCTION

Specific sludge production in wastewater treatment varies widely from 35 to 85 g dry solids per population equivalent per day (gTS PE^{-1} d^{-1}). It is considered as typical a daily sludge production pro capita of about 60–80 gTS PE^{-1} d^{-1} (data from Germany and France, Ginestet, 2007a), 65 gTS PE^{-1} d^{-1} (data from Italy, Battistoni *et al.*, 2002), which corresponds to around 250 g of wet sludge PE^{-1} d^{-1}.

2.3.1 Primary sludge production

The production of primary sludge is related to the amount of settleable solids in raw wastewater whose solids content is typically of 50–60 gTSS PE^{-1} d^{-1} or 110–170 gTSS/m^3 of treated wastewater (Tchobanoglous *et al.*, 2003). The most common approach for calculating the primary sludge production is to consider the quantity of TSS in the raw wastewater (typically 90–120 gTSS PE^{-1} d^{-1}) and assuming a TSS removal rate usually in the range 50–65% (Turovskiy and Mathai, 2006).

2.3.2 Biological excess sludge production

Heterotrophic biomass present in activated sludge grows on organic biodegradable substrate, both soluble and particulate, entering with the influent wastewater (raw or pre-settled). As indicated above, the autotrophic biomass is often neglected in the mass balances and therefore active biomass is considered to be mainly composed of heterotrophic organisms, X_{BH}.

Organic matter is oxidised by heterotrophic microorganisms to produce H_2O and CO_2 in the process known as *catabolism*. This process requires the availability of an electron acceptor – which may be oxygen or nitrate – and lead to the production of energy as ATP. This energy is then used by microorganisms to grow forming new cells and to guarantee maintenance functions (such as the renewal of cellular constituents, maintenance of osmotic pressure, nutrient transport, motility, etc...) in the process called *anabolism*. The ratio between the organic matter forming new cells and the organic matter oxidised in the process is known as *maximum growth yield*. In aerobic conditions the growth yield can reach 0.60–0.70, which means that an amount of 60–70% of organic biodegradable matter removed in the biological treatment is converted into new cellular biomass.

Simultaneously biological decay of X_{BH} occurs, which creates two fractions:

- biodegradable particulate COD (X_S);
- endogenous residue considered as inert particulate COD (X_P), which accumulates in the system.

The X_S fraction is subjected to hydrolysis process and is further oxidised to generate new cellular biomass (*cryptic growth*), while the endogenous residue (8–20%) remains and accumulates in the sludge.

A simplified scheme of these processes leading to sludge accumulation in a biological treatment of influent wastewater is indicated in Figure 2.4.

2.3.3 Calculation of biological excess sludge production

Biological excess sludge production is due to the growth and decay of heterotrophic biomass, the accumulation of endogenous residue and the accumulation of inert solids entering the plant with the influent wastewater.

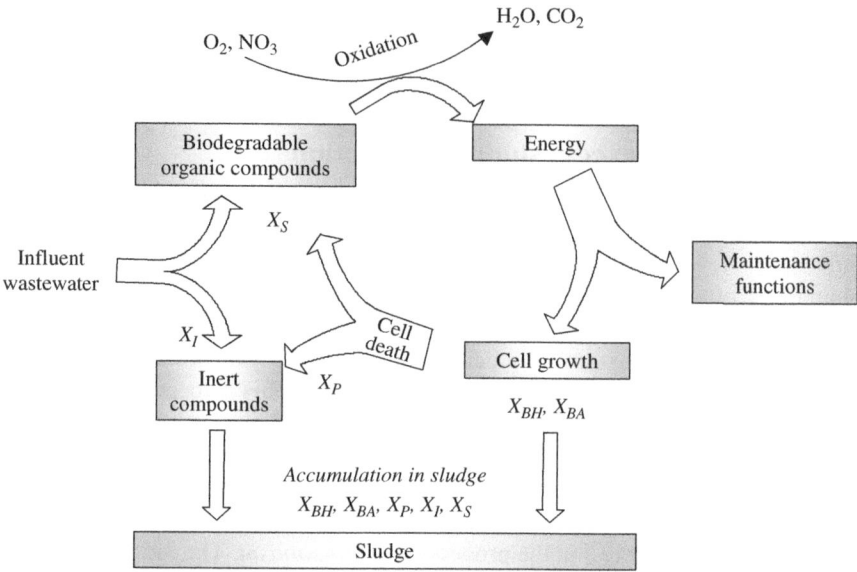

Figure 2.4. Simplified scheme of the processes leading to sludge production in the biological treatment of influent wastewater.

The following expression indicates the sum of these contribution, expressed as VSS:

$$X(kgSSV/d) = \frac{1}{SRT} \cdot \left[\underbrace{X_{BH}}_{\substack{\text{term} \\ \text{related to} \\ \text{heterotrophic} \\ \text{biomass}}} + \underbrace{X_P}_{\substack{\text{term} \\ \text{related to} \\ \text{endogenous} \\ \text{residue}}} + \underbrace{X_I}_{\substack{\text{term} \\ \text{related to} \\ \text{inert} \\ \text{solids}}} \right]$$

where each term is calculated using the following expressions:

$$X_{BH} = \frac{(COD_{b,in} - COD_{b,out}) \cdot Q_{in} \cdot Y_H \cdot SRT}{1 + b_{H,T} \cdot SRT}$$

$$X_P = f \cdot b_{H,T} \cdot X_{BH} \cdot SRT$$

Sludge composition and production in full-plants

$$X_I = Q_{in} \cdot \frac{COD_{nbp,in}}{f_{cv}} \cdot SRT$$

where parameters are as follows:

$COD_{b,in}$ = concentration of biodegradable COD in the influent wastewater (mg/L);

$COD_{b,out}$ = concentration of biodegradable COD in the effluent wastewater (mg/L);

$COD_{nbp,in}$ = concentration of inert particulate COD in the influent wastewater (mg/L);

Q_{in} = flow rate of influent wastewater (m³/d);

Y_H = maximum growth yield of heterotrophic bacteria, 0.67 gCOD/gCOD or 0.45 gVSS/gCOD;

SRT = sludge retention time (d);

$b_{H,T}$ = decay rate of heterotrophic bacteria at the operative temperature (d⁻¹);

f = endogenous fraction, 0.08–0.2 (-);

f_{cv} = conversion factor of VSS into COD, 1.42–1.48 gCOD/gVSS.

The sludge production, X expressed as VSS, can be converted to TSS, dividing by the ratio VSS/TSS of sludge (f_v):

$$X(kgTSS/d) = \frac{X(kgVSS/d)}{f_v}$$

If the autotrophic (nitrifying) biomass is also to be taken into account, the term X_{BA} has to be added:

$$X(kgVSS/d) = \frac{1}{SRT} \cdot \left[\underbrace{X_{BH}}_{\substack{\text{term} \\ \text{related to} \\ \text{heterotrophic} \\ \text{biomass}}} + \underbrace{X_{BA}}_{\substack{\text{term} \\ \text{related to} \\ \text{autotrophic} \\ \text{biomass}}} + \underbrace{X_P}_{\substack{\text{term} \\ \text{related to} \\ \text{endogenous} \\ \text{residue}}} + \underbrace{X_I}_{\substack{\text{term} \\ \text{related to} \\ \text{inert} \\ \text{solids}}} \right]$$

in which X_{BA} is expressed as:

$$X_{BA} = \frac{Q_{in} \cdot Y_A \cdot \Delta N - NO_3 \cdot SRT}{1 + b_{A,T} \cdot SRT}$$

where:

Y_A = maximum growth yield of autotrophic (nitrifying) bacteria, 0.16 gVSS/gN;

$\Delta N{-}NO_3$ = concentration of NO_3-N produced during nitrification (mgN/L);

$b_{A,T}$ = decay rate of autotrophic (nitrifying) bacteria at the operative temperature (d^{-1}).

Decay rates of heterotrophic and autotrophic bacteria assume the range of values indicated in Table 2.1.

Table 2.1. Values of decay rate at 20°C and dependence on temperature.

Parameter	units	value at 20°C	dependence from T	α
$b_{H,20}$	d^{-1}	0.06 – 0.24 d^{-1} (0.12)	$b_{HT} = b_{H20} \cdot \alpha^{(T-20)}$	1.03 – 1.08 (1.04)
$b_{A,20}$	d^{-1}	0.05 – 0.15 d^{-1} (0.08)	$b_{AT} = b_{A20} \cdot \alpha^{(T-20)}$	1.03 – 1.08 (1.04)

The knowledge of biodegradable COD and inert particulate COD in the influent wastewater is needed to calculate the COD fractions in sludge, on the basis of the expression indicated above. Conversely, methods for estimating sludge production based only on total COD and BOD_5 are less accurate.

The biodegradable ($COD_{b,in}$) and inert particulate COD ($COD_{nbp,in}$) can be estimated through the fractionation of total COD in the influent wastewater (Figure 2.5) by using a respirometric approach.

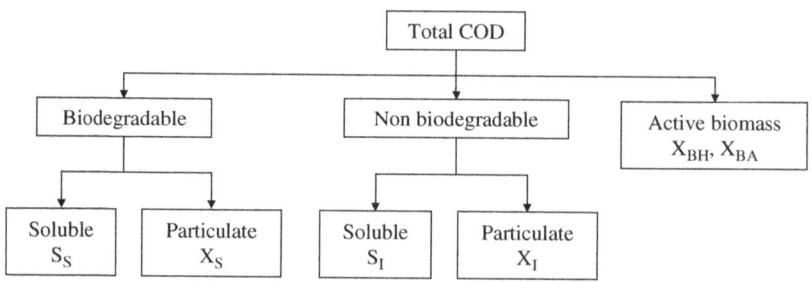

Figure 2.5. Fractionation of total COD in the influent wastewater.

Table 2.2 shows some typical data of COD fractionation in influent raw wastewater, obtained according to the subdivision in Figure 2.5 above.

Table 2.2. Data of COD fractionation in influent raw wastewater (own data) (fractions are expressed as percentage of total COD; mean value±st.dev. is indicated).

Soluble non-biodegradable (S_I)	Soluble biodegradable (S_S)	Particulate biodegradable (X_S)	Particulate non-biodegradable (X_I)
3.1±1.7	32.3±8.7	47.2±8.7	17.4±4.6
	$COD_{b,in}$		$COD_{nbp,in}$

The terms X_{BH} and X_{BA} can be considered as negligible in influent wastewater and however, their quantification is not easy. One example calculated on the basis of our own data is the following: assuming a concentration of influent COD of 466 mg/L and the mean values of COD fractionation indicated in Table 2.2, the concentration of $COD_{b,in}$ is $466 \times (0.323 + 0.472) = 369$ mg/L, while the concentration of $COD_{nbp,in}$ is $466 \times 0.174 = 81$ mg/L. The values of Y_H, $b_{H,T}$ and f_{cv} are assumed as indicated above and mean temperature of 16°C is chosen. The coefficient f is considered equal to 0.20. An average value of $COD_{b,out}$ of 5 mg/L is assumed. The variations of the VSS fractions in sludge as a function of SRT are indicated in Figure 2.6 (the calculation is independent on Q_{in}).

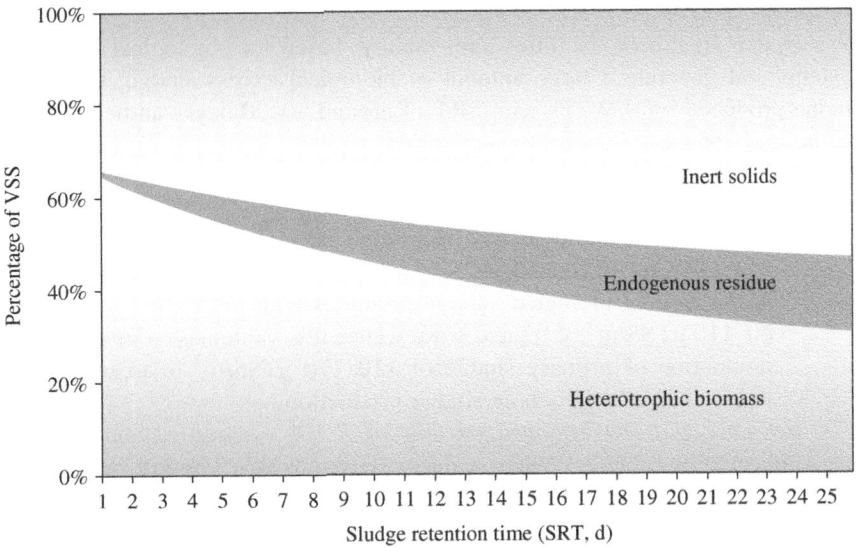

Figure 2.6. An example of variations of VSS fractions in sludge as a function of SRT, based on data described in the text.

In this example the heterotrophic biomass is only a limited part of the VSS of sludge, around 35% at SRT of 20 days (which corresponds to 25% of TSS of sludge). An increasing mass of the inert fractions X_I and X_P will accumulate in the sludge for increasing values of SRT.

From these estimates, the TSS in sludge can be considered in general made up of 20–30% of inorganic compounds, 10–30% of bacterial biomass, 40–50% of organic matter (not bacterial).

The sludge fractionation is important to correctly evaluate the potential of a sludge reduction route, because to obtain high sludge reduction it is necessary to act on inert fractions which often account for the majority of sludge mass, especially in the case of biological processes with long SRT.

This observation was also confirmed in a simulation of the sludge production in biological processes combined with sludge disintegration, by mathematical models (Yoon and Lee, 2005). The authors highlighted that the entity of the conversion of inert particulates to a biodegradable substrate significantly affects sludge reduction efficiency. If this conversion process is zero in a certain technique, complete sludge reduction is obviously not possible.

2.4 TYPICAL SLUDGE PRODUCTION DATA

Wastewater treatment facilities are mainly based on biological treatment systems and generate a large amount of biological excess sludge; the excess sludge produced by WWTPs with SRT of around 10–20 days can be estimated on the basis of the following typical values:

- *per unit of treated wastewater*:
  - in the case of biological sludge produced with raw wastewater: up to 250 gTSS/m^3 of treated wastewater;
  - in the case of biological sludge produced with pre-settled wastewater: 60–110 gTSS/m^3 of treated wastewater; this value has to be added to a production of primary sludge of 110–170 gTSS/m^3 of treated wastewater to obtain the whole sludge production;
- *per unit of COD removed or unit of BOD$_5$ removed*: typical data is indicated in Table 2.3, where values referred to WWTPs with and without digestion of sludge are indicated.

For increasing values of SRT a reduction of sludge production is expected. In Figure 2.7 the specific production of excess sludge over SRT is indicated – as kgTSS/kgCOD$_{removed}$ and as kgVSS/kgCOD$_{removed}$ – referring to data reported by Ginestet and Camacho (2007) and research data produced by the authors of this book (own data) during long-term monitoring of WWTPs located in the province of Trento (Italy).

Table 2.3. Specific production of excess sludge in WWTPs with and without digestion.

	WWTPs without sludge digestion	WWTPs with digestion	References
kgTSS/kgCOD$_{removed}$	0.31±0.12	0.27±0.07	Own data
	0.32±0.08	–	Ginestet and Camacho (2007)
	0.35	–	Henze (1992)
kgVSS/kgCOD$_{removed}$	0.23±0.13	0.18±0.10	Own data
	0.25±0.06	–	Ginestet and Camacho (2007)
kgCOD/kgCOD$_{removed}$	0.34±0.13	0.27±0.10	Own data
kgTSS/kgBOD$_5$ $_{removed}$	0.69±0.29	0.57±0.15	Own data
kgVSS/kgBOD$_5$ $_{removed}$	0.51±0.22	0.38±0.14	Own data

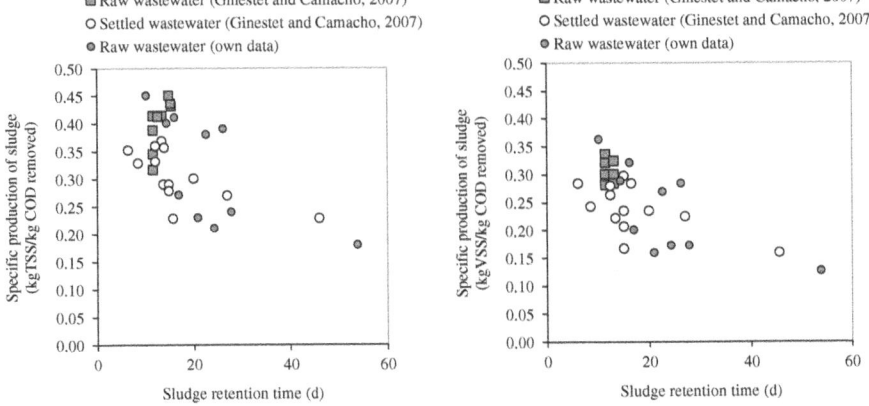

Figure 2.7. Specific production of excess sludge as a function of sludge retention time (SRT).

3
Current sludge disposal alternatives and costs in critical areas

> *Historically, it was common to see schematics that showed the water treatment scheme in detail [...] and an arrow at the end that simply said "sludge to disposal"*
> *(Neyens et al., 2004)*

3.1 INTRODUCTION

The type of final disposal of sludge is strongly influenced by highly variable conditions such as climatic conditions, land resources, need for fertilisers for agriculture, cost of land, transport costs, distance to the final disposal location, capacity of treatment facilities, local regulations, the local economy and other conditions.

Land use and landfilling – traditionally the cheaper alternatives – are often the most widely used in many countries. Today, composting is also the first step for sludge reuse in many countries of the world and it is a widely adopted alternative, but standards required for compost are more or less stringent in

© 2010 IWA Publishing. *Sludge Reduction Technologies in Wastewater Treatment Plants.* By Paola Foladori, Gianni Andreottola and Giuliano Ziglio. ISBN: 9781789065305. Published by IWA Publishing, London, UK.

different countries, with relative fertiliser legislation to control heavy metal or other contaminant content.

In some areas, sludge management is in rapid evolution due to (i) the continuous increase in sludge production, as a consequence of the fast growing number of WWTPs, which mostly implement biological treatment, and (ii) the limited choice of options for sludge disposal.

An overview of the current status and future prospects has recently been published by Spinosa (2007a). In Table 3.1 the disposal options in some different countries, currently adopted and planned for future years, are summarised.

Table 3.1. General situation of sludge disposal options in different areas, currently adopted and planned for future years (information from Spinosa, 2007a).

	On-site storage	Ponds, lagoons	Landfill	Agriculture/Land application	Composting	Incineration	Industrial reuse
Western Europe							
Eastern Europe							
North America							
Japan							
Australasia							
China							
South-East Asia							
Central and Latin America							
Africa							

In Western European countries the percentage of sludge used in agriculture varies widely among areas, in the range 10–70% (Müller, 2007). The future view, as indicated by Müller (2007) is a reduction of landfill for sewage sludge in response to the issues such as gas emissions – which may enhance global warming – hazardous contaminants and the loss of nutrients to the natural recycling process. Future consequence may include changes to the disposal choices, from landfill to land application, or from land application to incineration, or directly from landfill to incineration.

In Eastern Europe, where agricultural use and landfill are the most frequent disposal options (38% and 39% respectively, in the areas of Czech Republic,

Slovakia, Poland, Hungary and Slovenia), the amount of landfilled sludge will decrease, while a slow increase in the choice of incineration or other thermal treatment methods can be expected (Jenicek, 2007). The agricultural use of sludge with strictly controlled quality remains a sustainable solution (Jenicek, 2007). Meanwhile in Turkey sludge is stored in landfill areas or used for agricultural purposes, while incineration as a final disposal method is not so common (Jenicek, 2007).

Throughout North America in recent decades, technologies have become focused on means of reducing sludge masses and the stabilisation for the production of biosolids with Class A or equivalent properties (Dentel, 2007). Traditional composting sometimes appears problematic in big cities due to operating costs and odour production, while alternatives for meeting Class A requirements, including aerobic thermophilic digestion processes or drying and palletising technologies, could be able to produce a usable product without the operating complexities of composting (Dentel, 2007).

In Japan the percentage of incinerated sludge reaches 72%, and the final destination of incinerated ash includes landfill and industrial reuse. Incineration has been commonly used as a pre-treatment before landfill for 30 years in Japan (Okuno, 2007). Another important destination for sludge is Portland cement manufacturing plants; incinerated ash are used as a substitute for the raw material in production of Portland cement (Okuno, 2007). On the contrary, agricultural reuse is limited to 14% and is currently decreasing, while composting is not widely applied because of strict regulations for heavy metal levels and the need to operate with odour-free composting plants which have higher operating costs for the enclosed buildings and the off-gas deodorisation (Okuno, 2007).

In Australasian areas, there is a strong emphasis on recycling biosolids to land, either directly or as composted or blended product. In many urban areas (Sydney, Melbourne, Adelaide, Perth, Auckland, for a total population of about 11 million) disposal regards mainly recycling to land, as compost, for site rehabilitation, stockpiled, etc., or disposed in landfill, with the exception of Canberra where 100% of sludge is incinerated (Dixon and Anderson, 2007).

In many zones of South-East Asia disposal of sewage sludge to landfill and for soil conditioning is the main disposal route. In Singapore the majority of dewatered sludge is landfilled in an offshore site and in recent years in Taiwan all sewage sludge has been landfilled, but co-incineration and composting are planned for the near future (Lee, 2007).

The level of production and disposal of the sewage sludge produced in China is not known exactly, but a sharp increase of sludge production in future years is expected. It is estimated that sludge production in China is increasing at about 20% per year (Lee, 2007; Wei and Liu, 2006). Large amounts of the sewage sludge produced are directly used for agriculture (Wei and Liu, 2006). Where land is limited (big cities and some developed areas), thermal drying will be an alternative – even though it is more expensive – with the aim of also using the sludge treated by thermal drying as fertiliser, soil improver or construction material (Lee, 2007).

In Latin America, agriculture is the favoured route for sludge applications, due to the positive effects of the organic matter and nutrients (Barrios, 2007). In this context, suitable processes for sludge treatment aimed at microbial reduction are the main priority because this aspect is related to public health issues. In contrast, few attempts are aimed at reducing the mass of sludge since that option would reduce the beneficial effect of its reuse (Barrios, 2007).

In many developing countries sludge treatment and disposal are not high priorities and thus low cost solutions are often favoured. The management and disposal of the solids from on-site sanitation systems are often neglected during the planning of wastewater treatment systems.

Information about sludge treatment and disposal, for example, from African countries, is very limited. Isolated information available locally on sewage sludge treatment and disposal can be located – some examples are in Egypt and South Africa – but the fact that information is not readily available does not mean that innovative local management practices are not implemented (Snyman, 2007). The management of sludge is typically limited to on-site storage, because the treatment plants often consist of pond systems that have never been de-sludged (Snyman, 2007). However, in regions with a distinct wet season, such as Central Africa or South-East Asia, long-term storage is not feasible. Conversely, hot dry climates with an evaporation rate greater than the precipitation rate offer suitable conditions for natural dewatering (solar sludge drying, conventional drying beds) or sludge lagoons (Bauerfeld *et al.*, 2008). These processes are feasible on a large scale, especially where inexpensive land is available.

In South Africa the sludge generated from about 900 WWTPs is disposed of both dewatered or without employing dewatering technology in direct land application. The most widely employed sludge disposal options are agriculture and other land applications, while the remaining sludge is in part landfilled (Snyman, 2007).

In general, the options for sludge treatment/disposal depend on (1) physico-geographical, technical and economic factors, and on (2) ethical factors (values and priorities) related to the acceptability of specific practices or technologies (Bauerfeld *et al.*, 2008; Dentel, 2004). These very different options lead to very different costs for the sludge disposal, which are very high when the treatment/disposal options are limited or areas for their placement are scarce. The costs for sludge treatment and disposal in Europe, where the costs are becoming critical in some areas, are indicated in the following sections.

3.2 TOTAL COSTS FOR SLUDGE TREATMENT AND DISPOSAL

While wastewater treatment requires some hours, the treatment of the sludge produced requires several days or weeks and the costs for sludge management and disposal may reach 40–60% of the total costs for managing the entire WWTP, including personnel, maintenance, energy and sludge treatment and disposal.

There is a great variability in the cost of the alternatives for sludge disposal and treatment in the various countries and the costs for treatment + disposal can vary according to local conditions, widely ranging from 250 to more than 1000 €/t TS (Ginestet, 2007b).

Taking into account the costs for treatment + disposal + transport a mean value of 470±280 €/t TS is indicated in European countries (Paul *et al.*, 2006b). Huysmans *et al.* (2001, in Belgium) indicated a total cost of 480 €/t TS for sludge dewatering + incineration + transport of ash to landfill.

In Denmark, an additional tax of approx. 45 € per tonne TS leaving the treatment plant has been imposed (cited by Recktenwald *et al.*, 2008).

In the evaluation of the economic viability of sludge reduction techniques, the costs for sludge reduction have to be compared to the savings on the conventionally used treatment + disposal of sludge.

In the following sections the separate costs are indicated for European countries as costs for sludge treatment and for sludge disposal.

3.2.1 Sludge treatment costs in Europe

Sludge treatment costs are very variable depending on the units involved in the WWTP configuration: thickening, aerobic or anaerobic digestion, dewatering, pumping, conditioning, etc... All expenditure on these steps, such as chemicals for sludge conditioning or energy for digestion, are associated with sludge treatment. On the contrary, all additional steps in sludge treatment after dewatering (thermal drying, composting on site of the plant, application of sludge to agriculture by employees of the plant, etc...) are regarded as part of sludge disposal.

The cost for sludge treatment may vary from 150 €/t TS from extraction to dewatering (Frost & Sullivan, 2003), to 180 €/t TS for sludge dewatering only, taking into account the chemicals used, labour costs and depreciation (Huysmans *et al.*, 2001).

Paul *et al.* (2006b) referred to a range of costs for sludge treatment of 61–298 €/t TS, with a mean value of 160±80 €/t TS, considering some literature data on sludge treatment costs published in the last decade.

The costs for sludge treatment have to be added to the disposal costs, in order to obtain the overall cost of sludge processing.

3.2.2 Sludge disposal costs in Europe

The costs of the main alternatives for sludge disposal have been evaluated by Frost & Sullivan (2003), in a study on the subject based on a survey covering the whole European zone, and are summarised in Table 3.2 and Table 3.3, which indicates the costs per tonne of wet sludge and per tonne of dry solids respectively (Ginestet, 2007b).

Table 3.2. Costs for disposal of wet sludge (Frost & Sullivan, 2003; Ginestet, 2007b).

Sludge disposal options	Percentage of this option in EU zone	Wet sludge costs (€/t wet sludge)	
		average	min–max
Landfill	21.0%	71.7	35–120
Agriculture	47.0%	27.8	0–50
Composting	6.5%	41.2	35–70
Incineration	19.9%	74.5	38–98
Other	5.6%	26.0	15–50
Transport	100%	16.3	0–50
Average value (transport included)	–	*63.4*	*0–120*

Table 3.3. Costs for disposal of dry solids (Frost & Sullivan, 2003; Ginestet, 2007b).

↓ Sludge disposal options		Dry solid costs (€/t dry solids)		
	Dry content level→	10%	20%	30%
Landfill		717	359	239
Agriculture		278	139	93
Composting		412	206	137
Incineration		745	373	248
Other		260	130	87
Transport		163	82	54
Average value (transport included)		*634*	*317*	*211*

In the 2nd column of Table 3.2 the transport is indicated as 100% because all the disposal options indicated (landfill, agriculture, composting, incineration or other) require transport from the WWTP where the sludge is produced.

As can be observed in Table 3.2, the preferred alternative is land application (47%) – sometimes coupled with composting (6.5%) – and secondly landfill disposal (21%) and incineration (19.9%). The higher costs for the disposal of wet sludge are associated with incineration (38–98 €/t wet sludge) and landfill (35–120 €/t wet sludge), but a further increase of this cost is expected in the future.

Considering all the options for sludge disposal including transport, the average cost is 63.4 €/t wet sludge (range 0–120 €/t wet sludge).

As early as 2000, Novak (2000) indicated the costs for sludge disposal reported from Germany as higher than 80 €/t wet sludge, at least twice as high as those found for Austria in the same period, in which landfill and agriculture were the main options.

These costs can be converted per unit of dry solids (expressed as TS or TSS), considering 3 levels of sludge dry content: 10%, 20%, 30% (Table 3.3).

The average cost per tonne of dry solids (variable in the range 211–634 €/t TS) decreases progressively for increasing values of dry content in dewatered sludge. The higher cost is associated with incineration, which requires 248–745 €/t TS, depending on dry content, and landfill.

Huysmans et al. (2001, in Belgium) reported values in agreement with the foregoing, indicating a cost of 250 €/t TS for residual sludge incineration (running costs, depreciation, ash disposal) and transport cost of 50 €/t TS (for transport from WWTP to incineration and final ash disposal in landfill).

Also some French studies focused on agricultural recycling only, indicated wide ranging costs of 42–545 €/t TS, with average values of 133 and 178 €/t TS (Ginestet, 2007b), in agreement with the data indicated in Table 3.3 for sludge dewatered to a dry content around 20%.

Considering a typical dry content for dewatered sludge of 20%, and all the alternatives for disposal and transport, the mean cost up to now is around 317 €/t TS.

Paul et al. (2006b) referred to costs for sludge disposal in the range of 110–457 €/t TS, with a mean value of 240±110 €/t TS, considering literature data published over the last decade.

In 2006 the costs for sludge disposal in the Province of Trento (Italy) – the Italian region where the authors carry out their research – was 74 €/t wet sludge, excluding transport. These costs are in agreement with the values indicated in Table 3.2. Due to the alpine location of this area, it is difficult to use the sludge in agriculture. Thermal drying is required for long distance transport to final disposal in landfill sites or agricultural recycling in more suitable areas. Taking transport into account, a total cost of 108 €/t wet sludge is reached. Considering the mean dry content of dewatered sludge of 18.7%, the cost for sludge disposal is 394 €/t TS without transport and 577 €/t TS with transport.

4
Principles of sludge reduction techniques integrated in wastewater treatment plants

4.1 INTRODUCTION

Many studies have been produced since the '90s on alternative technologies for direct on-site reduction of sludge production (as dry mass and not only in volume).

The proposed methods are based on physical, mechanical, chemical, thermal and biological treatments. Most of them are aimed at solids solubilisation and disintegration of bacterial cells in sludge, with the objectives of:

© 2010 IWA Publishing. *Sludge Reduction Technologies in Wastewater Treatment Plants.* By Paola Foladori, Gianni Andreottola and Giuliano Ziglio. ISBN: 9781789065305. Published by IWA Publishing, London, UK.

- reducing sludge production directly in the wastewater handling units;
- reducing sludge mass in the sludge handling units and simultaneously improving biogas production in anaerobic digestion or, in some cases, dewaterability;
- (in few cases) producing an additional carbon source to support denitrification and phosphorous removal in the wastewater handling units.

In this chapter the main mechanisms of sludge reduction techniques integrated in wastewater or sludge handling units are presented.

The mechanisms for sludge reduction techniques can be identified as follows:

(1) **cell lysis and cryptic growth** (§ 4.2): basis of most of the proposed techniques and most widely discussed in this book;
(2) **uncoupled metabolism** (§ 4.3);
(3) **endogenous metabolism** (§ 4.4): basis of the self-digestion process which commonly occurs in the aerobic and anaerobic digestion units;
(4) **microbial predation** (§ 4.5);
(5) **hydrothermal oxidation** (§ 4.7).

A further mechanism, which is linked to biodegradability increase of inert solids, could be added, even though it can not be considered a separate mechanism because it takes place in combination with other mechanisms, as further described in § 4.6.

Figure 4.1 summarises these 5 mechanisms indicating the main technologies which activate each mechanism, divided according to their integration in wastewater or sludge handling units.

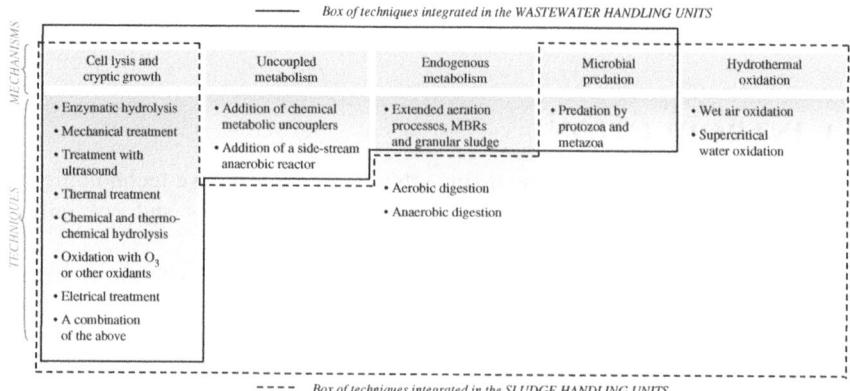

Figure 4.1. Scheme of the main mechanisms for sludge reduction, classified according to their integration in wastewater or sludge handling units.

Principles of sludge reduction techniques

These two categories can be identified on the basis of the mechanisms described:

(1) integration in the wastewater handling units (unbroken line in Figure 4.1): the objective is to reduce sludge production directly in the wastewater treatment units instead of realising post-treatments of sludge after its production;
(2) integration in the sludge handling units (broken line in Figure 4.1): the reduction of sludge mass occurs after its production in the wastewater handling units.

Figure 4.2 presents the processes commonly used to treat wastewater and sludge in a WWTP, divided into the two paths, as further considered in this book:

(1) wastewater handling units (known also as the water line)
(2) sludge handling units (known also as the sludge line).

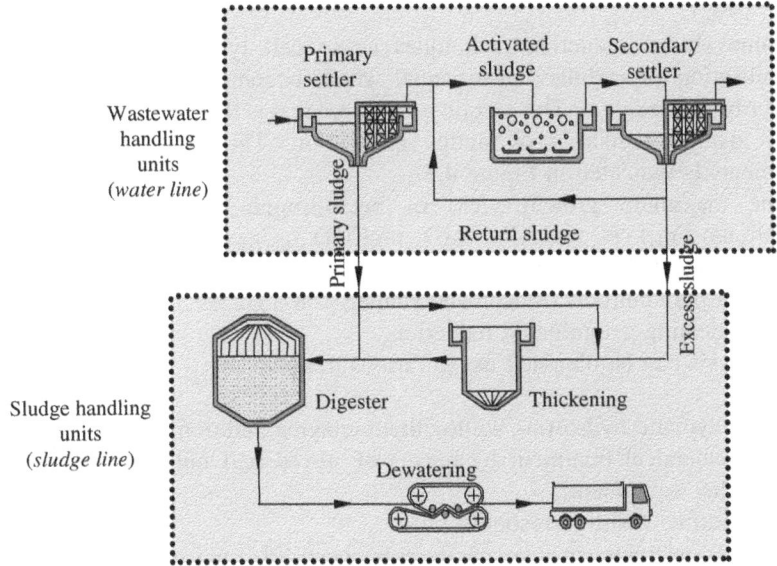

Figure 4.2. The common lay-out adopted to define the wastewater handling units (water line) and the sludge handling units (sludge line) in a WWTP.

These various processes, which can be applied for sludge reduction, are presented in specific sheets in Chapters 5 and 6 where the techniques integrated in the wastewater handling units and in the sludge handling units are described respectively.

4.2 CELL LYSIS AND CRYPTIC GROWTH

The term "cryptic growth" was introduced to indicate the reutilisation of intracellular compounds (both carbonaceous compounds and nutrients) released from cell lysis, for the growth of viable cells of the same population. Historically, the terms cryptic growth, endogenous activity, and autodigestion have often been used in an analogous manner in the literature (Ryan, 1959; Gaudy *et al.*, 1971; Hamer, 1985; Mason and Hamer, 1987; Canales *et al.*, 1994).

The occurrence of cryptic growth in the production of biomass in activated sludge systems was first demonstrated by Gaudy *et al.* (1971), who tested sonication as a cell-lysing technique and evaluated the solubilisation rate of biomass. Since the mid '80s the contributions investigating cell death, autolysis, and subsequent cryptic growth have increased widely, carried out both in lab-scale activated sludge reactors and in full-scale plants (*inter alia* Hamer, 1985; Sakai *et al.*, 1992).

Some sludge reduction techniques cause cell lysis with the consequent solubilisation of cellular constituents, which become substrate available for further biodegradation. The cryptic growth process is thus induced which results in an overall reduction of sludge production. The effect of these sludge treatments is indicated in Figure 4.3.

The maximum growth yield of heterotrophic bacteria under aerobic conditions (Y_H) is typically 0.67 $mgCOD_{synthetised}/mgCOD_{removed}$. In the presence of cell lysis-cryptic growth, Y_H can reach for example 0.43 gCOD/gCOD for pure cultures (Mason and Hamer, 1987; Hamer, 1985; Canales *et al.*, 1994) indicating a significant reduction.

Cell lysis can be obtained using various treatments:

(1) enzymatic hydrolysis with/without enzyme addition;
(2) mechanical treatment by means of stirred ball mills, homogenisers or other equipment;
(3) treatment with ultrasound;
(4) thermal treatment at temperatures between 40°C and approximately 220°C;
(5) chemical and thermo-chemical hydrolysis adding acid or alkaline reagents, sometimes coupled with temperature increase;
(6) oxidation with ozone, H_2O_2 or chlorination;
(7) electrical treatment;
(8) a combination of the above.

Principles of sludge reduction techniques

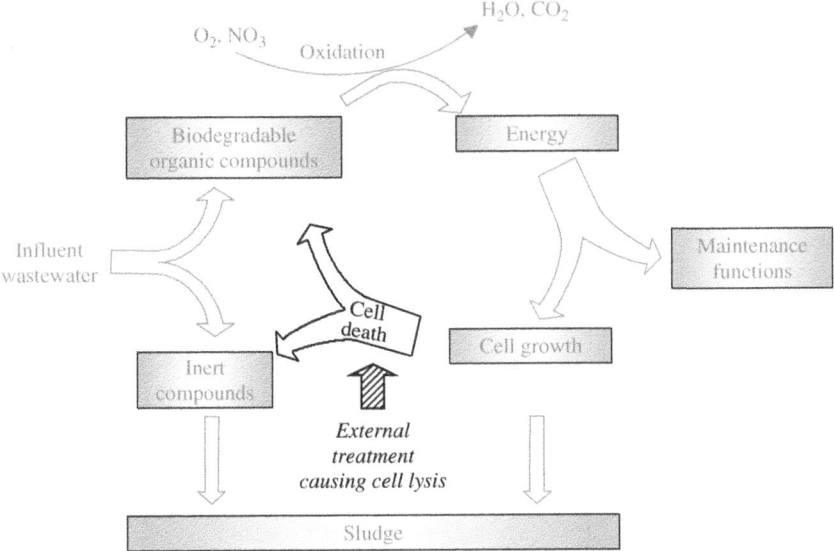

Figure 4.3. Effect of an external treatment on cell death and lysis, affecting sludge production.

Furthermore, many of these treatments cause the dispersion of sludge flocs reducing the size of the larger aggregates into smaller fragments and increasing the total surface area available for hydrolysis (Figure 4.4). Some models introduced to describe the hydrolysis process by activated sludge demonstrated the importance of the particle size and the particle surface on the hydrolysis rate (Dimock and Morgenroth, 2006).

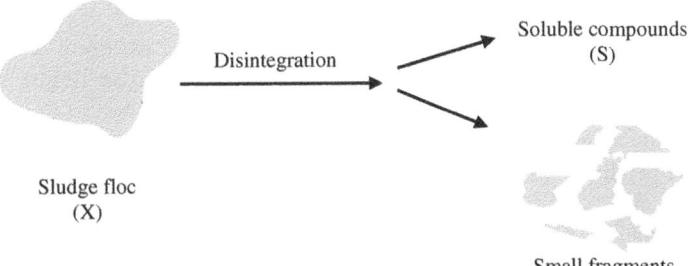

Figure 4.4. Scheme of the breakup of sludge flocs after a treatment for sludge disintegration and cell lysis.

These treatments can be integrated in the wastewater handling units or in the sludge handling units, according to the options indicated in Figure 4.5. The lysate produced by a treatment undergoes biodegradation in the activated sludge stages (integrated in the wastewater handling units) or in the digesters (sludge handling units).

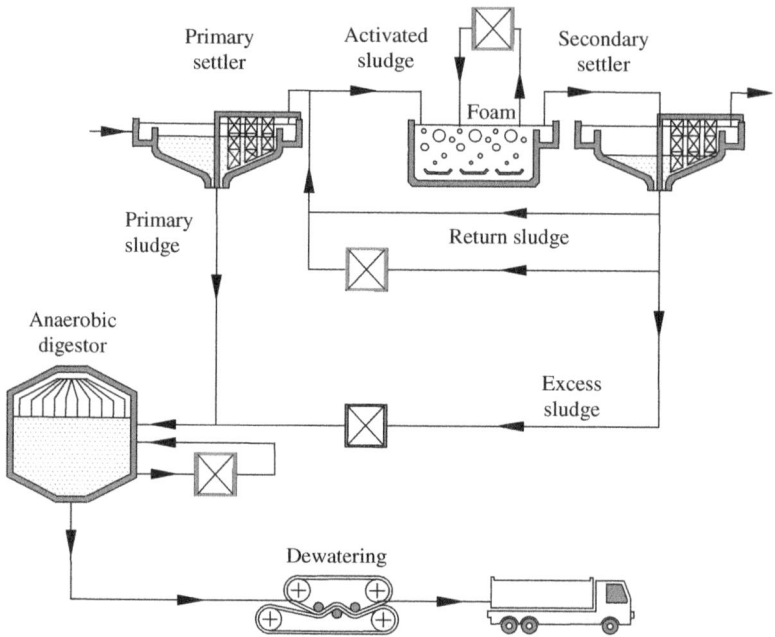

Figure 4.5. Alternatives for the integration of sludge reduction techniques in the wastewater handling units or in the sludge handling units. The symbol ⊠ indicates the possible location of the necessary equipment.

From an economic point of view, some techniques are more advantageous when applied to sludge with high solid content, such as return sludge or thickened sludge.

When the sludge reduction technique is applied to the return sludge, the treatment produces biodegradable carbonaceous matter supporting denitrification in activated sludge stages. When it is integrated in the sludge handling units, improvements in biogas production and sludge stabilisation in anaerobic digesters are obtained.

In general, the application of sludge reduction technologies to primary sludge (generated in primary settling) is not advisable, because solids in primary sludge are easily and spontaneously hydrolysable in biological reactors or digesters. Therefore, an additional treatment may be superfluous in enhancing the hydrolysis process.

An additional benefit can be the reduction of foam caused by filamentous bacteria in biological reactors.

4.3 UNCOUPLED METABOLISM

Microorganisms use organic matter in wastewater as a carbon source to obtain energy and produce new cells. The catabolic process transforms organic matter into energy and metabolites. This energy is then used to meet maintenance requirements and to support the anabolic process in which metabolites are converted into new biomass.

Adenosine triphosphate (ATP) plays an important role in the process of substrate oxidation (catabolism) and cell synthesis (anabolism) as briefly indicated in Figure 4.6. In the case of many aerobic bacteria, ATP is generated by oxidative phosphorylation, in which electrons are transported from an electron donor (substrate) to a final electron acceptor (O_2). The energy released during the conversion of ATP back to ADP + P is used for cell anabolism, growth and maintenance functions. The growth yield is directly proportional to the quantity of energy (ATP) produced.

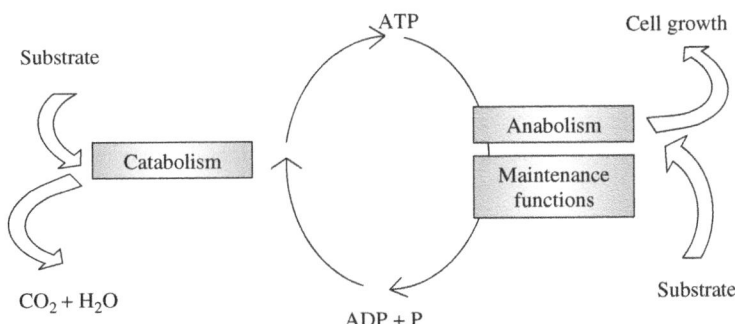

Figure 4.6. Relationship between catabolism and anabolism.

Russel and Cook (1995) introduced the term *"uncoupling"* to define the inability of oxidative phosphorylation to generate the maximum theoretical quantity of metabolic energy in the form of ATP. The ATP lost in reactions not associated with the growth is called *"ATP spilling"*. The uncoupled metabolism increases the discrepancy in the energy level between catabolism and anabolism, limiting the energy available for anabolism. As a consequence, bacteria first satisfy their maintenance functions before spending energy on growth and consequently there is less energy available for the synthesis of new cells. The biomass yield is thus reduced and the sludge production decreases.

The phenomenon of uncoupled metabolism is encountered in conditions such as: presence of inhibitory compounds or some heavy metals (Zn, Ni, Cu, Cr), not optimal temperatures, nutrient limitations, during transition periods in which cells are adjusting to changes in their environment, such as under alternating aerobic-anaerobic phases (Chudoba *et al.*, 1992a; Mayhew and Stephenson, 1998; Low and Chase, 1999a; Liu, 2000). In anoxic or anaerobic catabolism (in the absence of oxygen as the terminal electron acceptor) the overall ATP generation is much than is lower than in aerobic processes. Consequently, anoxic or anaerobic metabolism is considerably less efficient than aerobic metabolism, resulting in much lower biomass yields (Low and Chase, 1999a; see § 8.3).

The biomass grown per gram of ATP consumed (Y_{ATP}) depends on cell composition, on growth rate and on maintenance requirements. Because these factors vary from species to species, it is expected that Y_{ATP} demonstrates diverse values according to the type of microorganisms. The uncoupled metabolism in mixed cultures will favour the species which is more efficient in the production and exploitation of ATP, which are the species characterised by higher Y_{ATP} (Low and Chase, 1999a).

The process of uncoupled metabolism affects the sludge production as indicated in Figure 4.7. The aim of this process is to uncouple energy intended for anabolism, in order to reduce the growth yield, without reducing the efficiency of organic matter removal from wastewater.

From a technological point of view, uncoupled metabolism can be obtained by:

(1) adding chemical uncoupling compounds, such as: 2,4-dinitrophenol, para-nitrophenol, pentachlorophenol and 3,3',4',5-tetrachlorosalicyl-anilide, etc...; this process is described in § 8.8;
(2) subjecting activated sludge to cyclic aerobic and anaerobic conditions by means of side-stream anaerobic reactors. By inserting an anaerobic stage the most energy efficient electron acceptors (such as oxygen and nitrate) are no longer available. An example is the OSA process (*Oxic-Settling-Anaerobic*), which is made up of a conventional activated sludge stage integrated with an anaerobic digester supplied by the return sludge. This process is described in more detail in § 8.4.

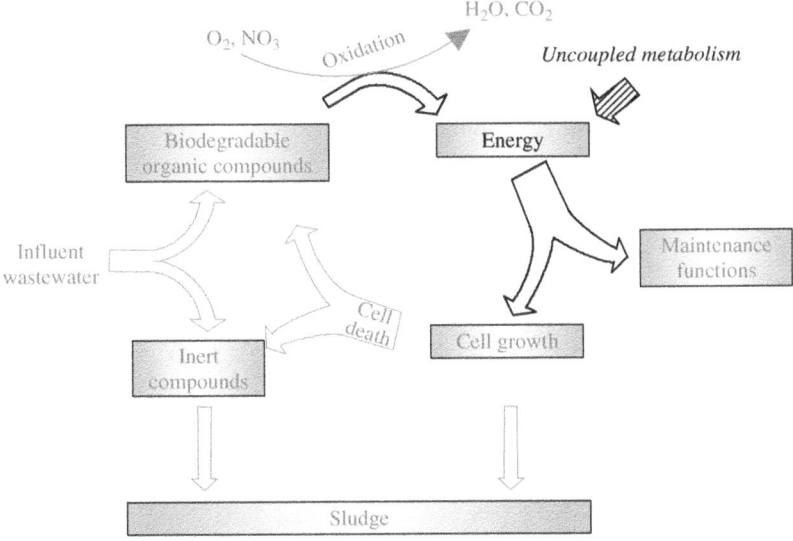

Figure 4.7. Effect of uncoupled metabolism in the scheme of sludge production.

4.4 ENDOGENOUS METABOLISM

When external substrate is available, energy obtained from the substrate biodegradation is used for maintenance requirements of bacteria (basic metabolic energy requirements such as membrane potential, renewal of proteins, maintenance of osmotic pressure, motility, etc...- see van Loosdrecht and Henze, 1999) as well as for the synthesis of new cellular biomass (Figure 4.8).

In the absence of external substrate, only a part of the cellular constituents (e.g. internally stored substrate such as glycogen or polyhydroxy-alkanoate, PHA) can be oxidised to carbon dioxide and water to produce the energy needed for cell maintenance requirements.

The concept of the endogenous metabolism was introduced to describe the observation that storage compounds are used for maintenance purposes when the external substrate is completely depleted. In other words, endogenous metabolism should be defined as a state when no net growth is possible, but cells consume energy to remain viable.

The fact that aerobic sludge consumes oxygen in the absence of external substrates has been observed since the mid '50s, introducing various hypotheses to explain the phenomenon of endogeneous respiration in the absence of external substrate, such as:

- oxidation of the cellular tissues;
- conversion of intracellular reserve material such as glycogen or polyhydroxy-alkanoate;
- decay of cells and subsequent consumption of the dead cells by the viable ones, to synthesise new biomass (cryptic growth).

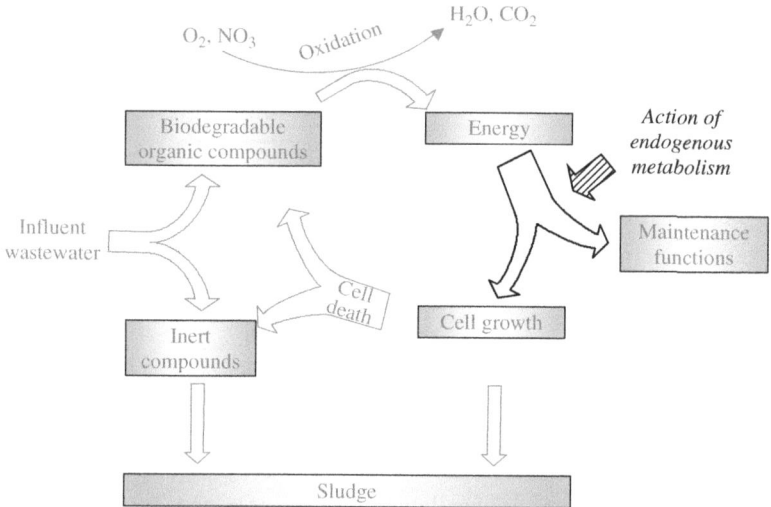

Figure 4.8. Effect on endogenous metabolism in the scheme of sludge production.

On the basis of the definition given by van Loosdrecht and Henze (1999) endogenous respiration is "the respiration with oxygen or nitrate using cell internal components". From the equations proposed by van Loosdrecht and Henze (1999; see § 7.9) it can be seen that the maintenance and the endogenous respiration concept are mathematically equivalent, and these two concepts cannot easily be distinguished from each other under experimental conditions.

By increasing energy requirements for non-growth activities, in particular maintenance functions, the amount of energy available for the growth of biomass decreases. Therefore a significant reduction of sludge production can be achieved by maximising the energy used for maintenance requirements rather than for cellular synthesis (Low and Chase, 1999b).

It is well known that sludge production is lower in activated sludge plants with long SRT and operating at low applied loads or low F/M ratios (*Food/ Microorganisms*). For example, endogenous metabolism allows biomass production to be reduced in aerobic reactors by 12% when biomass concentration (M) is increased from 3 to 6 g/L. The reduction reaches 44% when biomass concentration is increased from 1.7 to 10.3 g/L by increasing the SRT (Low and Chase, 1999a). However, there is a limit to the potential increase of sludge concentration in activated sludge systems equipped with conventional settlers and only systems based on membrane filtration (MBR) or biofilm processes (granular sludge) can overcome this limit.

The endogenous metabolism acts as indicated in the scheme of Figure 4.8. Endogenous metabolism is exploited in the following processes:

(1) stabilisation of sludge in aerobic and anaerobic digestion (§ 8.1, and § 8.2);
(2) low-loaded activated sludge plants: extended aeration processes (§ 8.10);
(3) MBR reactors operating with high concentrations of solids (up to 8–15 gTSS/L), long SRT and low organic loads. In these systems it would be theoretically possible to reach a balance, in which the energy obtained from substrate biodegradation is equal to the energy required for maintenance. As a consequence, the production of new biomass could theoretically reach zero (Ghyooy and Verstraete 2000; Wagner and Rosenwinkel, 2000). In reality, zero-sludge production is not feasible in MBR supplied with municipal wastewater as described in § 8.11, also because of the continuous feed of inert solids entering with the influent wastewater;
(4) granular sludge systems (§ 8.12), which are based on a self-immobilisation of microorganisms treating wastewater, and are characterised by good settleability, strong microbial structure, high biomass retention and low sludge production (only 0.15 kgTSS/kgCOD$_{removed}$ compared to conventional activated sludge systems which produce typically 0.27–0.35 kg TSS/kgCOD, see § 2.4).

4.5 MICROBIAL PREDATION

The biological processes involved in wastewater treatment represent a complex ecosystem (constituted by bacteria and predators) and sludge production could be reduced by increasing the microbial predation.

Both viable and dead bacteria are a food source for higher organisms, such as protozoa and metazoa. When one organism eats another the total amount of biomass decreases and the transfer to a higher trophic level of the food chain occurs. In this way part of the biomass and the potential energy is lost as heat and excretory products, which causes a reduced growth of biomass and lower sludge production (Figure 4.9).

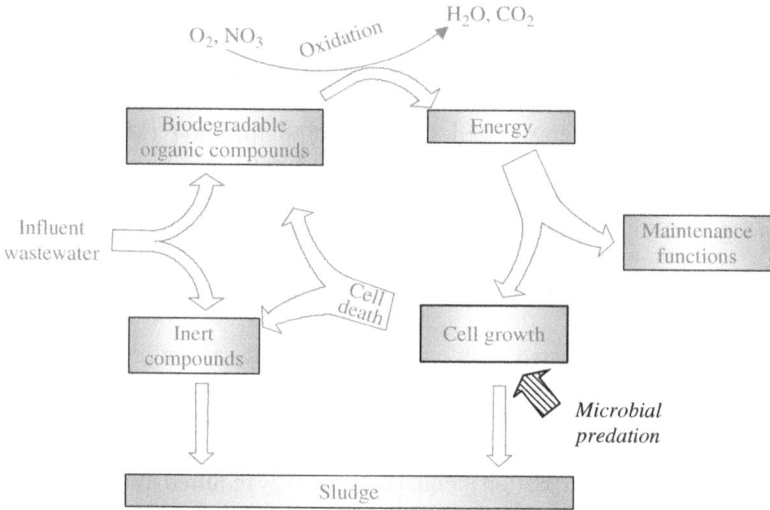

Figure 4.9. Effect of microbial predation in the scheme of sludge production.

Aquatic oligochaetes may be used for treatment and reduction of excess sludge as described in the review of Ratsak and Verkuijlen (2006), entitled "Sludge reduction by predatory activity of aquatic oligochaetes in WWTPs: science or fiction?". Predation by oligochaetes can be exploited either in the wastewater hangling units or in the sludge handling units. The oligochaetes used for the treatment of excess activated sludge can be divided into two groups, which occupy different niches (Ratsak et al., 2006):

– the large aquatic worms such as Tubificidae, Lumbriculidae;
– the small aquatic worms such as Naididae, Aeolosomatidae.

The main drawback of the predation process is the difficulty to ensure stable, long-term, favourable conditions for predator development and reproduction.

4.6 BIODEGRADABILITY INCREASE IN INERT SOLIDS

As described in § 2.2.1, sludge contains inert solids which reach the biological reactors with influent wastewater and accumulate in the sludge. To achieve the reduction of sludge mass one alternative would be to act on the inert solids, to convert – at least partly – into biodegradable compounds.

How a treatment affects the increase of biodegradability of inert solids and the reduction of sludge production is indicated in Figure 4.10.

Principles of sludge reduction techniques

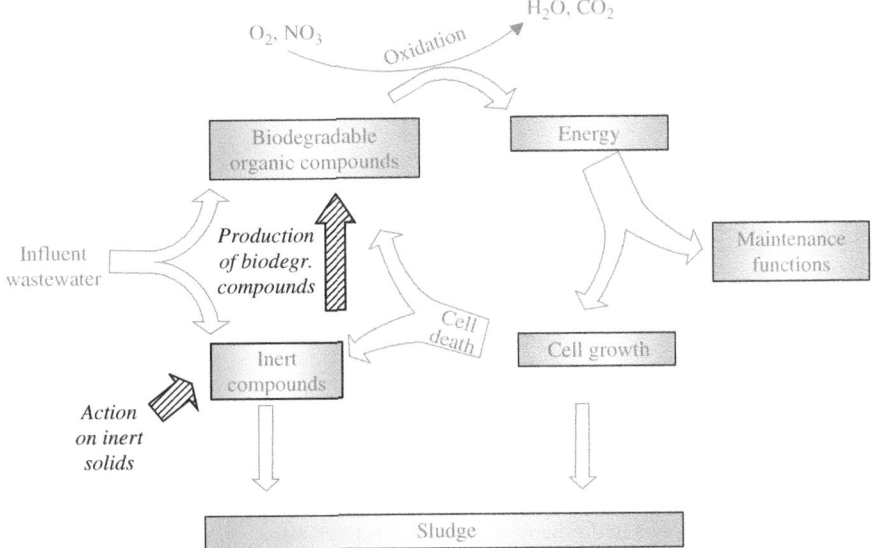

Figure 4.10. Effect of increase of biodegradability of inert solids in the scheme of sludge production.

To increase the biodegradability of inert solids there is no specific technology. For example, ozonation is expected to be a strong oxidant on all sludge components: whether cellular biomass or inert compounds. This treatment on cellular biomass induces cell lysis-cryptic growth, while its effect on inert solids is an enhancement of solubilisation with a potential increase of biodegradability of the compounds released.

4.7 HYDROTHERMAL OXIDATION

Once produced, the sludge can be treated using hydrothermal oxidation processes to reduce the amount requiring disposal (Figure 4.11).

The sludge mineralisation using oxygen can be achieved at high temperature (>850°C) in the gaseous phase (incineration) or at a relatively low temperature (in the range 150–320°C or >374°C) in the liquid phase: water (subcritical or supercritical) is used as a reaction medium. Two hydrothermal oxidation processes are available:

(1) subcritical water oxidation, including wet air oxidation (WAO);
(2) supercritical water oxidation (SCWO).

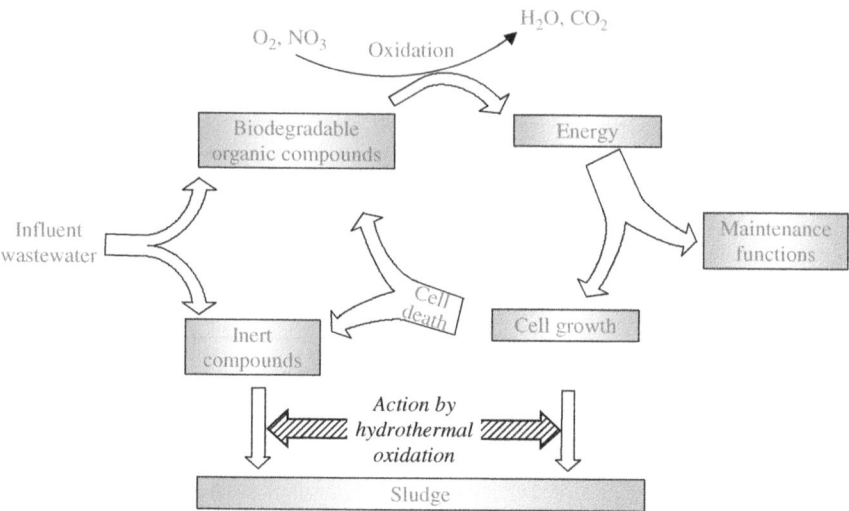

Figure 4.11. Effect of hydrothermal oxidation in the scheme of sludge production.

Operating at high pressures and temperatures with the addition of strong oxidants, the particulate organic compounds in sludge are mostly oxidised or solubilised.

In particular, hydrothermal oxidation affects bacteria biomass and organic solids even if they are inert.

5
Overview of the sludge reduction techniques integrated in the *wastewater handling units*

The aim of sludge reduction techniques integrated in the wastewater handling units is to reduce sludge production directly during wastewater treatment, rather than performing post-treatments of sludge after production.

This may be achieved via processes based on the following mechanisms (Figure 5.1): (a) cell lysis-cryptic growth, (b) endogenous metabolism, (c) uncoupled metabolism, (d) microbial predation.

The most widely adopted techniques in sludge reduction are based on *cell lysis-cryptic growth* and include the following solutions:

- enzymatic hydrolysis with/without added enzymes (§ 5.1, § 5.2);
- mechanical treatment (§ 5.3);

© 2010 IWA Publishing. *Sludge Reduction Technologies in Wastewater Treatment Plants.* By Paola Foladori, Gianni Andreottola and Giuliano Ziglio. ISBN: 9781789065305. Published by IWA Publishing, London, UK.

- treatment with ultrasound (§ 5.4);
- thermal treatment (§ 5.5);
- chemical and thermo-chemical hydrolysis (§ 5.6);
- oxidation with ozone or other strong oxidants (§ 5.7, § 5.8);
- electrical treatment (§ 5.9);
- a combination of the above.

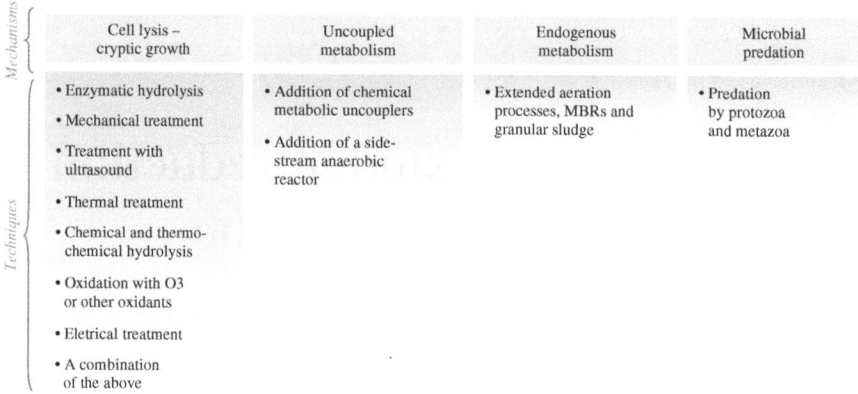

Figure 5.1. Scheme of the main sludge reduction techniques integrated in the wastewater handling units.

After undergoing one of these treatments, the sludge is recirculated into the activated sludge stage where the degradation of the lysate produced during treatment occurs with consequent reduction in overall sludge production.

Techniques based on *uncoupled metabolism* are the following:

- addition of chemical metabolic uncouplers (§ 5.10);
- addition of a side-stream anaerobic reactor at ambient temperature (§ 5.11).

Sludge reduction techniques based on *endogenous metabolism* are basically:

- extended aeration activated sludge processes (§ 5.12);
- Membrane Biological Reactors (§ 5.13);
- granular sludge (§ 5.14).

These systems can not be easily integrated in an existing wastewater plant, since their introduction implies the renovation of most of the biological units.

The process based on *microbial predation* by organisms such as metazoa and protozoa is described in § 5.15.

In the following sections for each alternative technique presented above a short description, a flow diagram and a table of advantages and drawbacks are given. Techniques which feature the widest applicability, will be described in more detail in the various chapters of the book, where experimental results, efficiency and operational conditions, etc... will be fully indicated.

5.1 ENZYMATIC HYDROLYSIS WITH ADDED ENZYMES
(See also Chapter 8, Section 8.7)

Hydrolytic enzymes adsorb to the sludge-substrate and attack the polymeric substances leading to solid solubilisation and biodegradation enhancement. Enzymatic treatment of sludge is based on the mechanisms of solubilisation, cell lysis and cryptic growth. Considering the high presence of proteins, carbohydrates and lipids in the composition of excess sludge, the addition of enzymes such as protease, lipase, cellulase, emicellulase and amylase could be advisable. Commercial enzymes for the hydrolysis of organic matter has been already used for the improvement of sludge biodegradation and reduction or to enhance organic waste degradation, but not always the expected performances are predictable. Many of these enzymatic products are patented and their exact composition is generally confidential; in some cases these mixtures contain other stimulatory nutrients as well as enzymes.

Cations such as Ca^{2+}, Mg^{2+} or Fe^{3+} have a role in the enzymatic treatment of sludge. The removal of these cations by means of cation-binding agents leads to the disruption of flocs and solid solubilisation, resulting in an increase of the specific surface area available for enzymatic hydrolysis. However, at the moment the dosages indicated for these cation-binding agents are very high and not yet recommended for practical applications at large/industrial scale.

The installation of an enzyme dosage system represents the only investment cost (Figure 5.2). To date, promising results in sludge reduction are referred often to high enzyme dosages and the viability at full-scale has not been yet completely investigated.

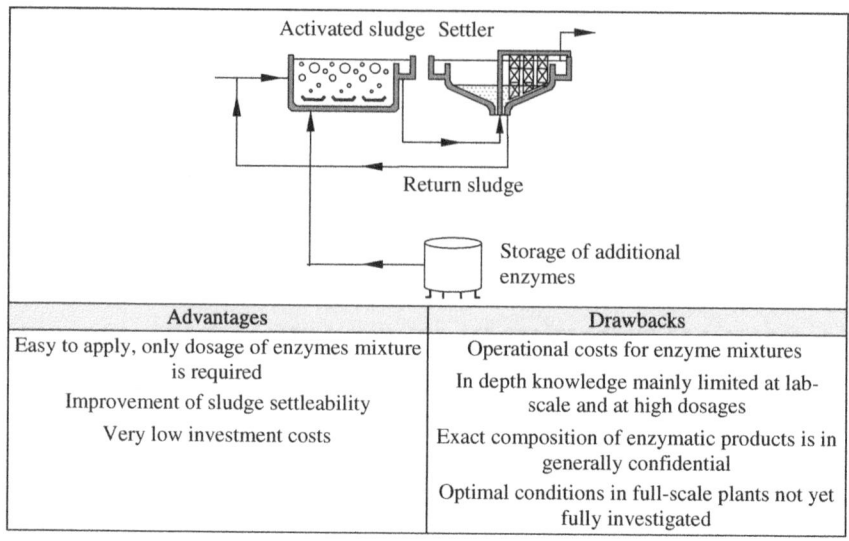

Advantages	Drawbacks
Easy to apply, only dosage of enzymes mixture is required	Operational costs for enzyme mixtures
Improvement of sludge settleability	In depth knowledge mainly limited at lab-scale and at high dosages
Very low investment costs	Exact composition of enzymatic products is in generally confidential
	Optimal conditions in full-scale plants not yet fully investigated

Figure 5.2. Enzyme addition in wastewater handling units.

5.2 ENZYMATIC HYDROLYSIS BY THERMOPHILIC BACTERIA (THERMOPHILIC AEROBIC REACTOR)

(See also Chapter 8, Section 8.6)

The process is based on a thermal action and an enzymatic attack occurring at thermophilic temperatures in an aerobic reactor treating part of the return sludge. The lysated sludge is then recirculated in the activated sludge stages where cryptic growth occurs (Figure 5.3).

Before entering the thermophilic reactor, the sludge can be mechanically thickened in order to save energy when heating the sludge. In the thermophilic reactor, at temperatures of 55–70°C (HRT = 1–3 d), thermophilic bacteria – such as *Bacillus stearothermophilus* – produce hydrolytic enzymes such as thermostable extracellular protease, responsible of the enhancement of sludge solubilisation compared to the thermal action alone. Thermophilic bacteria then pass from the thermophilic aerobic reactor to the activated sludge stage, where they become inactive and form spores which again return to the thermophilic aerobic reactor. This process occurs under thermophilic temperatures, and the thermal effect also render the cells in sludge more susceptible to enzyme attack. VSS solubilisation is 30–40% and causes a high strength lysate, recirculated in the activated sludge stage. This process requires the additional reactor and heating to thermophilic temperatures, with resulting additional costs which are however modest due to the compact volume of the reactor (due to the short retention time).

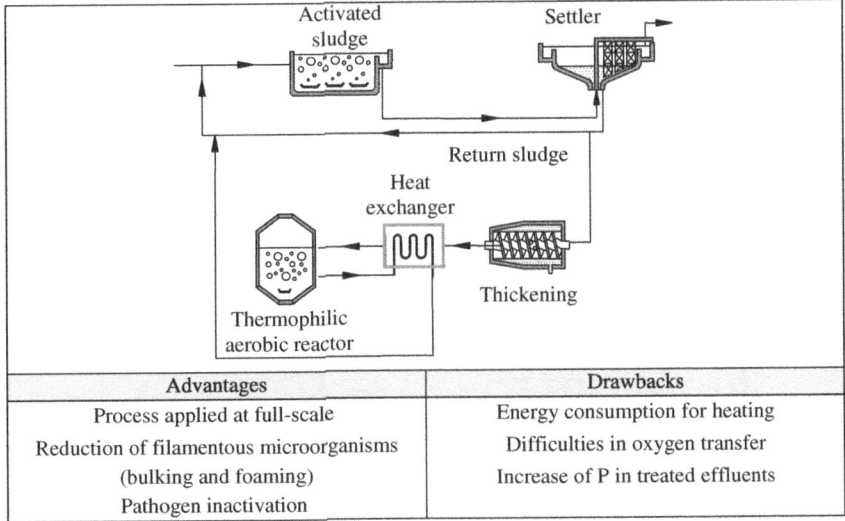

Figure 5.3. Thermophilic aerobic reactor integrated in wastewater handling units.

S-TE PROCESS® (acronym of Solubilisation by Thermophilic Enzyme) and Biolysis®E process are technologies applied at full-scale proposed on the market.

5.3 MECHANICAL DISINTEGRATION
(See also Chapter 9)

Several systems have been proposed for the mechanical disintegration of sludge, in which energy is supplied as pressure or rotational/translation movement. The aim is to enhance sludge solubilisation as a consequence of the bacteria cell disintegration and the disaggregation of biological flocs. In general, at low applied energy only floc disintegration is observed, while high energy is required to damage microbial cells.

The mechanical disintegration treatment is integrated in the activated sludge process through the addition of an appropriate unit treating part of the return sludge. Sludge is disintegrated and the lysate obtained is recirculated into the activated sludge reactors (Figure 5.4). Thus the process is based on the cell lysis-cryptic growth mechanism. The systems proposed for mechanical disintegration are:

- lysis-thickening centrifuge;
- stirred ball mills;
- high pressure homogenisers;
- high pressure jet and collision
- rotor-stator disintegration systems.

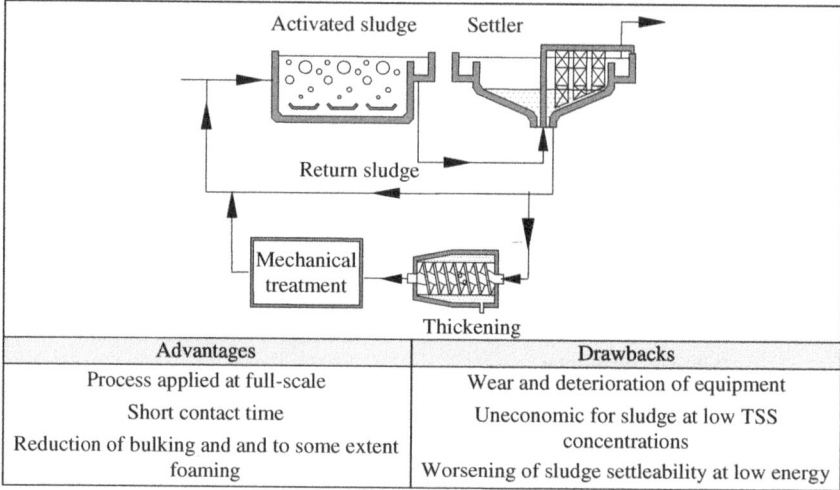

Advantages	Drawbacks
Process applied at full-scale	Wear and deterioration of equipment
Short contact time	Uneconomic for sludge at low TSS concentrations
Reduction of bulking and and to some extent foaming	Worsening of sludge settleability at low energy

Figure 5.4. Mechanical disintegration integrated in wastewater handling units.

These systems differ widely with regards to configuration, operational conditions, level of sludge solubilisation and energy consumption. In the case of low energy applications the floc disaggregation is moderate and this may cause a worsening of sludge settleability, but not in all the cited techniques. In the case of sludge with a high presence of filamentous bacteria (bulking phenomena and foaming) the settleability may be improved due to the separation of structures and bridges among filaments.

5.4 ULTRASONIC DISINTEGRATION
(See also Chapter 10)

The ultrasonic disintegration treatment consists of an ultrasound generator operating at frequencies of 20–40 kHz and in a device, which usually is a sonotrode, to transmit mechanical impulses to the bulk liquid. In the application of ultrasounds, pressure waves lead to cavitation bubbles forming in the liquid phase, which grow and then implode releasing localised high energy (local heating and high pressure), which cause sludge disintegration and, at high energy, the rupture of microbial cells.

The basic mechanism of ultrasonic disintegration is cell lysis-cryptic growth. Since the most important mechanism of ultrasonic disintegration is ultrasonic cavitation, it is advantageous to apply ultrasounds at low frequencies and at high energy levels.

In the scheme of ultrasonic disintegration integrated in the wastewater handling units, a part of the return sludge is treated continuously or in batch mode in a contact reactor equipped with sonotrodes (Figure 5.5). The subsequent biodegradation of lysate is completed in the activated sludge stage. Among the mechanical disintegration systems, sonication is the most energy hungry. A pre-thickening unit before the ultrasonic disintegration is advisable, to operate at higher solid concentration, which allows energy consumption to be reduced. Although several full-scale applications already exist of ultrasonic disintegration integrated in the sludge handling units, the application in activated sludge systems is rarer, due to economic reasons.

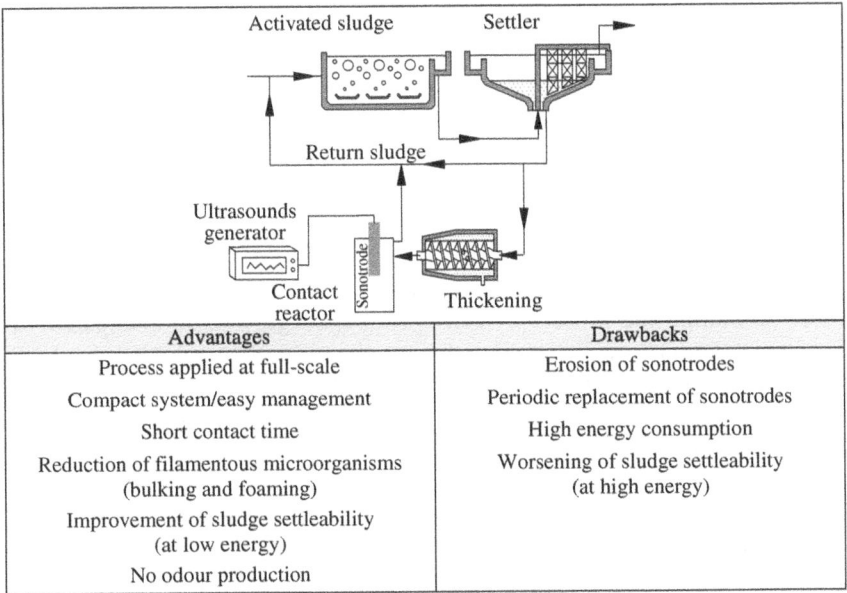

Advantages	Drawbacks
Process applied at full-scale	Erosion of sonotrodes
Compact system/easy management	Periodic replacement of sonotrodes
Short contact time	High energy consumption
Reduction of filamentous microorganisms (bulking and foaming)	Worsening of sludge settleability (at high energy)
Improvement of sludge settleability (at low energy)	
No odour production	

Figure 5.5. Ultrasonic disintegration integrated in wastewater handling units.

5.5 THERMAL TREATMENT
(See also Chapter 11)

The application of a thermal treatment of sludge (by heating sludge) produces disaggregation of sludge flocs, high level of solubilisation, cell lysis and release of intracellular bound water. The main parameter for thermal treatment is temperature, whilst the duration of treatment has generally less influence, expecially when high temperatures (>100°C) are applied. Several investigations confirmed that the highest sludge solubilisation is obtained around 180°C and higher temperatures do not causes appreciable increase of sludge biodegradability which even may decrease, due to the formation of refractory compounds linked to Maillard reactions. However, also the thermal treatment at T<100°C, integrated in the activated sludge stages, causes a significant reduction of excess sludge production, directly linked to an immediate decrease of biological activity and an increase of maintenance requirement. For sludge reduction by thermal treatment, the sludge is heated by steam and/or by heat exchangers prior to enter a contact reactor; then the lysated sludge is recirculated in the activated sludge system (Figure 5.6). Thickening before thermal treatment is advisable because higher solid concentration in thickened sludge means a save in energy and contact reactor volumes. SVI decreases with the rise in temperature, since the solubilisation of EPS (hydrated compounds able to absorb huge quantity of water) causes the release of a part of linked water.

5.6 CHEMICAL AND THERMO-CHEMICAL HYDROLYSIS
(See also Chapter 12)

Chemical or thermo-chemical treatments are based on alkaline or acid reagents. Coupling an increase in temperature with a strong change in pH, cell breakage occurs promoting the process of cell lysis-cryptic growth. As compared with the simple thermal treatment, the thermo-chemical treatment has an higher efficiency in sludge solubilisation when applied at the same temperature, however giving additional costs for reagents. Alkaline reagents, such as NaOH, are considered to be more efficient than the acids (HCl or H_2SO_4) and NaOH is effectively the most use reagent. Optimal conditions to induce sludge solubilisation and reduce costs are pH>10, temperature>50–60°C, contact time less than 1 h, since longer time do not improve solubilisation effectively. After the hydrolysis in a contact reactor, the lysate is recirculated in the activated sludge stages for further biodegradation (Figure 5.7).

Overview: techniques integrated in the *wastewater handling units* 51

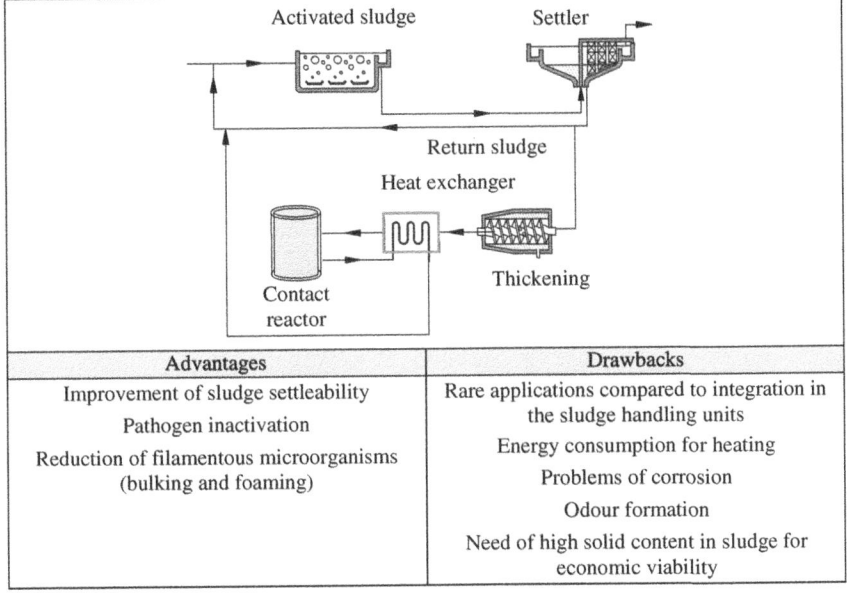

Advantages	Drawbacks
Improvement of sludge settleability	Rare applications compared to integration in the sludge handling units
Pathogen inactivation	Energy consumption for heating
Reduction of filamentous microorganisms (bulking and foaming)	Problems of corrosion
	Odour formation
	Need of high solid content in sludge for economic viability

Figure 5.6. Thermal treatment integrated in wastewater handling units.

The integration of a thermo-chemical treatment in the wastewater handling units at full-scale is rare and as far as we are aware, it is difficult to find successful results in the literature. The reason is the unfavourable economic balance of this application due to the high energy requirement for heating a low-concentrated sludge and the increased reagent dosage.

5.7 OXIDATION WITH OZONE (OZONATION)
(See also Chapter 13)

The advanced oxidation processes consist of the use of ozone (this section), hydroxide peroxide or chlorine (see § 5.8) and the combination of various oxidants. These oxidative treatments combined with biological degradation have been demonstrated to be very efficient in sludge reduction, but, generally, the main limitation is their economic feasibility.

The treatment based on ozone (ozonation) for sludge reduction has been proposed since the mid '90s, initially integrated in the wastewater handling units. To date, sludge ozonation has been successfully applied at full-scale both in industrial and municipal WWTPs. Sludge ozonation causes floc disintegration, cell lysis, organic matter solubilisation and, to a lesser extent, a partial subsequent oxidation of solubilised organics to carbon dioxide (mineralisation).

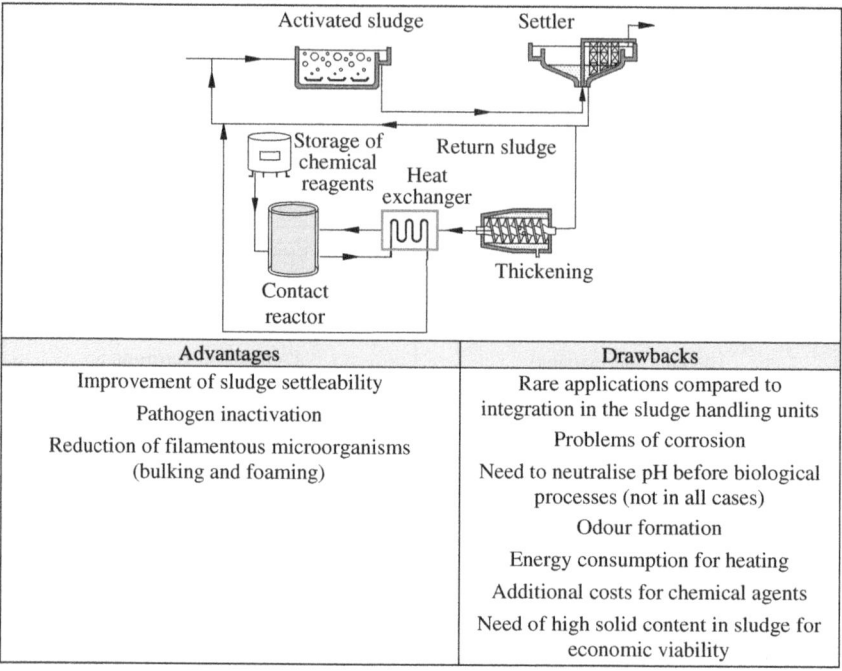

Figure 5.7. Thermo-chemical hydrolysis integrated in wastewater handling units.

Ozonation can be applied to a part of the return sludge or directly to the sludge taken from the activated sludge tanks. The ozonated sludge is then recirculated in the activated sludge stages where cryptic growth occurs at the expense of the released biodegradable organic matter (Figure 5.8).

The recommended ozone dosage is in the range 0.03–0.05 $gO_3/gTSS_{produced}$, which is appropriate to achieve a balance between sludge reduction efficiency and cost. At the moment, sludge ozonation results economically sustainable for WWTPs with large capacity or in the areas where sludge disposal costs are very high, or in the case of operational problems such as sludge foaming and bulking.

Overview: techniques integrated in the *wastewater handling units* 53

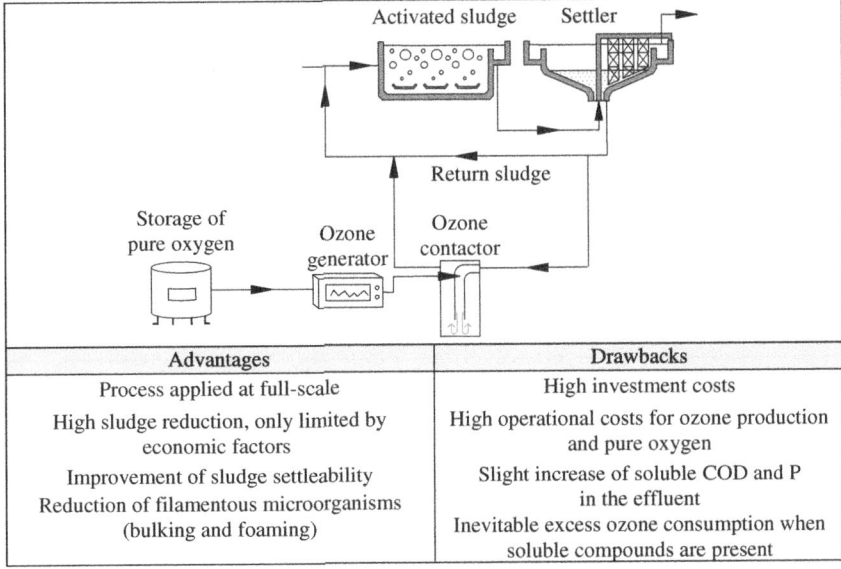

Figure 5.8. Ozonation integrated in wastewater handling units.

5.8 OXIDATION WITH STRONG OXIDANTS (DIFFERENT FROM OZONE)

Beyond ozonation (see § 5.7), the advanced oxidation processes include also the use or other strong oxidants, such as hydroxide peroxide or chlorine (this section). An advantage is that hydrogen peroxide and chlorine requires less expensive equipment and less qualified personnel compared to ozonation.

The combination of a thermal treatment allows a higher efficiency of peroxidation to be reached, due to the synergistic effect between temperature and H_2O_2. In fact, catalase – which is a terminal respiratory enzyme present in all aerobic living cells, able to break down H_2O_2 to protect cells from damage caused by reactive oxygen species – is active at low temperature (15–45°C) and gradually lost activity beyond 60°C (Wang et al., 2009). In the configuration of peroxidation, a portion of the return sludge undergoes heating (up to 95°C) and oxidation in a contact reactor with hydrogen peroxide (Paul and Sahli, 2006). H_2O_2 (aqueous solution, 6%) was added continuously for a contact time of 1 h, to reach dosages of up to 0.5 $gH_2O_2/gTSS$ (Camacho et al., 2002b). Then the lysate is recirculated to the activated sludge reactors and the process is based on the cell lysis-cryptic growth mechanism (Figure 5.9). TOC solubilisation increased linearly with H_2O_2 dosage (specific ratio of 20 mgTOC$_{released}$/molH$_2$O$_{2,consumed}$. Mineralisation of organic matter was observed above 0.34 gH$_2$O$_2$/gTSS (Camacho et al., 2002b). The peroxidation integrated in an activated sludge process fed with real wastewater (pilot-scale) using dosages of 0.12 gH$_2$O$_2$/gTSS for a contact time of 150 min at 93°C (stress frequency 0.17/d) led to a TSS reduction of 50% (Paul et al., 2006b). Wang et al. (2009), coupling H_2O_2 + microwave treatment (to heating sludge to 80°C) indicated a range of 0.1–1.0 gH$_2$O$_2$/gCOD of sludge to optimise sludge disintegration and costs of H_2O_2.

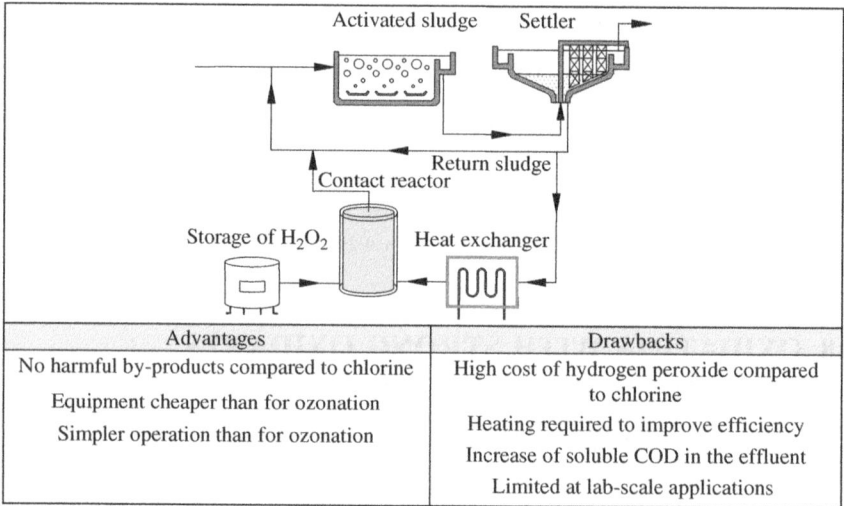

Figure 5.9. Peroxidation integrated in wastewater handling units.

Conventional inorganic oxidising agents such as chlorine and hypochlorite can be theoretically used for sludge solubilisation, cell lysis and thus sludge reduction.

In the chlorination treatment, a portion of the return sludge is treated with Cl_2 and the equipment needed is minimal (Figure 5.10). Chlorination was applied at lab-scale at dosages of 0.066 gCl$_2$/gTSS$_{produced}$ per 1 min (per day) to an activated sludge process fed with synthetic wastewater, obtaining a sludge reduction of 65% (Saby et al., 2002). No mineralisation is observed. At high dosages chlorination may worsen the biological process efficiency, and especially nitrification due to disinfection effect. Chlorine is a weak oxidant as

Overview: techniques integrated in the *wastewater handling units* 55

compared to ozone, requiring higher dosages, but, although the chlorination may be a cost-effective treatment, its application in sludge reduction is limited by the formation of harmful by-products (trihalomethanes, THMs). A concentration of THMs less than 200 ppb was detected in experiments at lab-scale by Chen *et al.* (2001b) using 0.066 $gCl_2/gTSS_{produced}$, but probably volatilisation of THMs during chlorine treatment occurred. Soluble COD concentration in the effluent from the biological stage resulted higher than that in the effluent of the reference system without the sludge chlorination treatment. Up to now, the use of chlorination for sludge reduction remains affected by these operational problems, which cause difficulties for the application in conventional WWTPs. As far as we know, full-scale applications are not reported in the literature.

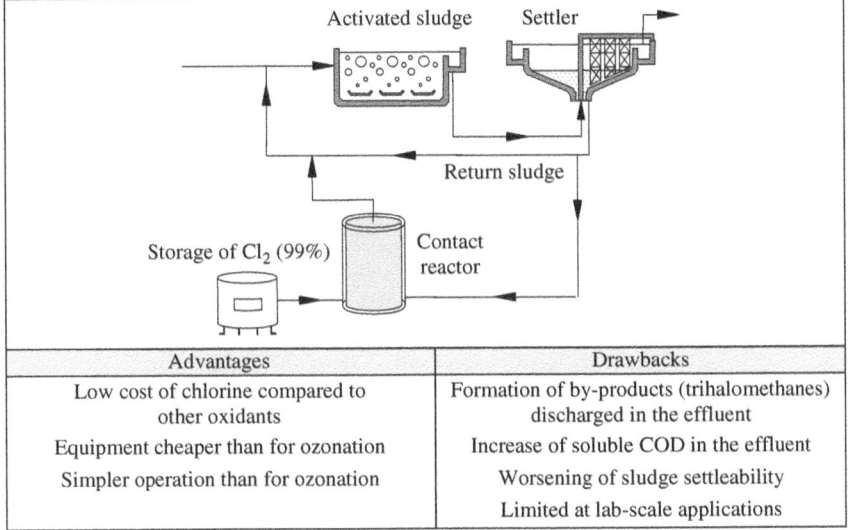

Advantages	Drawbacks
Low cost of chlorine compared to other oxidants	Formation of by-products (trihalomethanes) discharged in the effluent
Equipment cheaper than for ozonation	Increase of soluble COD in the effluent
Simpler operation than for ozonation	Worsening of sludge settleability
	Limited at lab-scale applications

Figure 5.10. Chlorination integrated in wastewater handling units.

5.9 ELECTRICAL TREATMENT

The pulsed electric field (PEF) technology has been widely used in biology, medicine and food applications for years. Recently, it has been proposed for treating sludge, due to the breakage effect on microbial cells. A PEF device sends high-voltage (>20 kV) electrical pulses thousands of times per second across the medium (sludge). The strong electrical field attacks the phospholipids and the peptidoglycan – the main constituents of cell membranes and walls – which are negatively charged and thus susceptible to electrical field action. The consequent pore openings in the cell membranes and cell walls causes rupture of the cells and lysis. As well as cell lysis, PEF breaks down large aggregates, and may reduce complex organic molecules to simpler forms. Due to the very strong electric field, which pulses on and off, the treatment time is in milliseconds.

The PEF treatment of sludge in the wastewater handling units was applied by Heinz (2007), treating part of the return sludge which is then recirculated in activated sludge stages. In this configuration, thickening of sludge before PEF is advisable in order to increase the solid concentration up to 60 gTSS/L, while maintaining pumpability. The addition of a heat exchanger prior to the PEF treatment allows the sludge temperature to be increased up to 35°C, which enhances the effect of the electrical pulses. The PEF treatment causes a further temperature increase of about 20°C in the treated sludge and this surplus energy can be used in the heat exchanger to heat inlet sludge from ambient temperature towards 35°C. Applying a specific energy of 100,000 kJ/m^3 the TSS reduction was 27–45% depending on the type of sludge. However, this application is still at pilot-scale and further investigation is needed to fully understand the process.

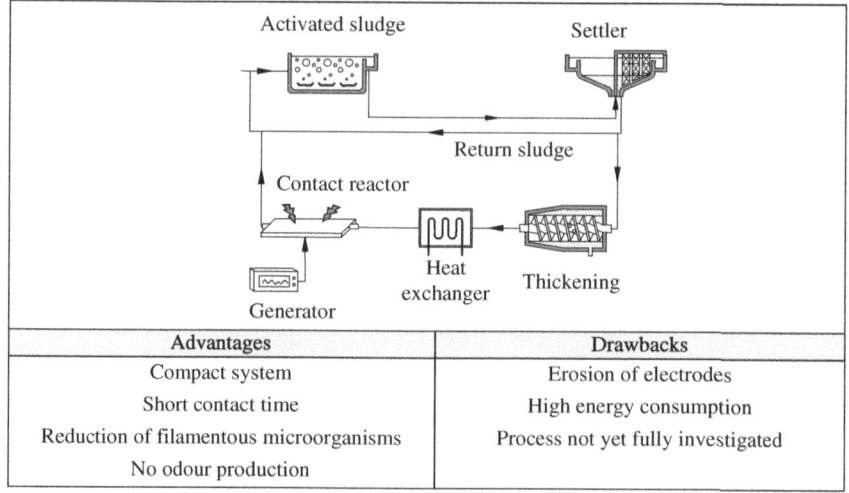

Advantages	Drawbacks
Compact system	Erosion of electrodes
Short contact time	High energy consumption
Reduction of filamentous microorganisms	Process not yet fully investigated
No odour production	

Figure 5.11. Electrical treatment integrated in wastewater handling units.

5.10 ADDITION OF CHEMICAL METABOLIC UNCOUPLERS

(See also Chapter 8, Section 8.8)

The process of uncoupled metabolism has been studied since the '90s for the reduction of sludge production. It can be obtained by using chemical metabolic uncouplers such as chlorinated and nitrated phenols, or 3,3',4',5-tetrachlorosalicylanilide (TCS). These molecules diffuse relatively freely through the phospholipid bilayer with a transport rate proportional to the concentration gradient across the cell membrane. Once inside the membrane, the phenolic hydroxyl dissipate the proton gradient which is a driving force for ATP production, resulting in the dissociation between anabolism and catabolism. High concentrations of these metabolic uncouplers are needed to favour a higher energy dissociation and the consequent cell growth reduction. The addition of metabolic uncouplers does not block electron transport along the respiratory chain to oxygen, and therefore the efficiency of substrate removal may remain good in most cases; however some compounds cause a reduction of substrate removal up to 26%.

The process is based on the simple addition of the chemical metabolic uncouplers to the wastewater handling units; the effect of the compounds added occurs directly during contact with activated sludge (Figure 5.12). Efficiency in the reduction of sludge production depends on the type of compound added and on the dosage.

However, little is known about the effect over long periods in which acclimatisation could play a role, or about the optimal conditions for the process or on the potential negative side-effects caused by these compounds.

5.11 SIDE-STREAM ANAEROBIC REACTOR (AT AMBIENT TEMPERATURE)

(See also Chapter 8, Section 8.4)

The integration of an anaerobic reactor (operating at ambient temperature) fed with part of the return sludge in an activated sludge process originates the *Oxic-Settling-Anaerobic* (OSA) process (Figure 5.13) in which a significant sludge reduction has been demonstrated since the beginning of the '90s. The cyclic alternation of aerobic/anaerobic conditions uncouples catabolism and anabolism, which causes a decrease in growth yield favouring sludge reduction. However, the mechanisms causing sludge reduction with the OSA system are not yet fully understood and also other explanations based on cell lysis-cryptic growth have been proposed.

58   Sludge Reduction Technologies in Wastewater Treatment Plants

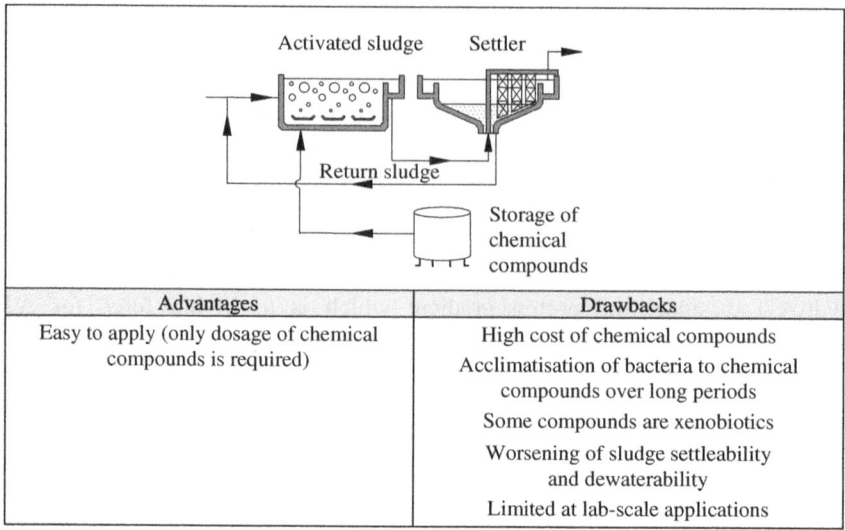

Figure 5.12. Addition of chemical metabolic uncouplers in wastewater handling units.

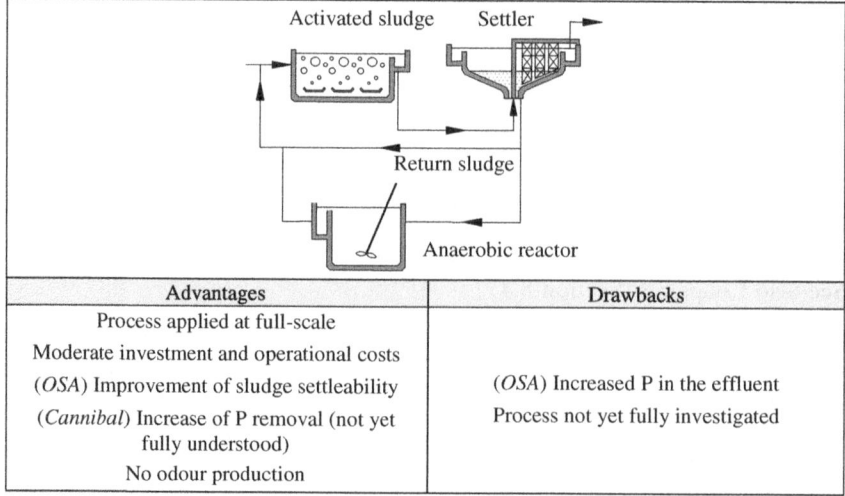

Figure 5.13. Side-stream anaerobic reactor integrated in wastewater handling units.

The process known as Cannibal®system is based on physical treatments (screening, hydrocyclones to separate inert solids) and a biological anaerobic reactor (interchange reactor). The introduction of this process in an existing WWTP requires the addition of a mixed tank, which has to operate without oxygen (only short periodic aeration), at a high biomass concentration, a sufficiently long retention time (SRT of 8–15 d), an interchange rate between 4% and 7%, without any wastewater feeding and maintaining a low redox potential (ORP), set approximately at −250 mV. A reduction of the observed sludge yield of 60% was demonstrated in pilot-scale plants fed with synthetic wastewater. Field operations indicate that the Cannibal process allows a significant reduction of sludge, but only a few studies have been conducted to fully understand and explain the basic mechanisms causing sludge reduction. Although the OSA system led to an increase of P in the effluent, the Cannibal system seems to contribute to P removal, but up to now the fate of P in the system remains enigmatic.

5.12 EXTENDED AERATION PROCESS
(See also Chapter 8, Section 8.10)

When an activated sludge system operates at a sufficiently long sludge age and low F/M ratio, such as in extended aeration processes, the excess sludge production is generally reduced, due to lower observed biomass yield which depends by SRT. Other processes occurs at long SRT and can affect sludge production: maintenance energy requirements, endogenous respiration, cell decay and grazing by predators.

Extended aeration processes require long aeration times and low applied organic loads, around 0.04–0.1 $kgBOD_5$ $kgTSS^{-1}$ d^{-1} compared to 0.08–0.15 $kgBOD_5$ $kgTSS^{-1}$ d^{-1} in conventional activated sludge processes. Therefore the oxidation volume results increased with respect to conventional activated sludge processes (Figure 5.14) working at F/M ratios and SRT in the typical ranges. Theoretical zero sludge production is not achievable even for very long SRT, because sludge contains always a part of inert solids, entering the system with influent wastewater, which accumulate in the sludge. However, in practice, a significantly lower quantity of excess sludge is produced compared to conventional activated sludge processes.

Drawbacks of the extended aeration processes – applied to reduce sludge production – may be the formation of pinpoint flocs characterised by poor settleability and the excessive costs for aerating larger oxidation tanks.

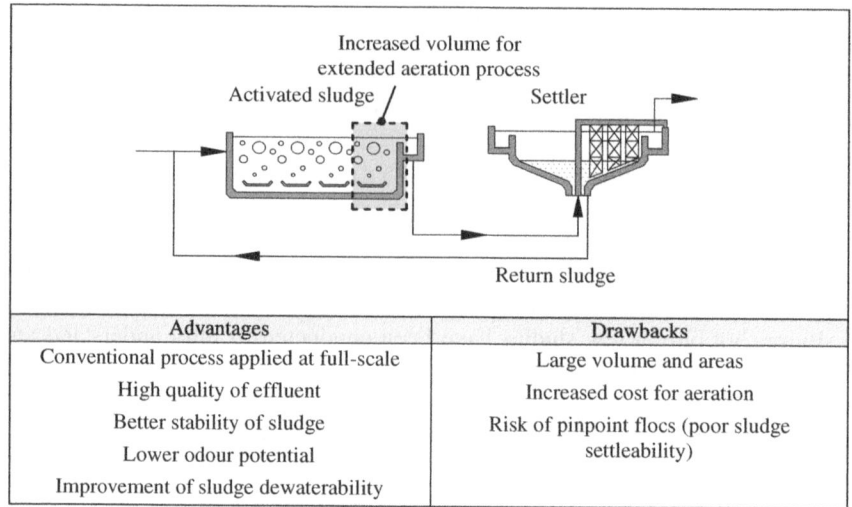

Advantages	Drawbacks
Conventional process applied at full-scale	Large volume and areas
High quality of effluent	Increased cost for aeration
Better stability of sludge	Risk of pinpoint flocs (poor sludge settleability)
Lower odour potential	
Improvement of sludge dewaterability	

Figure 5.14. Extended aeration process in wastewater handling units.

5.13 MEMBRANE BIOLOGICAL REACTORS
(See also Chapter 8, Section 8.11)

In MBR systems the separation of sludge and effluent takes place in a highly efficient membrane module rather than in a conventional gravity settler (Figure 5.15). The sludge production in the MBRs is usually expected to be lower than in conventional activated sludge processes, due to higher sludge concentration (7–20 gTSS/L), lower F/M ratio and longer SRT. This conditions do not favour cell growth causing increased energy maintenance requirements. A significant sludge reduction is expected especially at organic loads lower than 0.1 kgCOD kgSSV^{-1} d^{-1}, without worsening in COD removal efficiency and nitrification.

The feasibility of a MBR process with complete sludge retention was demonstrated and high removal efficiency and very limited sludge production was confirmed. However, in the practice, high TSS concentrations may have potential adverse effects on membrane, such as fouling, increased cleaning requirements, oxygen transfer limitations, increased sludge viscosity, worsening in sludge filterability and reduction of biological activity. Although MBRs permit to obtain high removal efficiency and very limited sludge production, complete sludge retention is less feasible in practice.

Overview: techniques integrated in the *wastewater handling units* 61

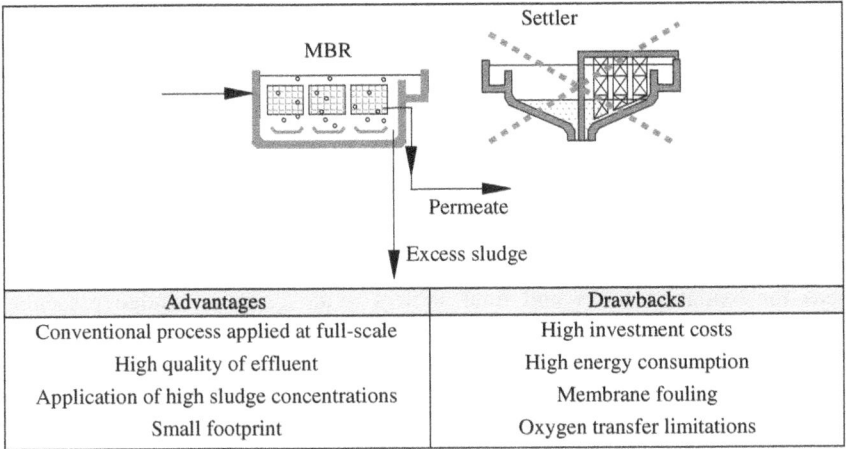

Advantages	Drawbacks
Conventional process applied at full-scale	High investment costs
High quality of effluent	High energy consumption
Application of high sludge concentrations	Membrane fouling
Small footprint	Oxygen transfer limitations

Figure 5.15. Membrane biological reactors in wastewater handling units.

A long SRT in MBRs may also be theoretically suitable for favouring the abundant growth of bacterial predators, enhancing mineralisation and further contributing to reduce sludge production, but opinions are divided. In some experiences the worm population introduced in the MBR was unstable, and worm growth and disappearance alternated. The integration of MBRs with some disintegration techniques (alkaline treatment, ozonation, ultrasonic disintegration) have also been proposed, with the aim to increase the decay rate of biomass, maintaining a relatively low TSS concentration in MBRs.

5.14 GRANULAR SLUDGE
(See also Chapter 8, Section 8.12)

Granular sludge systems, which are based on a self-immobilisation of microorganisms treating wastewater, permit to obtain some advantages compared to conventional activated sludge: (1) very high biomass concentration in the reactor (15–60 kgTSS/m^3); (2) very high treated organic loads, up to 10 kgCOD m^{-3} d^{-1}; (3) very low sludge production.

Up to now, the mechanisms which describe the formation of granules are not fully understood. The compact and dense aerobic granules present a strong microbial structure and generally good settleability. The observed sludge yield in granular sludge reaches only 0.07–0.15 kgTSS/kgCOD$_{removed}$ compared to conventional activated sludge systems which presents typical values of 0.27–0.35 kgTSS/kgCOD. The low sludge production is probably due to the endogenous metabolism and the high maintenance requirements. Denitrification occurs in the internal part of the granules and the energy available for anabolism is very low, resulting in a limited bacterial growth. In the configuration of the *Sequencing Batch Biofilter Granular Reactor* (SBBGR) shown in Figure 5.16. A submerged upflow filter is filled with aerobic granules and equipped with an external recirculation flow to obtain a homogeneous distribution of substrate and oxygen through the bed (Ramadori *et al.*, 2006). In the treatment of municipal wastewater the maximum organic load compatible with nitrification was 5.7 kgCOD m^{-3} d^{-1}. With respect to activated sludge – which require large areas for oxidation tanks and final settlers – the granular sludge presents a smaller foot-print.

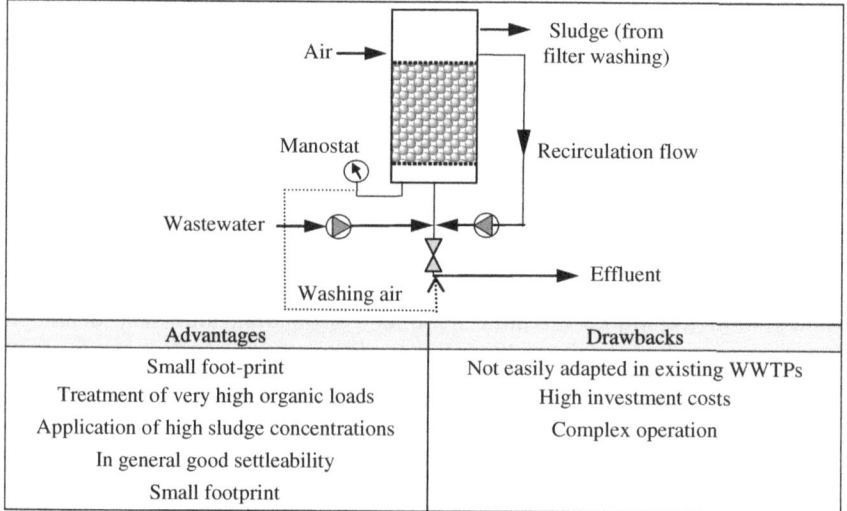

Advantages	Drawbacks
Small foot-print	Not easily adapted in existing WWTPs
Treatment of very high organic loads	High investment costs
Application of high sludge concentrations	Complex operation
In general good settleability	
Small footprint	

Figure 5.16. Granular sludge system in wastewater handling units.

5.15 MICROBIAL PREDATION
(See also Chapter 8, Section 8.9)

Protozoa and metazoa play an important role in activated sludge processes, thanks to their grazing effect resulting in a clearer effluent. Sludge reduction by predation is based on the loss of energy in the food chain and by the change of part of the sludge from a solid to a liquid or gas. The performance of metazoa in sludge reduction has attracted more attention compared to protozoa.

Despite efforts to control the growth and reproduction of predators in the biological systems, the conclusion is that it is very difficult to manage predators directly within activated sludge. Thus some experiences developed innovative predator-reactors separate from activated sludge stages in order to favour the growth of predators (Figure 5.17).

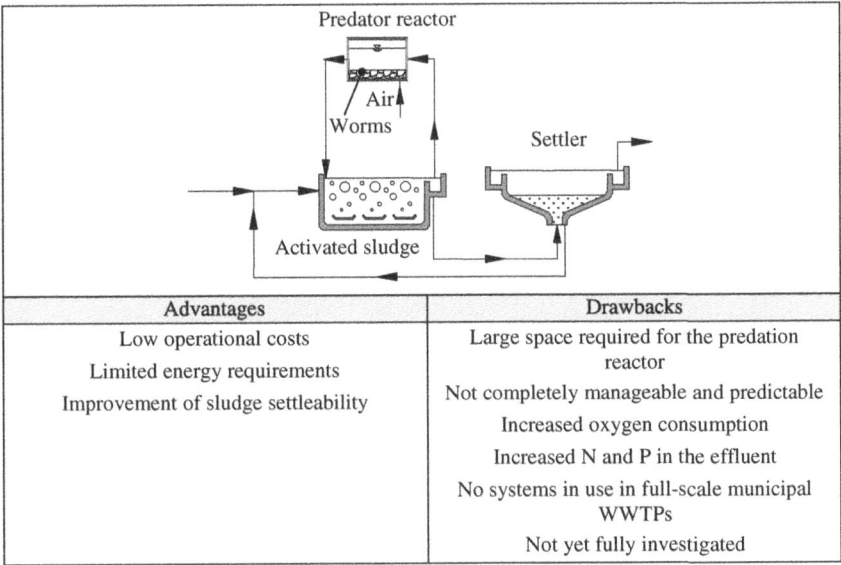

Figure 5.17. Predator-reactor integrated in wastewater handling units.

Predation can be achieved in a two-stage system: (1) a first aerobic stage (chemostat) with short HRT, which favours fast-growing dispersed bacteria, and (2) a second stage with a longer SRT to favour the growth of predators (activated sludge reactor, biofilm system or MBR). In practice, this two-stage system greatly increases the biological volume and its operational costs, and thus it is generally not feasible to apply for municipal wastewater. Alternatively, an additional specialised predation-reactor integrated in the wastewater handling units can be applied, suitable for the growth of predators. The feasibility at full-scale has not been fully tested in municipal wastewater with nutrient removal.

6
Overview of the sludge reduction techniques integrated in the *sludge handling units*

With the integration of sludge reduction techniques in the sludge handling units the reduction of sludge production is generally achieved while increasing the efficiency of sludge digestion. The aerobic or anaerobic digestion processes are mainly limited by hydrolysis which, since it is based on the endogenous metabolism, makes sludge biodegradation particularly slow. This limit can be overcome by enhancing the initial stage of hydrolysis through an additional treatment of sludge integrated in the aerobic or anaerobic digestion.

This may be achieved via processes based on the following mechanisms (Figure 6.1): (a) cell lysis, (b) endogenous metabolism, (c) microbial predation, (d) hydrothermal oxidation.

© 2010 IWA Publishing. *Sludge Reduction Technologies in Wastewater Treatment Plants.* By Paola Foladori, Gianni Andreottola and Giuliano Ziglio. ISBN: 9781789065305. Published by IWA Publishing, London, UK.

66   Sludge Reduction Technologies in Wastewater Treatment Plants

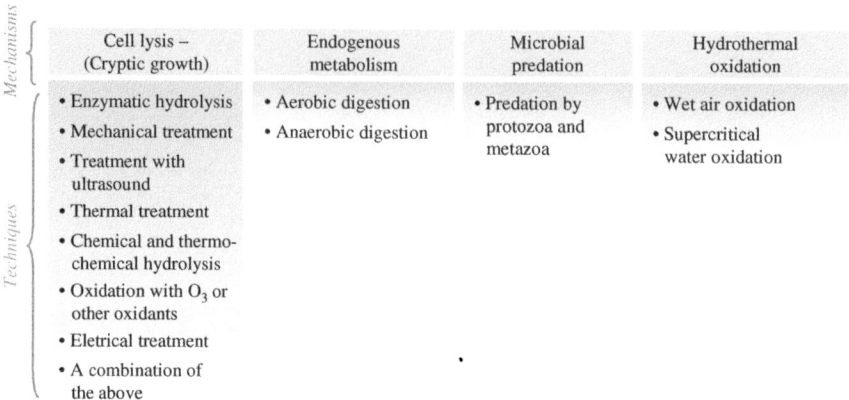

Figure 6.1. Scheme of the main sludge reduction techniques integrated in the sludge handling units.

The alternatives for carrying out the pre-treatment based on the *cell lysis* mechanisms are the following:

- enzymatic hydrolysis with/without added enzymes (§ 6.1 and to a less extent § 6.12, § 6.13);
- mechanical treatment (§ 6.2);
- treatment with ultrasound (§ 6.3);
- thermal treatment (§ 6.4, § 6.5);
- chemical and thermo-chemical hydrolysis (§ 6.6);
- oxidation with ozone or other strong oxidants (§ 6.7, § 6.8);
- electrical treatment (§ 6.9);
- a combination of the above.

The processes based on *endogenous metabolism* integrated in the sludge handling units are basically:

- aerobic digestion (§ 6.10);
- anaerobic digestion (§ 6.14).

We will not dwell on these two since the processes are both well known; reference will be made to them in § 8.2.1 and § 8.2.2 merely as regards certain specific details. Possible modifications to these units are:

– the application of alternating aerobic/anoxic/anaerobic conditions in conventional aerobic digestors (§ 6.11);

Overview: techniques integrated in the *sludge handling units* 67

- the application of thermophilic conditions (§ 6.12, § 6.13, § 6.15) to the digestor in order to enhance the endogenous metabolism or favour enzymatic hydrolysis.

The process based on *microbial predation* refers to the use of predators such as worms (§ 6.16).

The treatments referred to under *hydrothermal oxidation* are the following:

- wet air oxidation (§ 6.17);
- supercritical water oxidation (§ 6.18).

In the following sections, for each alternative techniques a short description, a flow diagram and a table indicating the advantages and drawbacks are given. A more detailed discussion of the most relevant technologies will be presented in the following chapters, where experimental results, efficiency and operational conditions of the various techniques will be addressed.

6.1 ENZYMATIC HYDROLYSIS WITH ADDED ENZYMES
(See also Chapter 8, Section 8.7)

Sludge solubilisation by enzymes is more intense the nearer the temperature is to the 50°C, considered optimal for hydrolitic enzyme activity. For this reason, the enzymatic treatment is best applied with anaerobic digestion in mesophilic or thermophilic conditions (Figure 6.2). Hydrolytical enzymes have the role to improve the hydrolysis step prior to acidogenesis, which is well known to be the rate-limiting step for the anaerobic digestion.

The treatment of sludge with various combinations of commercial protease, lipase, cellulase, hemicellulase, glycosidic enzymes has been tested for the enzymatic treatment + anaerobic digestion, with the aim to improve sludge reduction and enhance biogas production. The hydrolytic enzymes added in anaerobic reactors at lab-scale, demonstrated effectively good results in floc disintegration, significant reductions in EPS, better filterability and higher biogas production. Dewatering properties can be also improved, leading to a reduction of total sludge volume and a minor polymer dosage for dewatering. Although proteases are very effective in reducing sludge solids and in improving settling, when used glycosidic enzymes, they resulted in better solubilisation of the sludge than proteases and lipases.

Some types of cation binding agents can have a role to remove cations as Ca^{2+}, Mg^{2+} or Fe^{3+} from flocs, favouring the disruption of floc structure end enhancing the enzymatic activity during methanogenesys.

68   Sludge Reduction Technologies in Wastewater Treatment Plants

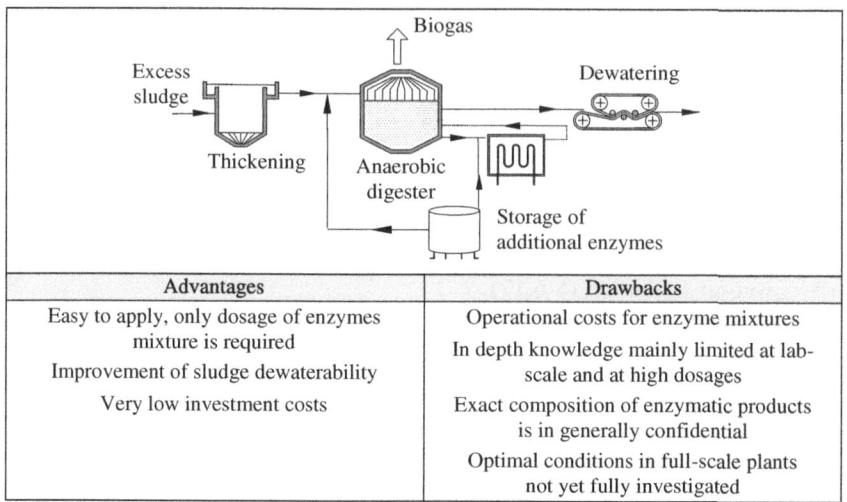

Figure 6.2. Enzyme addition in sludge handling units.

The installation of a dosage system for the enzymes represents the only investment cost, but the cost of enzyme mixtures has to be considered as the most important and sometimes limiting aspect for their use at full-scale.

6.2 MECHANICAL DISINTEGRATION
(See also Chapter 9)

Mechanical disintegration is often proposed as integration in the sludge handling units, which is a preferable configuration rather than the integration in the wastewater handling units. The reason is that it is more economic to operate with high concentrated sludge (thickened), in order to reduce the energy needed for disintegration. It is possible to insert a pre-treatment of sludge before anaerobic digestors or the treatment on a part of the sludge return flow (Figure 6.3).

The mechanical disintegration, due to the reduction of size and compactness of biological flocs, enhances the contact among bacteria, substrates and enzymes. Therefore the biodegradability of sludge is enhanced with a consequent increase in biogas production in anaerobic digestion.

With a high disintegration level the sludge dewatering performance is enhanced and a higher dry content is obtained respect to the untreated sludge. An increase of the flocculant dosage used for sludge conditioning and dewatering

Overview: techniques integrated in the *sludge handling units* 69

is observed, probably required for the neutralisation of superficial charges of colloidal particles increased in number after the mechanical treatment.

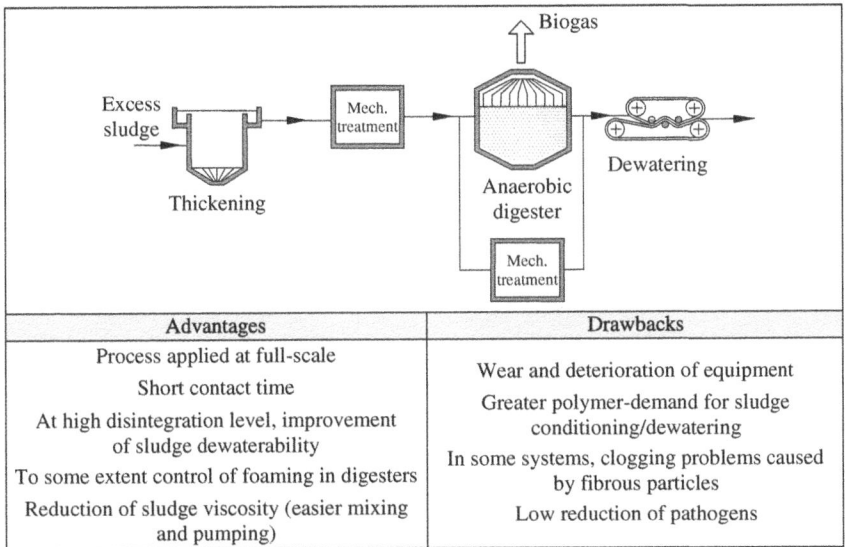

Advantages	Drawbacks
Process applied at full-scale	Wear and deterioration of equipment
Short contact time	Greater polymer-demand for sludge conditioning/dewatering
At high disintegration level, improvement of sludge dewaterability	In some systems, clogging problems caused by fibrous particles
To some extent control of foaming in digesters	Low reduction of pathogens
Reduction of sludge viscosity (easier mixing and pumping)	

Figure 6.3. Mechanical disintegration integrated in sludge handling units.

In some experiences, it was observed that foaming problems in anaerobic digesters – caused by filamentous microorganisms – can be reduced by adopting mechanical disintegration as pre-treatment.

6.3 ULTRASONIC DISINTEGRATION
(See also Chapter 10)

Ultrasonic disintegration is integrated in the sludge handling units as a pre-treatment before anaerobic digestion to obtain sludge reduction and an increase in biogas production (Figure 6.4). The application in the sludge handling units is more effective because the higher TS concentrations in treated sludge. In the presence of a higher TS concentration in the bulk liquid, a higher number of cavitation sites occurs and the probability that the solids come into contact with exploding cavitation bubbles increases. However, in practice, it is not be feasible to use too high TS concentrations, because there is a risk of overheating, sonotrode erosion and plant breakdown.

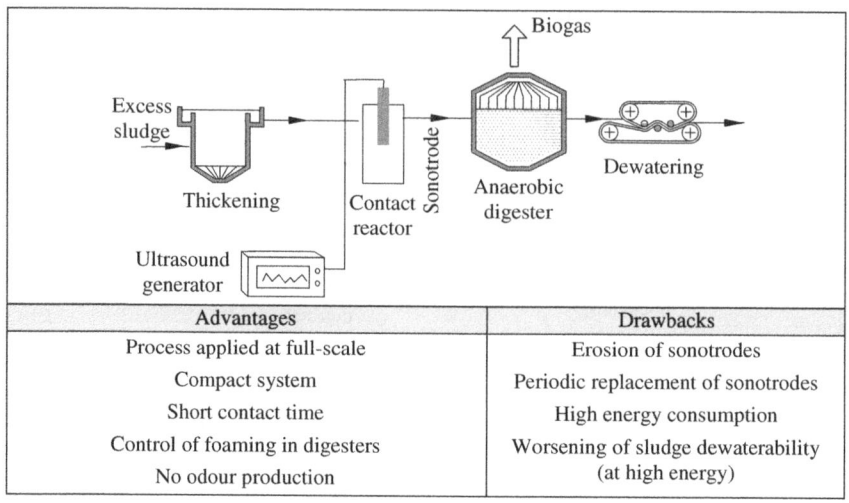

Figure 6.4. Ultrasonic disintegration integrated in sludge handling units.

At the energy levels usually applied in practice, sonication causes floc size reduction without damaging cells, because damage/death of bacteria requires too much energy (>60.000 kJ/kgTSS). However, the increase in the specific surface of sludge flocs favours contact among bacteria, substrates and enzymes, enhancing the overall sludge biodegradability and biogas production, as demonstrated by several full-scale applications as a pre-treatment before anaerobic digestion. The ultrasonic disintegration requires very high energy, but this drawback is offset by the ease of installation, the simple management and the compactness.

Reducing the specific energy applied in the pre-treatment before anaerobic digestion to more sustainable values, the benefit related to the VS mass reduction is reduced, but the increase in biogas production is always observed.

6.4 THERMAL TREATMENT
(See also Chapter 11)

The thermal treatment is integrated in the sludge handling units with the aim to (a) reduce sludge production, (b) enhance biogas production in anaerobic digesters and (c) obtain pathogen inactivation, (d) improve sludge dewaterability (Figure 6.5).

The integration of thermal treatment in the sludge handling units is generally preferred with respect to the integration in the wastewater handling units, due to the higher solid concentration (5–7% TS) in thickened sludge, resulting in a energy saving for heating and in a reducing reactor volume. The energy consumption for sludge heating can be covered by biogas production when thermal treatment is used as a pre-treatment before anaerobic digestion. The thermal treatment applied at full-scale is in general performed at temperatures in the range 160–180°C with short contact time around 0.5–1 h. Higher temperatures seem to limit the improvement of methane production due to the formation of slowly or hardly biodegradabile products. Full-scale plants of the combination of thermal pre-treatment + anaerobic digestion are Cambi process and BioThelys® process.

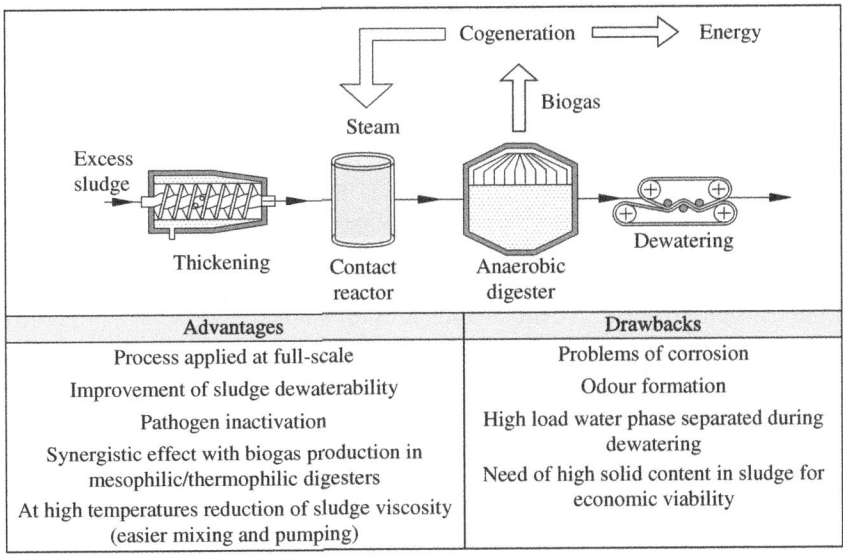

Figure 6.5. Thermal treatment integrated in sludge handling units.

Thermally hydrolysed sludge has a better dewaterability (resulting in a dry content of sludge cake up to 35%), lower viscosity and appears as a liquid even at solid content around 12%.

6.5 MICROWAVE TREATMENT
(See also Chapter 11, Section 11.8)

Microwave irradiation (2450 MHz) has been recently proposed in the field of WWTPs for heating sludge instead of conventional thermal methods. Interest in the use of microwaves is due to the significant reduction in reaction times and the expected reduction of energy requirements (as a result of much lower thermal losses in transferring energy). Microwave treatment has been proposed as a pretreatment before anaerobic digestion, aimed at the enhancement of digestion performance, the improvement of dewaterability and pathogen inactivation (Figure 6.6). Analogously to conventional heating, microwave application to sludge has the following effects: floc disintegration, cell lysis, EPS solubilisation. Disruption of the polymeric network of sludge flocs and cell lysis are caused by two effects: (1) thermal effect (heating), (2) athermal effect (orientation), also known as nonthermal or microwave effect. Due to the potential athermal effect, cells undergoing exposure to microwave treatment may be theoretically subjected to greater damage compared to cells conventionally heated to the same temperature. However, these two effects are not always distinguishable from each other, and up to now it is not yet clear which of them is prevalent. Applying 50–120°C, similar COD solubilisation was observed for the two types of treatment, resulting in no discernable microwave athermal effect, while for higher temperatures the conventional heating was more efficient. However, higher solubilisation with conventional heating may be due to the longer exposure times required to reach the desired temperature. However, the improvement of biogas production for microwave treated sludge seems to be related to an athermal effect on the mesophilic anaerobic biodegradability of sludge.

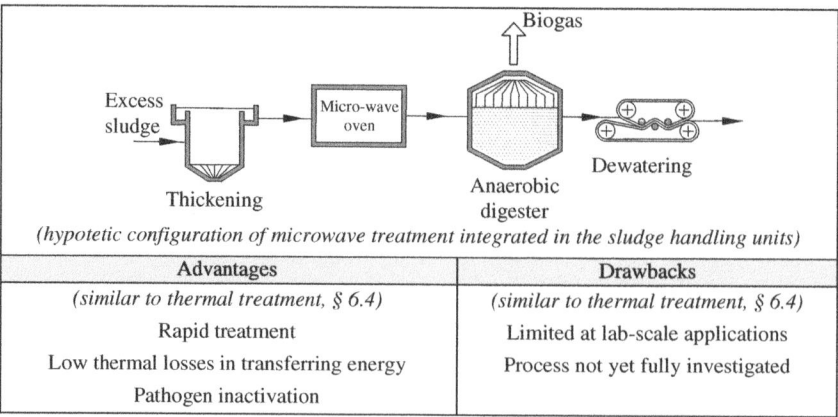

Figure 6.6. Microwave treatment integrated in sludge handling units.

Although data is available in the literature on sludge solubilisation, as far as we know, microwave technology has not yet been successfully applied at full-scale scale.

6.6 CHEMICAL AND THERMO-CHEMICAL HYDROLYSIS

(See also Chapter 12)

Acid or alkaline thermal hydrolysis is applied as: (1) pre-treatment prior to anaerobic digestion, to improve sludge biodegradation, reduce digester volume and enhance biogas production (Figure 6.7); (2) treatment of thickened sludge before dewatering, to reduce solids to be disposed of and increase solid content in dewatered cake. As in the thermal treatment, part of the heat used for the thermo-chemical pre-treatment can be recover for heating the mesophilic or thermophilic anaerobic process.

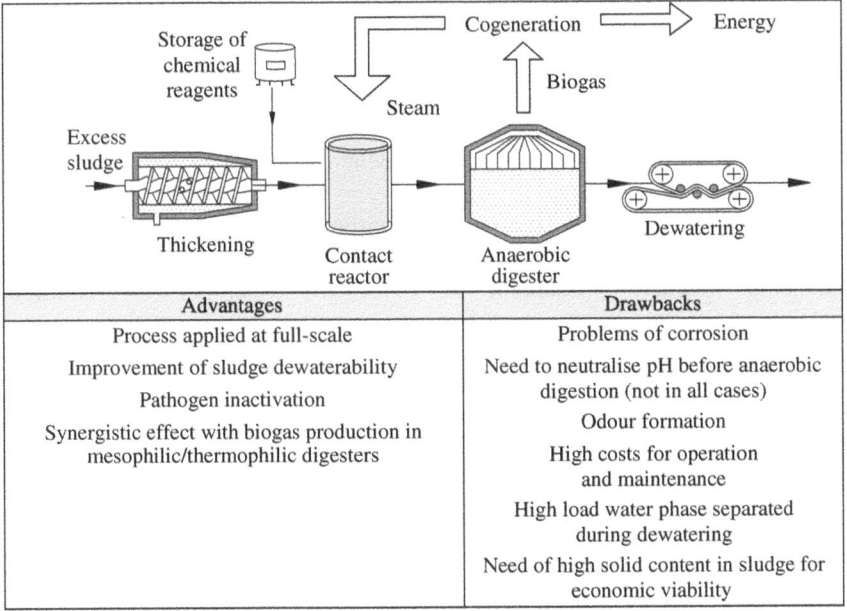

Advantages	Drawbacks
Process applied at full-scale	Problems of corrosion
Improvement of sludge dewaterability	Need to neutralise pH before anaerobic digestion (not in all cases)
Pathogen inactivation	Odour formation
Synergistic effect with biogas production in mesophilic/thermophilic digesters	High costs for operation and maintenance
	High load water phase separated during dewatering
	Need of high solid content in sludge for economic viability

Figure 6.7. Thermo-chemical hydrolysis integrated in sludge handling units.

In the configuration (1), since methanogenic bacteria activity and thus the production of methane is strongly affected by pH, which should be maintained in the optimal range of 6.6–7.6, pH neutralisation after strong acid/alkaline conditions may be needed. However, after sludge hydrolysis pH changes spontaneously and may becomes again suitable for biological processes.

Digestion of sludge pre-treated at temperatures up to 170–175°C resulted in an effective increased methane production, while higher temperatures (>180°C) do not lead to a further increase. Treatment at 100–120°C with acids or alkalis improves dewaterability of thickened sludge (5–6% TS content), leads to a reduction of the sludge mass to be dewatered (inducing solubilisation) and increases the dry content of sludge cake. The water phase separated from dewatering, rich in solubilised organic compounds, N and P, is recirculated to the wastewater handling units.

6.7 OXIDATION WITH OZONE (OZONATION)

The advanced oxidation processes applied to sludge reduction and integrated in the sludge handling units consists of the use of ozone (this section), hydroxide peroxide (see § 6.8) or oxygen at high temperature and high pressure (see § 6.17, § 6.18).

The efficiency of both anaerobic and aerobic stabilisation processes can be enhanced by integrating sludge ozonation. The partial solubilisation of sludge favours the hydrolysis of organic matter which is the main limiting factor of digestion processes.

The most widely used configuration is the integration of ozonation with the anaerobic mesophilic digesters with the aim of reducing sludge mass and enhancing methane production to cover the costs of the additional ozonation treatment. In this configuration, ozonation can be applied as: (1) a pre-treatment prior to the digester or (2) a post-treatment, operating on the return flow of digested sludge (Figure 6.8).

Although sludge filterability is generally deteriorated by ozonation, in some experiences it was observed that dewaterability was significantly enhanced after ozonation + anaerobic digestion, comparable with the dewaterability of untreated sludge. In this case a significantly lower production of dewatered sludge cake with lower water content can be obtained, contributing to savings in sludge disposal costs.

6.8 OXIDATION WITH STRONG OXIDANTS (DIFFERENT FROM OZONE)

Beyond ozonation (see § 6.7), another oxidative treatment (different from hydrothermal oxidation, see § 6.17, § 6.18) is peroxidation. The treatment with H_2O_2 require the combination with a thermal treatment to improve the efficiency. In this configuration, a portion of the digested sludge (taken from the return flow) passes through a heat exchanger before the oxidation in a contact reactor (Figure 6.9). The VSS reduction in the anaerobic digester (HRT = 30 d; T = 37°C) was 77% by applying peroxidation (2 $gH_2O_2/gSSV_{treated}$) + thermal treatment (90°C), compared to 52.8% when peroxidation is applied without thermal treatment, and 50.1% without any pre-treatments (Cacho Rivero et al., 2005). The increase of 26.9% in the VSS reduction demonstrates the efficacy of the combination with a thermal treatment. A 2 log reduction of fecal coliforms was also observed.

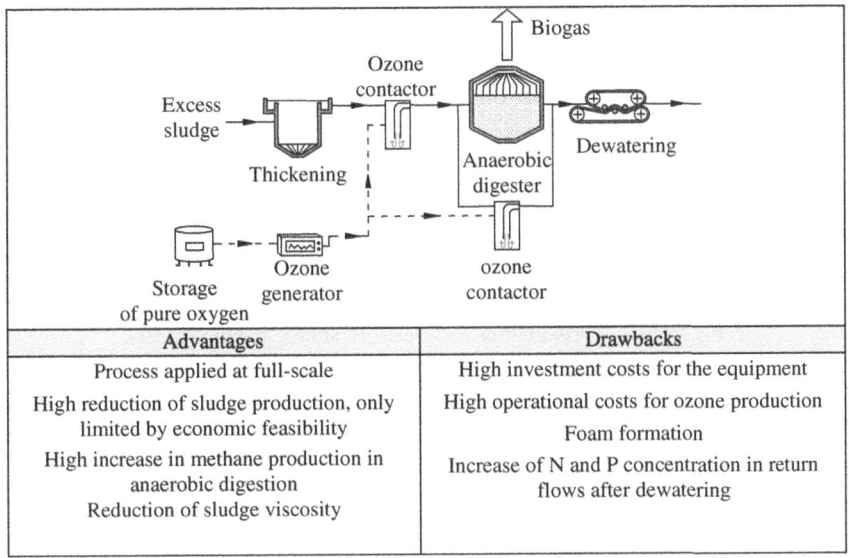

Figure 6.8. Ozonation integrated in sludge handling units.

H_2O_2 is also applied in combination with transition metal salts. The reaction obtained by using Fe(II)-salts in combination with H_2O_2 at very low pH (about 3) and ambient temperature and pressure, is known as Fenton's peroxidation. Although the basic mechanism of the Fenton's peroxidation is well known, the mechanisms to enhance sludge treatment is not fully understood. The ferrous iron catalyses the formation of highly reactive hydroxyl radicals that attack and destroy organic compounds. The result is sludge solubilisation, biodegradation increase, partial oxidation and enhancement of flocculation and dewaterability. Using dosages of 0.025 gH_2O_2/gTS and 1.67 $gFe^{2+}/kgTS$ at contact time of 60–90 min and pH 3, the amount of dry solids to be dewatered after peroxidation was 40% of the initial untreated amount. The performance of dewatering (as CST) was improved and led to dry content of 45% in filter cake compared to

25% in untreated sludge (Neyens *et al.*, 2004). COD load in the return flow resulted increased due to the solubilisation of organic compounds in the filtrate after dewatering.

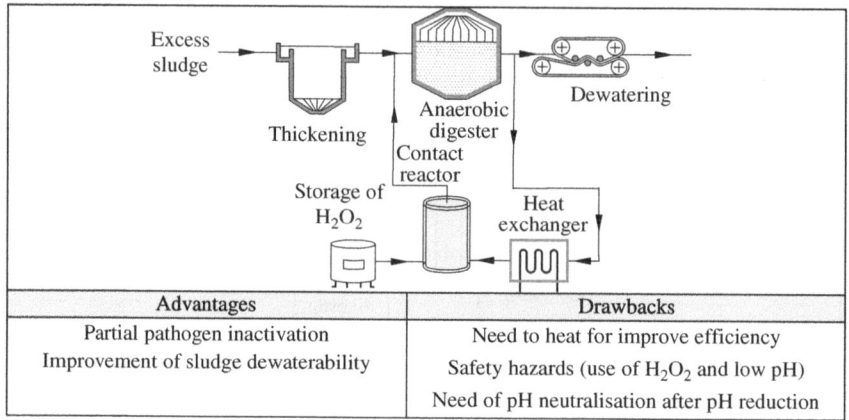

Figure 6.9. Peroxidation integrated in sludge handling units.

6.9 ELECTRICAL TREATMENT

The pulsed electric field (PEF), at 20–30 kV, has been proposed as a pre-treatment of sludge before anaerobic mesophilic digestion (Figure 6.10). Researchers have investigated PEF performance in this configuration on a small scale in recent years, and a full-scale application of a Focused Pulsed (FP) treatment – designed and adapted specifically for the treatment of organic solids (such as sludge) – was reported by Rittman *et al.* (2008). The objective of the PEF or FP pre-treatment is to improve VS reduction and biogas generation during the mesophilic anaerobic digestion.

The FP equipment consists of a cylindrical chamber through which the sludge flows. Electrodes at either end of the chamber are separated by an insulator that becomes an electric field when a voltage is applied. The focused pulsed unit stores energy in capacitors, then releases the energy across the electrodes in the form of an electrical pulse. The current flows from one electrode to the other via the sludge inside the chamber (Salerno *et al.*, 2009).

The treatment of waste activated sludge with FP technology causes:

- COD solubilisation and cell lysis
- disruption of a larger portion of the VSS, forming small colloids between 0.2 and 1.2 μm.

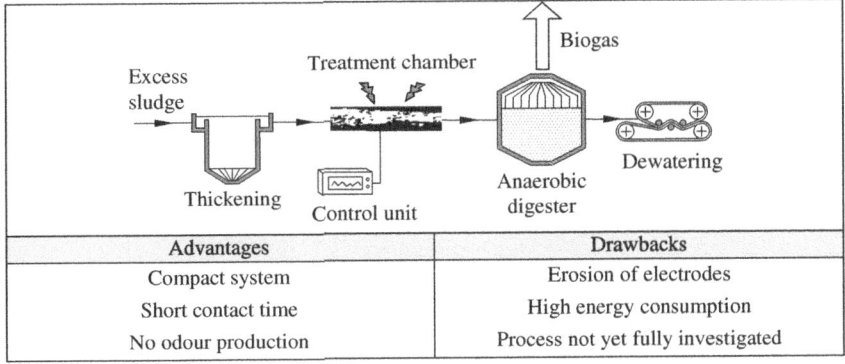

Figure 6.10. Electrical treatment integrated in sludge handling units.

This conversion of organic solids material produces soluble and colloidal particles which are more bioavailable during subsequent anaerobic digestion. One immediate benefit is an increase in methane production, up to 100% after 25–30 d of digestion, with an electric intensity up to 20 kWh/m^3 of reactor (residence time in the treatment chamber around 0.01 s). A threshold in the PF treatment of sludge may exist at approximately 10 kWh/m^3 of reactor.

The application of FP treatment at full-scale supports the conclusion that this type of pre-treatment before anaerobic digestion has the potential to significantly improve digester performance, potentially resulting in decreased residual biosolids, but this aspect has not yet been fully investigated.

6.10 AEROBIC DIGESTION
(See also Chapter 8, Section 8.2)

Conventional aerobic digestion with air is a widely applied process, especially in small-medium WWTPs. Aerobic digestion can be considered as an extension of the activated sludge process and the sludge reduction is originated by the endogenous metabolism. It is carried out normally at ambient temperature and the sludge reduction generally does not exceed around 30% of VSS even with long digestion time of up to 50 d. Aerobic digestion is managed in an open holding tank for a period ranging between 10 and 25 days, under aeration and mixing of the sludge, favouring the digestion of sludge to carbon dioxide and water (Figure 6.11).

Compared to anaerobic digestion, the aerobic digestion produces lower strength supernatant liquors and requires lower investment costs, but digested sludge has poorer mechanical dewatering characteristics and the process is significantly affected by temperature. Furthermore, no energy (biogas) can be recovered, while high energy is needed for air supply.

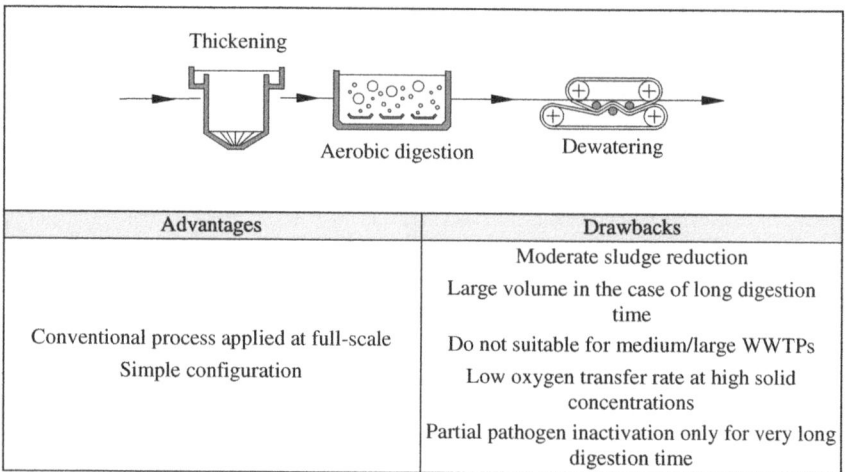

Advantages	Drawbacks
Conventional process applied at full-scale Simple configuration	Moderate sludge reduction Large volume in the case of long digestion time Do not suitable for medium/large WWTPs Low oxygen transfer rate at high solid concentrations Partial pathogen inactivation only for very long digestion time

Figure 6.11. Aerobic digestion in sludge handling units.

Improvements to conventional aerobic digestion include:

- the inclusion of alternating aerobic/anoxic/anaerobic phases to favour nitrogen removal (see § 6.11);
- modification of the process at mesophilic or thermophilic temperatures (see § 6.12, § 6.13).

6.11 DIGESTION WITH ALTERNATING AEROBIC/ ANOXIC/ANAEROBIC CONDITIONS

(See also Chapter 8, Section 8.3.2)

The conventional aerobic digestion process – in which the tank is continuously aerated – can be enhanced by including alternating aerobic/anoxic (and/or anaerobic) phases in order to favour nitrogen removal through the alternation of the nitrification/denitrification process (Figure 6.12). The aerobic/anoxic periods can be optimised by using probes for nitrogen forms (NH_4 and/or NO_3) or ORP/pH probes.

Theoretically the enhancement of sludge reduction is based on the fact that the heterotrophic growth yield assumes lower values under anoxic conditions, when compared to aerobic ones. Typically the anoxic growth yield (0.30–0.36 gVSS/gCOD) is 67–80% of the aerobic yield (0.45 gVSS/gCOD). The value of the growth yield influences the mass of bacterial biomass and therefore, from a theoretical point of view, it will affect sludge production. But the influence on sludge reduction is expected to be low, since bacterial biomass is only 15–30% of TSS of sludge.

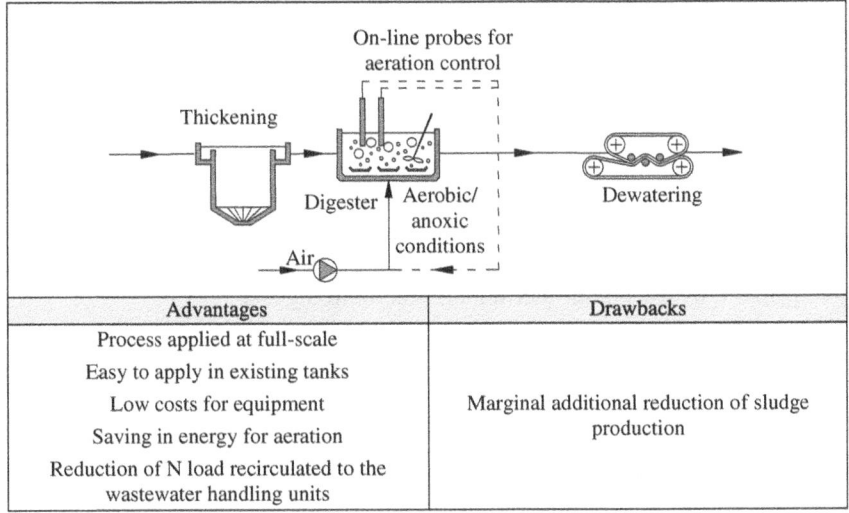

Figure 6.12. Digestion with alternating conditions in sludge handling units.

Several studies have investigated the performance of aerobic digestion under alternating aerobic/anoxic phases and a marginal additional reduction of sludge production is observed compared to continuously aerated digestion. Conversely, nitrogen removal is significantly improved, reducing the N loads recirculated to the wastewater handling units.

In conclusion, digestion under alternating aerobic/anoxic conditions produces similar sludge reduction rates but with lower operating costs.

6.12 DUAL DIGESTION
(See also Chapter 8, Section 8.6.2)

Aerobic digestion of sludge – usually at ambient temperature – can also be performed either at mesophilic (33–35°C) or thermophilic temperatures (55–65°C). Thermophilic aerobic processes have been proposed as:

- a pre-treatment before anaerobic digestion (this section)
- a part of the two-stage autothermal thermophilic aerobic digestion (see § 6.13).

In a thermophilic aerobic reactor, some strains of bacteria belonging to *Bacillus* spp. – facultative aerobes or microaerophiles – produce heat stable extracellular proteases which help to degrade the microbial cells in sludge by hydrolysing the cell wall compounds. Furthermore, this thermal treatment renders microbial cells in the sludge more susceptible to enzyme attack and favours their autolysis. Thermophilic treatment can be integrated in the sludge handling units as a preliminary stage before conventional mesophilic anaerobic digestion (Figure 6.13). In this configuration a high VFA accumulation – extremely useful to improve biogas formation in subsequent anaerobic digestion – can be achieved by keeping the sludge under microaerobic conditions. This combination is known as dual digestion. This process can be considered as based on the mechanism of endogenous metabolism in anaerobic digester and cell lysis by enzymatic hydrolysis in thermophilic aerobic digester.

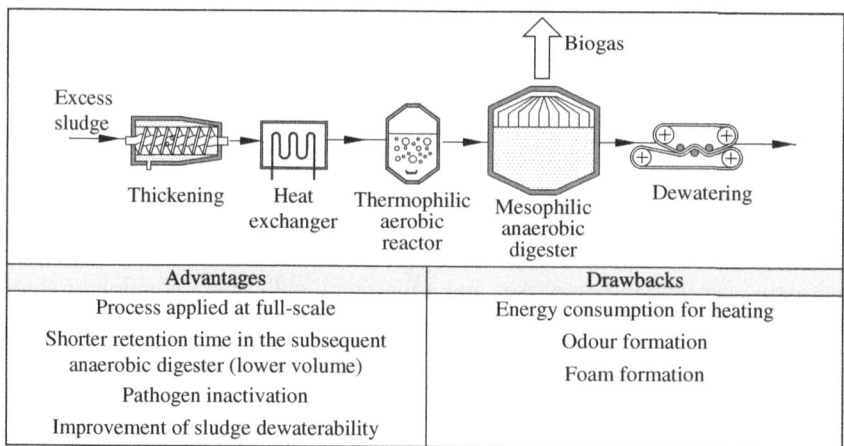

Advantages	Drawbacks
Process applied at full-scale	Energy consumption for heating
Shorter retention time in the subsequent anaerobic digester (lower volume)	Odour formation
Pathogen inactivation	Foam formation
Improvement of sludge dewaterability	

Figure 6.13. Dual digestion in sludge handling units.

The aerobic conditions in the thermophilic aerobic reactor can be reached by supplying air or pure oxygen. Temperatures are maintained in the range 55–65°C, which is optimal for thermophilic hydrolysis. The retention time in the thermophilic aerobic reactor is around (or less than) 1 d, while the retention time in the subsequent anaerobic digester is 10–12 d (shorter than typical retention time for non pre-treated sludge, which leads to a reduction of digester volume). This configuration also include increased level of pathogen inactivation.

6.13 AUTOTHERMAL THERMOPHILIC AEROBIC DIGESTION

(See also Chapter 8, Section 8.6.3)

The process known as autothermal thermophilic aerobic digestion (ATAD) can be considered as a variation of conventional aerobic digestion, but carried out at thermophilic temperatures (55–70°C) and with pure oxygen. These temperatures can be achieved by conserving the heat produced in the exothermal oxidation reactions in insulated reactors. Approximately 20,000 kJ of heat per 1 kg VS destroyed is produced by the process, and since no supplementary heat is provided, the process is "autothermal". The feed sludge is thickened (at high solid content, around 4–5%) to reduce reactor volume and sludge volume to be heated.

The configuration is composed by two-stages (Figure 6.14):

- the first stage operating under aerobic and mesophilic (or thermophilic) conditions, in the range 35–50°C (typical value 40°C);
- the second stage operating under aerobic and thermophilic conditions, in the range 50–70°C (typical value 55°C).

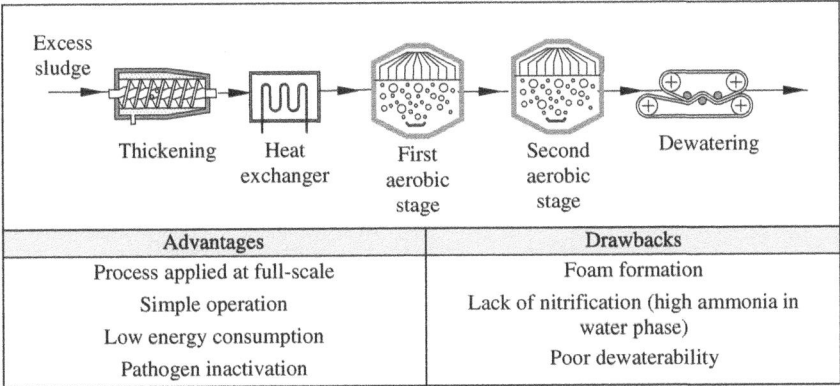

Advantages	Drawbacks
Process applied at full-scale	Foam formation
Simple operation	Lack of nitrification (high ammonia in water phase)
Low energy consumption	
Pathogen inactivation	Poor dewaterability

Figure 6.14. Autothermal thermophilic aerobic digestion in sludge handling units.

The two stages operate in series under microaeration and with retention time of 6–12 d, which is lower than the retention time in conventional aerobic and anaerobic digesters.

The ATAD process already has several applications in Europe and in the USA, installed in the last 2 decades, and is recommended for small to medium plants. An important operational factor is the aeration of the digesters, whose optimisation is crucial in order to avoid over-aeration (increase in costs, cooling effect) or under-aeration (low efficiency, odours). Automatic management of the aeration system using sophisticated devices is currently being introduced.

6.14 ANAEROBIC DIGESTION
(See also Chapter 8, Section 8.2)

Anaerobic digestion is the most widely used process to reduce both primary and biological excess sludge, preferred especially in larger WWTPs with capacity above 20,000–30,000 PE, in which the energy recovered by exploiting the methane gas produced becomes economically advantageous (Figure 6.15).

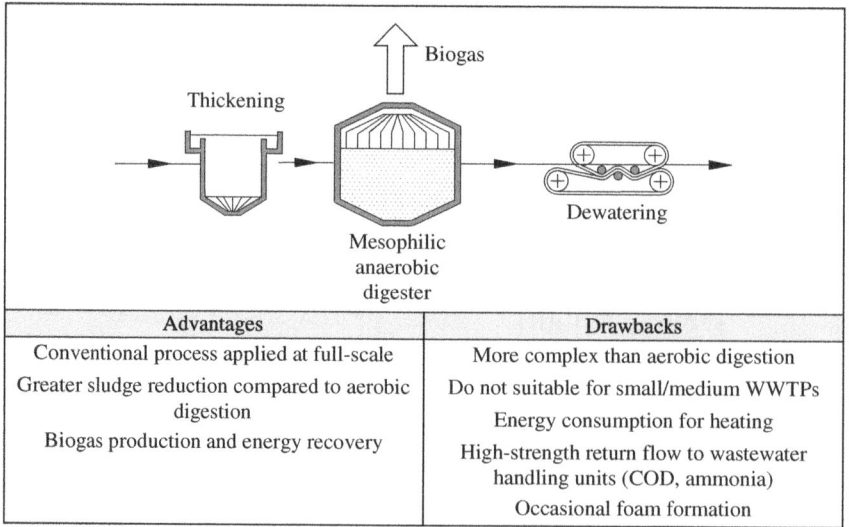

Advantages	Drawbacks
Conventional process applied at full-scale	More complex than aerobic digestion
Greater sludge reduction compared to aerobic digestion	Do not suitable for small/medium WWTPs
	Energy consumption for heating
Biogas production and energy recovery	High-strength return flow to wastewater handling units (COD, ammonia)
	Occasional foam formation

Figure 6.15. Anaerobic digestion in sludge handling units.

Anaerobic digestion is carried out under mesophilic conditions (33–35°C) or thermophilic conditions (53–55°C). Because the rate-limiting step is the enzymatic hydrolysis of solids, several efforts have been made to apply pretreatment methods (mechanical, thermal and/or chemical treatments, ozonation, etc...) to improve the hydrolysis rate, enhance sludge reduction and increase biogas production.

In conventional mesophilic anaerobic digestion about 40–45% of organic matter is degraded, usually leading to TSS reduction of around 30%.

Very high concentrations of COD and ammonia characterise the return flow to the wastewater handling units.

Recent research suggests the existence of floc destruction mechanisms which differ under aerobic and anaerobic conditions. Under anaerobic conditions, monovalent cations (such as potassium and ammonium) and a larger quantity of proteins are released compared to aerobic digestion. When sludge undergoes reducing anaerobic conditions, iron is reduced and solubilised, weakening the bond with proteins and causing a deflocculation of sludge, which contribute to the overall sludge reduction during digestion.

6.15 THERMOPHILIC ANAEROBIC DIGESTION
(See also Chapter 8, Section 8.5)

Thermophilic anaerobic digestion (Figure 6.16) enhances performance compared to mesophilic anaerobic digestion, obtaining an acceleration of biochemical reactions, higher biogas production rate, and a considerable pathogen inactivation. Methanogenic activity increases passing from mesophilic to thermophilic conditions, although the specific growth rate of thermophilic microorganisms is significantly lower than mesophilic ones, because of the greater maintenance energy required by thermophilic organisms.

Conventional mesophilic digesters are normally designed for a retention time of 20–30 d, while for thermophilic digestion a retention time of 10–12 d is sufficient. Thermophilic anaerobic digestion therefore uses more compact reactors. VSS reduction reaches 40–45%, corresponding to a total solid reduction of 30%, similar to mesophilic digestion, but obtained with shorter digestion time.

Despite the greater energy consumption, thermophilic digestion is a good option if the sludge solid content is higher than 3% (mechanically thickened sludge) and heat recovery from the effluent sludge is performed.

6.16 MICROBIAL PREDATION
(See also Chapter 8, Section 8.9)

Microbial predation by worms may lead to substantial and cost-effective sludge reduction, but up to now knowledge of the process is not fully complete and therefore the application of predators for sludge reduction at full-scale is still not entirely manageable and predictable. In the proposed configuration for treating excess sludge after production, an additional specialised predation-reactor can be integrated with the sludge handling units, using *Tubificidae* or *Lumbriculidae* for sludge reduction by transforming it into worm faeces (Figure 6.17). In new reactor concepts, the separation of the worms from the excess sludge was obtained by the immobilisation of *L. variegatus* in a carrier material, which allows the worms to be recovered as a valuable protein-rich product from the waste sludge.

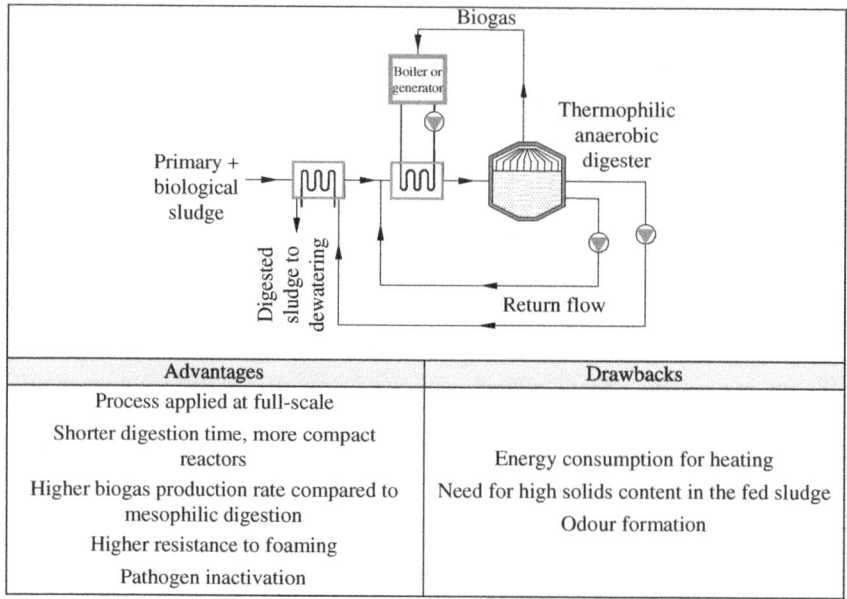

Advantages	Drawbacks
Process applied at full-scale	
Shorter digestion time, more compact reactors	Energy consumption for heating
Higher biogas production rate compared to mesophilic digestion	Need for high solids content in the fed sludge
Higher resistance to foaming	Odour formation
Pathogen inactivation	

Figure 6.16. Thermophilic anaerobic digestion in sludge handling units.

The application of this process is still at lab-scale; in full-scale applications, the need for large areas of the carrier material may affect economic feasibility.

6.17 WET AIR OXIDATION

Wet air oxidation (subcritical water oxidation) of sludge (Figure 6.18) performed at temperatures of 150–320°C and a pressure of 1–22 MPa is a technology able to drastically reduce the organic matter in sludge: over 80% of total COD can be oxidised, while the remaining part is mainly a soluble COD (both rapidly biodegradable and inert). Air or pure oxygen are supplied in the reactor. The oxidation reactions are exothermic. Typical retention times are around 1 h. The fractions produced are: (1) low impact exhaust gas; (2) mineralised solid fraction (carbon content <5%); (3) a high-strength liquid fraction, typically recirculated to the oxidation bioreactors of WWTP (containing 25% of COD in the sludge). Organic matter in the sludge (proteins, lipids, starch and fibres) is converted into simpler dissolved compounds (sugars, amino acids, fatty acids and ammonia). The organic nitrogen of sludge is converted into ammonia and thus has to be removed, for example, by stripping and conversion into atmospheric N_2 through a catalytic reactor. Wet air oxidation can treat sludge with a solid content between 1 and 6%. The wet air oxidation process is commercialised as the Zimpro® process (Siemens Water Technologies Corp.) or the ATHOS® process (Veolia Water).

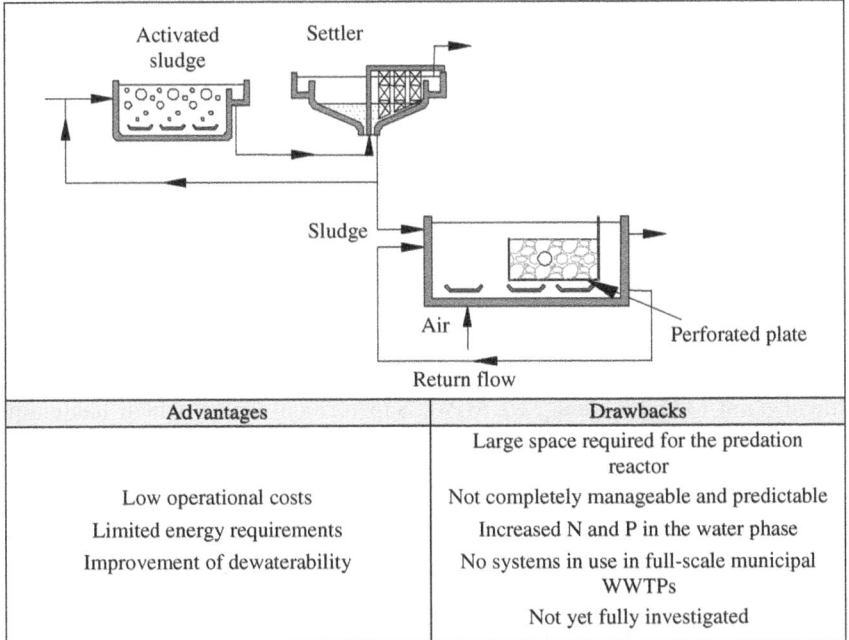

Advantages	Drawbacks
Low operational costs Limited energy requirements Improvement of dewaterability	Large space required for the predation reactor Not completely manageable and predictable Increased N and P in the water phase No systems in use in full-scale municipal WWTPs Not yet fully investigated

Figure 6.17. Predation-reactor integrated in sludge handling units.

Because these thermal processes are considered as final disposal routes since the organic matter is volatilised and an inert mineral residue is produced, they fall outside the scope of this book and therefore this technology will not be further described.

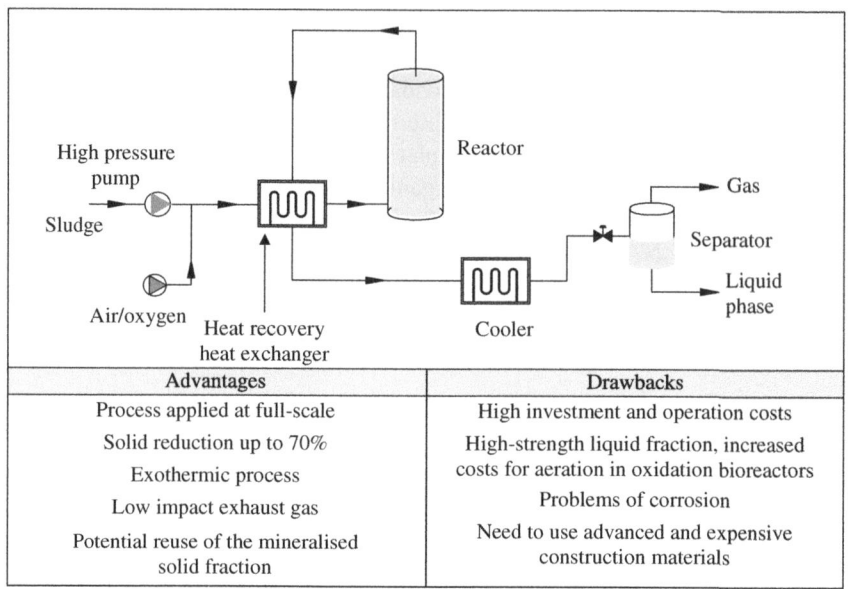

Figure 6.18. Wet air oxidation.

6.18 SUPERCRITICAL WATER OXIDATION

The state of supercritical fluid – which is a transition state between the liquid and gas phases – is reached by water at temperatures and pressures above the critical point (374.2°C and 22.1 MPa). Supercritical water, which has unique characteristics, some of which are highly useful for the hydrolysis and oxidation of complex organic waste streams, is an excellent solvent for organic molecules and reacts transforming them into small molecules. In the presence of oxidants (air or pure O_2), oxidation reactions occur readily, without diffusion limitations. When the supercritical water oxidation (SCWO) is applied to sludge (Figure 6.19), it undergoes an overall chemical transformation, forming H_2O, CO_2, and N_2 gas from organic compounds, and acids, salts and oxides from inorganic materials. SCWO can very effectively oxidize organics at temperatures of 400–650°C and pressures of 25–27 MPa.

SCWO is similar in principle to wet air oxidation, but in wet oxidation the liquid and gas phases remain separate and distinct, while reactions in SCWO occur in a single phase. In SCWO the complete oxidation of the organic matter in sludge can be achieved; conversely wet air oxidation can achieve effective hydrolysis of the sludge but incomplete oxidation. The investment cost is high, but off-set by the high sludge reduction. 80% of sludge can be oxidized and COD destruction efficiency reach 99%.

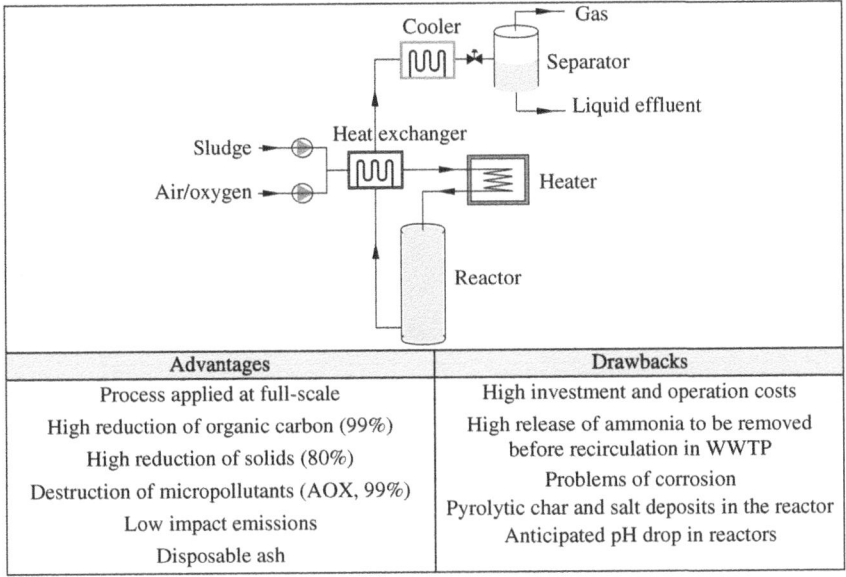

Advantages	Drawbacks
Process applied at full-scale	High investment and operation costs
High reduction of organic carbon (99%)	High release of ammonia to be removed before recirculation in WWTP
High reduction of solids (80%)	Problems of corrosion
Destruction of micropollutants (AOX, 99%)	Pyrolytic char and salt deposits in the reactor
Low impact emissions	Anticipated pH drop in reactors
Disposable ash	

Figure 6.19. Supercritical water oxidation.

Because these thermal processes are considered as final disposal routes, they fall outside the scope of this book and therefore this technology will not be further described.

7
Procedures for estimating the efficiency of sludge reduction technologies

> *"There is no simple batch test for predicting extent of sludge reduction"*
> *(Ginestet P., Camacho P., 2007)*

7.1 INTRODUCTION

One of the objectives in the European project EVK1-CT-2000-00050 (*W.I.R.E.S., Ways of innovation for the reduction of excess sludge*) was to determine whether and/or to what extent it is possible to predict sludge reduction from simple batch tests. The answer was that: *"There is no simple batch test for predicting extent of sludge reduction"*, confirming that long-term monitoring of pilot-plants is needed (Ginestet and Camacho, 2007).

The efficiency of sludge reduction techniques is often evaluated by experiments in pilot-plants and at lab-scale, under specific operational conditions, to simulate full-scale plant configurations and with the aim of

© 2010 IWA Publishing. *Sludge Reduction Technologies in Wastewater Treatment Plants.* By Paola Foladori, Gianni Andreottola and Giuliano Ziglio. ISBN: 9781789065305. Published by IWA Publishing, London, UK.

carrying out mass balances and assessing effective sludge reduction (or the observed sludge yield, Y_{obs}). The parameter Y_{obs} is the ratio between the amount of VSS (or TSS) produced and the amount of COD removed during the process. This parameter is widely used in assessment of sludge reduction.

Lab and pilot plants generally produce reliable, representative results, but they need long monitoring periods. This creates the need for quick tests to evaluate the efficiency of sludge reduction techniques, for example, by measuring COD solubilisation, inactivation of bacteria biomass, increase in sludge biodegradability, etc... The methods for determining these parameters are generally quick to perform – since they require at most a few hours to be completed – with resulting limited costs compared to long-term pilot-scale investigations.

It has to be underlined that a single batch test, such as COD solubilisation or bacteria inactivation, cannot – when considered alone – give a complete indication of the exact sludge reduction to be expected. For example, some treatments could produce a low COD solubilisation, but demonstrate a high VSS reduction when coupled with a biological treatment. Other treatments may give higher COD solubilisation, but moderate VSS reduction.

Certainly, the opportunity to use an array of simple and complementary batch tests could give more information on the expected sludge reduction and would also help to understand the mechanisms involved in the specific technology for sludge reduction.

For example:

(1) tests to evaluate COD and TSS solubilisation could give additional information about the transformation of particulate COD and solids (bacterial and non-bacterial fractions) in soluble COD (§ 7.2, § 7.3);
(2) tests to estimate sludge biodegradability, applied under aerobic, anoxic and anaerobic conditions, could predict the fate of the lysates in biological reactors (§ 7.4, § 7.5, § 7.6);
(3) tests to evaluate bacteria inactivation could give information about viability, death or damage to bacterial cells, but without contributing anything about the behaviour of the rest of the particulate COD which is not bacterial biomass (§ 7.7).

These tests, when considered singly, can not be exhaustive, but may provide useful information when used in appropriate combinations.

Simple batch tests are not available for all sludge reduction techniques; for example, in the evaluation of a side-stream anaerobic reactor integrated in the wastewater handling units, the above tests are not used, but lab- or pilot-plants are realised.

Procedures for estimating the efficiency of sludge reduction technologies 91

Furthermore, when pilot plants are used to evaluated sludge reduction, some operational parameters need to be considered or recalculated, such as:

- sludge retention time (§ 7.8)
- observed sludge yield (§ 7.9)
- sludge reduction in terms of solids (§ 7.10)
- treatment frequency (§ 7.11)
- physical properties of sludge, which affect sludge settleability and/or dewaterability (§ 7.12).

7.2 COD AND TSS SOLUBILISATION

The ability of a sludge reduction technique to solubilise the particulate fraction of sludge can be evaluated either through the degree of COD solubilisation or the degree of TSS solubilisation (or TSS disintegration).

The TSS disintegration is defined by the following expression, as a percentage:

$$\text{TSS disintegration}(\%) = \frac{\text{TSS}_0 - \text{TSS}_t}{\text{TSS}_0} \cdot 100$$

where:
TSS_0 = concentration of TSS in the untreated sludge;
TSS_t = concentration of TSS in the sludge after treatment.

The term TSS disintegration is more adequate to describe the expression above, because it takes into account both solubilisation + potential mineralisation.

The COD solubilisation (S_{COD}) is calculated with the following expression, as a percentage (*inter alia* Bougrier *et al.*, 2005; Cui and Jahng, 2006; Benabdallah El-Hadj *et al.*, 2007; Yan *et al.*, 2009):

$$\text{COD solubilisation} = S_{COD}\,(\%) = \frac{\text{SCOD}_t - \text{SCOD}_0}{\text{COD}_0 - \text{SCOD}_0} \times 100$$

where:
SCOD_0 = concentration of soluble COD in the untreated sludge;
SCOD_t = concentration of soluble COD in the sludge after treatment;
COD_0 = concentration of total COD in the untreated sludge.

The difference $\text{COD}_0 - \text{SCOD}_0$ as the denominator represents the particulate COD in the untreated sludge. The COD effectively solubilised after the treatment is thus compared with the particulate COD initially present in the untreated sludge.

Some authors use simplified expressions for calculating S_{COD}, considering negligible the term $SCOD_0$:

$$S_{COD}(\%) = \frac{SCOD_t - SCOD_0}{COD_0} \times 100 \quad \text{or} \quad S_{COD}(\%) = \frac{SCOD_t}{COD_0} \times 100$$

Another form takes into account the particulate COD, given by the difference between total COD and soluble COD (for example Saktaywin et al., 2005):

$$\text{solubilisation of particulate COD}(\%) = \frac{PCOD_0 - PCOD_t}{PCOD_0} \times 100$$

where:
$PCOD_0$ = concentration of particulate COD in the untreated sludge;
$PCOD_t$ = concentration of particulate COD in the sludge after treatment.

This expression coincides with S_{COD} only if total COD concentration remains constant before and after treatment ($COD_0 \equiv COD_t$). Conversely, for example, in the case in which mineralisation occurs during a treatment such as ozonation, this expression includes solubilisation of particulate matter + potential mineralisation.

From the various experiences referred to in the literature it can be seen that the analysis of soluble COD is not standardised: some authors use filtration on 0.45-μm-membrane, while other use 2-μm-membrane, etc...

COD and TSS solubilisation has been widely monitored in the literature as an approximate indicator of improvement in sludge reduction, expected when coupling the technique with a subsequent biological treatment, expecially in the case of techniques based on the mechanism of cell lysis-cryptic growth.

7.3 DEGREE OF DISINTEGRATION

The degree of disintegration is a widely applied parameter to evaluate the efficiency of physico/mechanical treatments for sludge reduction (see Chapters 9 and 10).

The degree of disintegration can be determined on the basis of COD solubilisation or oxygen consumption. The soluble COD allows us to understand how organic compounds are released in the bulk liquid, while oxygen consumption is due to the metabolism of bacterial biomass and the availability of biodegradable substrate. In the extreme situation where all the bacterial biomass is destroyed, there will be no oxygen consumption, while there will be a high concentration of soluble COD in the sludge.

7.3.1 Degree of disintegration based on COD solubilisation (DD_{COD})

To measure the degree of disintegration using COD, the maximum COD solubilisation has to be determined by an alkaline total fusion process using NaOH. The soluble COD concentration measured after this chemical treatment is indicated as $SCOD_{NaOH}$.

The degree of disintegration (DD_{COD}, *Disintegration Degree*, expressed as a percentage) is calculated through the following expression (Müller, 2000b):

$$DD_{COD}(\%) = \frac{SCOD_t - SCOD_0}{SCOD_{NaOH} - SCOD_0} \times 100$$

where:
$SCOD_0$ = concentration of soluble COD in the untreated sludge;
$SCOD_t$ = concentration of soluble COD in the sludge after treatment;
$SCOD_{NaOH}$ = maximum COD concentration that can be solubilised and corresponds to the soluble COD after alkaline hydrolysis.

The concentrations of $SCOD_0$ and $SCOD_t$ in sludge are determined after filtration at 0.45 µm (alternatively on the supernatant after centrifugation). Alkaline hydrolysis can be performed by using NaOH 0.5 mol/L or 1 mol/L for 22–24 h at room temperature (Tiehm *et al.*, 2001; Gonze *et al.*, 2003; Bougrier *et al.*, 2005; Nickel and Neis, 2007). The alkaline treatment is carried out on untreated sludge and $SCOD_{NaOH}$ is measured after filtration.

Some authors measure the maximum COD solubilisation after digestion with H_2SO_4, instead of NaOH (Braguglia *et al.*, 2006).

An attempt has been made to draw a correspondence between the parameters DD_{COD} and S_{COD}. $SCOD_{NaOH}$ is considered approximately one half of the total COD in the untreated sludge (Böhler and Siegrist, 2006):

$$SCOD_{NaOH} \approx 0.5 \cdot COD_0$$

Furthermore, considering that generally $SCOD_0 << SCOD_{NaOH}$ and $SCOD_0 << COD_0$, the following approximate expression is obtained:

$$\underbrace{\frac{SCOD_t - SCOD_0}{COD_0}}_{S_{COD}} \approx 0.5 \cdot \underbrace{\frac{SCOD_t - SCOD_0}{SCOD_{NaOH}}}_{DD_{COD}}$$

from which it appears that $S_{COD} \approx 0.5 \cdot DD_{COD}$.

This approximate relationship can be used when comparing results obtained in different experiments reported in the literature.

7.3.2 Degree of disintegration based on oxygen consumption

The disintegration degree can also be evaluated on the basis of the oxygen consumption by treated and untreated sludge (Kopp et al., 1997). According to the method proposed by Kopp et al. (1997) the disintegration degree based on oxygen consumption (indicated as DD_{O2}, *Disintegration Degree*, expressed as percentage) is calculated as follows:

$$DD_{O2}(\%) = \left(1 - \frac{OC_t}{OC_0}\right) \cdot 100$$

where:

- OC_0 is the oxygen consumption of the untreated sludge;
- OC_t is the oxygen consumption of the treated sludge.

In this type of test, oxygen consumption is measured in a short period and no respirograms are provided. Oxygen consumption is expected to decrease as a result of disintegration and the damage or death of cells. However, it should be considered that during the sludge disintegration two phenomena generally occur:

(1) the release of solubilised compounds (a part of which are biodegradable);
(2) the damage/death of a fraction of bacteria.

Consequently, the oxygen consumption of the treated sludge is the result of the respiration of the remaining bacteria which can oxidise the biodegradable compounds released by the treatment.

Therefore, to investigate separately bacterial damage and the release of biodegradable compounds, respirometry may be usefully adopted but the interpretation of the results is more complex.

The measurement of OUR (*Oxygen Uptake Rate*) by sequenced closed respirometers (see § 7.4) and the acquisition of respirograms for periods of several hours or one day, could help in this direction. A similar approach was also proposed by Camacho et al. (2002a) which estimated the heterotrophic biomass inactivation (I) on the basis of the maximum specific OUR after treatment (OUR_t) and the maximum specific OUR before treatment (OUR_{max}):

$$I(\%) = \left(1 - \frac{OUR_t}{OUR_{max}}\right) \cdot 100$$

This expression is similar to the one proposed above for DD_{O2} but the value OC is replaced with the more representative OUR value.

7.4 BIODEGRADABILITY EVALUATED BY RESPIROMETRY

Beyond the solubilisation of COD, widely applied for evaluating the potential of a sludge reduction technique, great attention should be focused on the biodegradability of the organic matter released, which plays a major role in sludge reduction processes (Paul *et al.*, 2006b).

To evaluate sludge biodegradability, respirometric tests can be carried out, according to the procedures developed for COD characterisation in wastewater by Spanjers and Vanrolleghem (1995), Vanrolleghem *et al.* (1999), Andreottola and Foladori (2007).

Various types of respirometers are available on the market with different configurations and operational procedures. For example, the type of respirometer proposed by Spérandio and Paul (2000), Ziglio *et al.* (2001), Andreottola and Foladori (2007), consists of a batch reactor where a sample of activated sludge is aerated continuously (option 1) or is aerated in an intermittent way on the basis of two set-points (option 2, Figure 7.1).

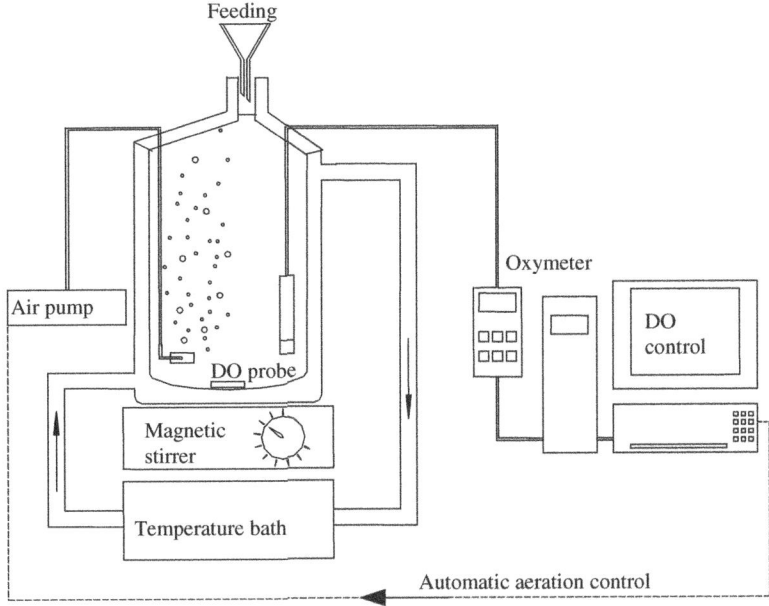

Figure 7.1. Scheme of a simple respirometer.

Dissolved oxygen (DO) is measured directly in the main reactor (option 2) or in a separate small measurement cell, where no oxygen is added (option 1). Depending on the available substrate, activated sludge can be in the growth phase or in the endogenous respiration phase. The sample to be tested for biodegradability is injected into the batch reactor filled with activated sludge and its biological degradation starts. The variations of DO concentration over time is monitored to calcolate the Oxygen Uptake Rate (OUR) which provides a curve known as a respirogram. An example of a respirogram is indicated in Figure 7.2, obtained using: activated sludge (V_0) + filtered sludge after ultrasonic disintegration treatment (V_t).

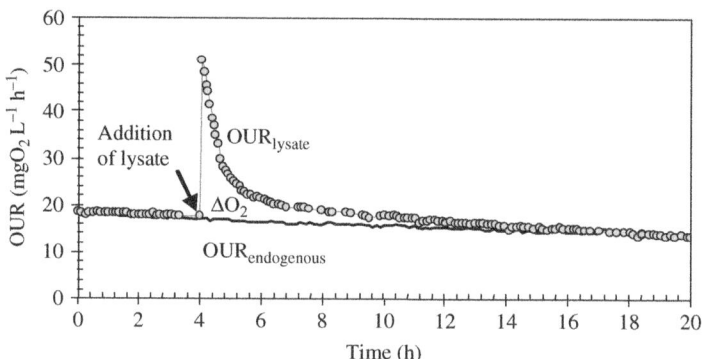

Figure 7.2. Respirograms of $OUR_{endogenous}$ and after addition of 0.45-μm-filtered sludge after sonication (OUR_{lysate}).

The integration of the respirogram gives the net amount of oxygen (ΔO_2) consumed during the biodegradation over a period of time (usually until the endogenous respiration rate has been restored, i.e., from a few hours to one day). Finally, ΔO_2 is converted into an equivalent amount of biodegradable COD (COD_b):

$$COD_b = \frac{\Delta O_2}{1 - Y_H} \cdot \frac{V_0 + V_t}{V_t} [mgCOD/L]$$

introducing the maximum growth yield, Y_H, equal to 0.67 gCOD/gCOD.

The concentration of COD_b represents the concentration of biodegradable COD in the filtered sludge, released by the treatment applied.

Procedures for estimating the efficiency of sludge reduction technologies 97

7.5 DENITRIFICATION RATE EVALUATED BY NUR TEST

When a technique for sludge reduction is integrated in the wastewater handling units and the mechanically, thermally or chemically treated sludge is recirculated in the pre-denitrification stage (according to the schemes of Figure 7.3), the released COD can be used to support nitrogen removal, but only when the COD/N ratio is sufficiently high (amount of COD released by the treatment per unit of nitrogen released).

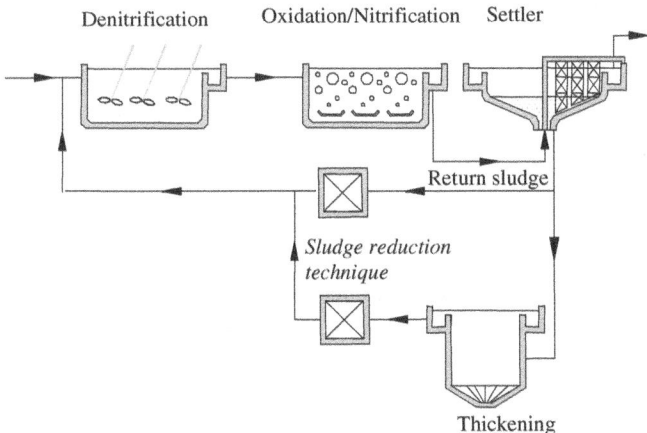

Figure 7.3. Options for the integration of a sludge reduction technique in the wastewater handling units to support denitrification.

The denitrification test is used to measure the denitrification rate and it is well known as a NUR test (*Nitrate Utilisation Rate*) (Kristensen *et al.*, 1992). Briefly, it consists of a batch test carried out in a lab-scale temperature-controlled reactor, in which activated sludge + treated sludge + nitrate are mixed without oxygen supply and the depletion of the NO_3-N concentration over time is monitored.

In a conventional NUR test carried out using wastewater as carbon source, three phases of nitrate reduction occur simultaneously (Figure 7.4):

(1) the first and highest denitrification rate ($V_{D,1}$) is determined by readily biodegradable COD;
(2) the second denitrification rate ($V_{D,2}$) is due to slowly biodegradable COD;

(3) the third and lowest denitrification rate ($V_{D,3}$) is related to the endogenous respiration.

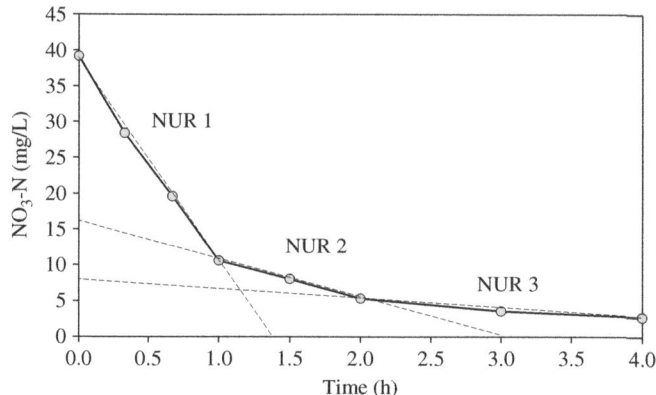

Figure 7.4. Example of a dynamic of NO_3-N concentration in a NUR test.

An example of the NUR test is indicated in Figure 7.4. From the graph, three slopes can be calculated, indicated as NUR with subscript 1,2,3 and expressed as $mgNO_3-N\ L^{-1}\ h^{-1}$:

$$NUR_1 = V_{D,1} + V_{D,2} + V_{D,3}$$

$$NUR_2 = V_{D,2} + V_{D,3}$$

$$NUR_3 = V_{D,3}$$

Finally the specific denitrification rate expressed as $mgNO_3-N\ gVSS^{-1}\ h^{-1}$ can be calculated, dividing by the VSS concentration in sludge.

The lysate derived from a sludge reduction technique, which is expected to be more biodegradable than the untreated sludge, may improve the denitrification rate, prolonging the duration of NUR_1 and NUR_2.

As an example, the specific denitrification rate measured by NUR tests and obtained for three different sludge reduction techniques is indicated Table 7.1, compared with the denitrification rates of some carbon sources commonly used for denitrification. The thermal treatment carried out at 110°C for 45 minutes gives the highest denitrification rate (5.5 $mgNO_3-N\ gVSS^{-1}\ h^{-1}$) compared to mechanical treatment (with stirred ball mills and high pressure homogenisers) and enzymatic hydrolysis, through enzyme addition at 35°C (Müller, 2000a).

Table 7.1. Denitrification rates obtained from three different technologies for sludge reduction compared to other carbon sources commonly used (*from* Müller, 2000a).

Sludge reduction technique	Specific denitrification rate (mg NO_3–N $gVSS^{-1}$ h^{-1})	Reference
Enzymatic hydrolysis with added enzymes (35°C, 3 d)	1.9	Müller (2000a)
Thermal treatment (110°C, 45 min)	5.9	Müller (2000a)
Mechanical disintegration (stirred ball mill and high pressure homogeniser)	2.2	Müller (2000a)
Wastewater	0.6–1	Kujawa and Klapwijk (1999)
Readily biodegradable COD	1–3	Kujawa and Klapwijk (1999)
Endogenous respiration	0.2–0.6	Kujawa and Klapwijk (1999)

7.6 ANAEROBIC BIODEGRADABILITY EVALUATED BY BIOGAS PRODUCTION

The biodegradability of sludge after mechanical, chemical or thermal treatment can be investigated in anaerobic batch tests. The more efficient the disintegration treatment is, the greater the biogas (or methane) production which will be expected.

Anaerobic batch tests can be carried out with different equipment, from the simplest to the most complex, maintained at constant pressure, or at constant volume and/or coupled with gas chromatography.

In its simplest configuration, the inoculum of sludge (untreated sludge) taken from a conventional digester is used in addition to the treated sludge to be tested. The lab reactors are completely mixed and if necessary buffered with nutrients or $NaHCO_3$, KH_2PO_4 to maintain the pH around optimal values for anaerobic digestion. For mesophilic digestion, the temperature has to be maintained around 33–35°C.

A blank test is usually conducted without the addition of treated sludge (only inoculum sludge) in order to obtain the baseline biogas production to be compared/substracted from the test with treated sludge. The enhancement of biodegradability of a sludge treated with a specific sludge reduction technique can be evaluated by comparing the volume of biogas produced by the treated sludge + inoculum with the inoculum itself.

Furthermore, it could be useful to carry out a parallel degradability test with a synthetic feed such as acetate or ethanol (completely biodegradable) in order to compare it to the treated sludge.

Measurements of biogas production (or methane, after CO_2 absorption on caustic material) were usually made for a period ranging from few days to several weeks. Specific biogas production (or specific methane production) is calculated by dividing the volume of biogas (or methane) produced, by the amount of COD added or can be expressed, for example, as $mLCH_4/gVSS$, or as gCOD/gVSS converting CH_4 in a corresponding amount of COD.

7.7 BACTERIA INACTIVATION

Most techniques proposed for the reduction of sludge production are aimed at the disintegration of the biological floc structure and the disruption of bacterial cells. In this way the cells undergo lysis and the intracellular compounds are released and further degraded in biological reactors, through the mechanisms of cell lysis-cryptic growth.

To investigate the efficiency of a technique to damage or lysate bacterial cells, different approaches have been proposed, both conventional and innovative.

Conventional approaches based on the cultivation of bacteria on selective media have been widely used to evaluate bacteria inactivation. But today it is well known that the culture-dependent analysis of microbial populations in activated sludge produces partial and heavily biased results (Wagner *et al.*, 1993), because only a small proportion (about 5%) are able to grow in nutrient media. In fact, most bacteria in activated sludge are in a physiological state known as *"viable-but-not-culturable"* (Nebe-von-Caron *et al.*, 2000; Foladori *et al.*, 2004).

Microscopic observations of biological flocs have also been carried out frequently, but the highly aggregated structure of activated sludge, in which bacteria are embedded, interferes with the microscopic images, limiting the efficiency of direct observations of single cells. However, the disaggregation or dispersion of flocs as a consequence of a sludge reduction treatment can be well appreciated by microscopic observations, as frequently shown in photos in the literature.

To obtain a more realistic view of bacteria populations, the application of advanced *in situ* techniques or direct molecular approaches is needed. In this way, a more complete description of bacterial populations in activated sludge and of their physiological states can be given.

The physiological status of cells (integrity, death, metabolic activity and reproductive growth) can be clarified taking into account the scheme of Figure 7.5 introduced by Nebe-von-Caron *et al.* (2000). Membrane integrity demonstrates the protection of constituents in intact cells and the potential capability of metabolic activity/repair and potentially reproductive growth. Cells without an intact membrane are considered as permeabilised and can be classified as dead cells. As their structures are freely exposed to the environment they will eventually decompose.

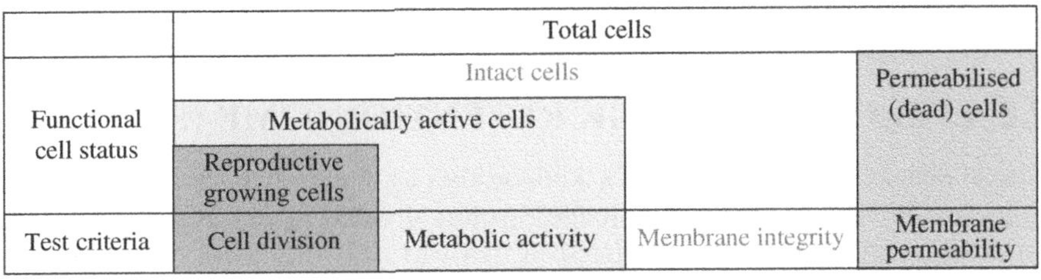

	Total cells			
Functional cell status	Intact cells			Permeabilised (dead) cells
	Metabolically active cells			
	Reproductive growing cells			
Test criteria	Cell division	Metabolic activity	Membrane integrity	Membrane permeability

Figure 7.5. Relationship of physiological states of cells according to Nebe-von-Caron et al. (2000).

Intact and dead cells can be identified simultaneously on the basis of their membrane integrity by applying a double fluorescent staining, such as Propidium Iodide (a dye which can enter only damaged or permeabilised cells, identified as dead) and SYBR-Green I (which can enter all cells, intact or damaged) (Ziglio et al., 2002; Foladori et al., 2007). Metabolic activity is a more restrictive condition, because it requires that cells be able to demonstrate one of the following functions: biosynthesis, pump activity, membrane potential or enzyme activity. Among these, enzymatic activity can be identified by using fluorogenic substrates such as BCECF-AM and FDA (Ziglio et al., 2002) or others, which release, after hydrolysis, fluorescein-based compounds, retained inside cells which become fluorescent.

The advantages of coupling fluorescent molecular probes followed by flow cytometry (FCM) for the rapid quantification of intact, active or dead cells in bacterial suspensions have been highlighted many times in the environmental field (inter alia Porter et al., 1997; Steen, 2000; Vives-Rego et al., 2000). FCM, able to count more than 1000 cells per second, is a powerful single-cell analysis that allows the rapid and precise quantification of free cells in a suspension to be obtained. With respect to the conventional observation of bacteria suspensions under microscope, FCM analysis is faster, results are more reliable and a greater number of samples can be analysed daily.

These approaches, and others not cited here, can be effectively used for the assessment of damage to microorganisms in sludge after the application of a sludge reduction technique. As a consequence of a mechanical, thermal or chemical treatment a reduction of intact cells is expected, as is an increase in dead cells or a net loss when disrupted.

For example, the PCR-DGGE fingerprinting – a powerful method for analysing the bacterial population in various environments – was used to evaluate the death of the bacteria and the destruction of the bacterial DNA contained in the sludge during an ozonation process (Yan et al., 2009). Foladori et al. (2007) applied fluorescent staining and FCM to evaluate integrity and death of bacteria in activated sludge during sonication.

7.8 EFFECT ON SLUDGE RETENTION TIME (SRT)

The sludge retention time (SRT), known also as sludge age, is defined as the mass of solids in the activated sludge reactors divided by solids removed daily and can be approximately calculated in two ways:

(1) when excess sludge is taken, as usual, from the secondary settler the expression is:

$$\mathrm{SRT(d)} = \frac{V \cdot x}{Q_s \cdot x_s}$$

(2) when excess sludge is taken from the reactor, SRT is calculated dividing V by Q_s, because $x = x_s$:

$$\mathrm{SRT(d)} = \frac{V \cdot x}{Q_s \cdot x} = \frac{V}{Q_s}$$

where:
  V = volume of biological reactors (m^3);
  x = TSS concentration in biological reactors ($kgTSS/m^3$);
  Q_s = daily excess sludge flow rate (m^3/d);
  x_s = TSS concentration in excess sludge flow ($kgTSS/m^3$).

The fact that excess sludge is reduced with increasing efficiency by a sludge reduction technique, results in an apparently higher SRT if the TSS concentration in the reactor remains constant. Therefore, the actual SRT obtained in the activated sludge system should be recalculated taking into account the excess sludge reduction rate (Böhler and Siegrist, 2004; Paul and Debellefontaine, 2007).

Due to the higher SRT, one might argue that the growth of the nitrifying biomass could be enhanced. However, the application of a sludge reduction technique (based on disintegration) may cause a significant reduction of nitrifiers due to partial inactivation or death, resulting in a shorter SRT of nitrifiers. The diminished SRT of the nitrifiers is therefore just about compensated by the increased apparent SRT due to the lower excess sludge production (Böhler and Siegrist, 2004).

The SRT of the nitrifiers in an activated sludge system integrated with a sludge reduction technique (treating $Q_{treated}$) can be calculated as follows, according to the expression proposed by Böhler and Siegrist (2007) in the case of ozonation:

$$SRT_{nitrifiers} = \frac{V}{Q_s + Q_{treated} \cdot \eta_{nitrifiers}}$$

where $\eta_{nitrifier}$ is the fraction of nitrifiers destroyed.

It is important to know the effective SRT of the nitrifiers to estimate the safety of nitrification in case of ammonium overloads (Böhler and Siegrist, 2007).

For example, in the case of ozonation of sludge, Böhler and Siegrist (2004) observed that the reduction of the nitrification capacity was similar to the entity of sludge reduction (see § 13.9.6). In this case, the reduction of the SRT of the nitrifiers due to ozonation is more or less compensated by the increased apparent SRT due to the lower excess sludge production (Böhler and Siegrist, 2007).

7.9 MAXIMUM GROWTH YIELD, OBSERVED BIOMASS YIELD, OBSERVED SLUDGE YIELD

Considering a mass balance of carbonaceous substrate in a chemostat system, Pirt (1965) proposed that a portion of the carbon source be used for maintenance and a portion used for anabolism. When the entire substrate is used for anabolism, the *maximum growth yield* (or synthesys yield; Y_H for heterotrophic bacteria), Y_H is theoretically obtained: this value is also termed "true" growth yield.

In a system containing a given amount of biomass, part of the carbon source is always used for maintenance. Therefore the *"observed biomass yield"* (Y_{obs}) is a function of the maximum growth yield (Y_H) and SRT through coefficient from maintenance (m_s) or endogenous respiration (k_e) as described by van Loosdrecht and Henze (1999), resulting in the following expressions:

$$Y_{obs} = Y_H \cdot \frac{1}{1 + SRT \cdot m_s \cdot Y_H}$$

$$Y_{obs} = Y_H \cdot \frac{1}{1 + SRT \cdot k_e}$$

where Y_H is 0.67 gCOD/gCOD (0.45 gVSS/gCOD) under aerobic conditions (see § 8.3). The observed biomass yield Y_{obs} is always lower than Y_H.

The forms of these equations indicate that the concepts of maintenance and endogenous respiration explained in the § 4.4 are mathematically equivalent.

Another expression to describe the Y_{obs} coefficient is the following (van Loosdrecht and Henze, 1999):

$$Y_{obs} = Y_H \cdot \frac{1}{1 + SRT \cdot b \cdot (1 - (1-f) \cdot Y_H)}$$

which is based on the concept of death-regeneration, including the decay/lysis of cells (indicated by decay rate, b) and the further conversion of decayed material into a substrate available to feed living cells. In the above expression the factor f is introduced, which represents the fraction of inert material formed during decay/lysis (endogenous residue).

In the pratical approaches in WWTPs, the parameter Y_{obs} is used in a more general way, to evaluate the specific sludge production, and not only to describe the net growth of bacterial biomass.

In this case, the parameter Y_{obs} is defined as "*observed sludge yield*" (or observed solid yield) and corresponds to the ratio $\Delta VSS/\Delta COD$. It is different from the observed biomass yield because it contains other organic solids from the wastewater that are measured as VSS, but are not biological (Tchobanoglous *et al.*, 2003).

It can be calculated as the total solid mass of sludge generated per unit of COD removed, and corresponds to the slope of the lines shown in Figure 7.6A.

A change (decrease) in the Y_{obs} coefficient is expected after the integration of a sludge reduction technique in an activated sludge process (see Figure 7.6B), demonstrating that the total amount of sludge can be effectively reduced.

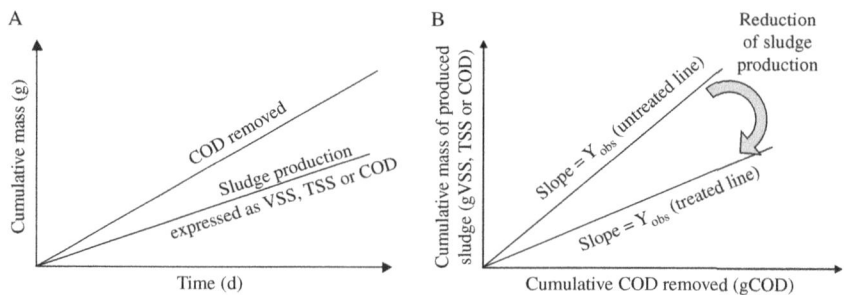

Figure 7.6. (A) Comparison between the mass of COD removed in the process and the production of excess sludge; (B) relations between the cumulative sludge increase and COD removed, for an untreated and a treated line.

Procedures for estimating the efficiency of sludge reduction technologies 105

7.10 EVALUATION OF SLUDGE REDUCTION

Although some of the methods described above allow us to obtain rapid information about the potential efficiency of techniques used in sludge reduction, generally, pilot-scale experiments are strongly recommended in order to perform more accurate evaluations of a process before full-scale applications. Long term experiments carried out at lab- or pilot-scale, using real wastewater, and running in parallel with a control treatment plant, should be performed to correctly evaluate the reduction of sludge production from a technical and economic point of view (Paul et al., 2006b).

The reason is the recognised importance of evaluating the performance of a new technique under the continuous variability of influent wastewater COD fractions and operational plant conditions.

In a lab- or pilot-plant, the reduction of sludge production should be evaluated by comparing two lines operating in parallel: (1) one line used as control, (2) the other treated line, with an identical configuration and identical operational characteristics to the control line, but integrated with the new technology for sludge reduction.

The quantification of the daily sludge production (X) for each line of an activated sludge plant should be calculated through an overall mass balance applied to the whole operational period, taking into account:

– the mass of solids in the excess sludge extracted daily;
– the mass of solids accumulated within the biological reactor (checking that the quantity of sludge accumulated in the settler is negligible);
– the mass of solids lost with treated effluent;
– (only in lab- or pilot-scale plants) the solids lost with sampling.

The values of sludge production in the treated line (X_t) and the control line (X_0) are used to calculate the reduction of sludge production (RSP):

$$\text{RSP}(\%) = \left(1 - \frac{X_t}{X_0}\right) \cdot 100$$

7.11 TREATMENT FREQUENCY

A parameter which is sensitive in the application of a sludge reduction technique, is the *treatment frequency* known also as *stress frequency* (SF, expressed as d^{-1}).

The treatment frequency is a parameter related to the time interval established for the sludge passing through the treatment unit (Camacho et al., 2002a; 2005). Considering, for example, a sludge reduction technique integrated in the activated sludge process as indicated in Figure 7.7, the treatment frequency is calculated as follows:

$$SF = \frac{\text{amount of sludge treated per day}}{\text{total amount of sludge in the system}} = \frac{Q_{treated} \cdot x_{treated}}{V \cdot x} \left[d^{-1} \right]$$

where:
$Q_{treated}$ = sludge flow rate fed to the treatment unit (m³/d);
$x_{treated}$ = TSS concentration in sludge flow fed in the treatment unit (kgTSS/m³);
V = volume of biological reactors (m³);
x = TSS concentration in biological reactors (kgTSS/m³).

For example a SF of 0.2 d^{-1} means that 20% of sludge present in the system undergoes the treatment daily (Figure 7.7).

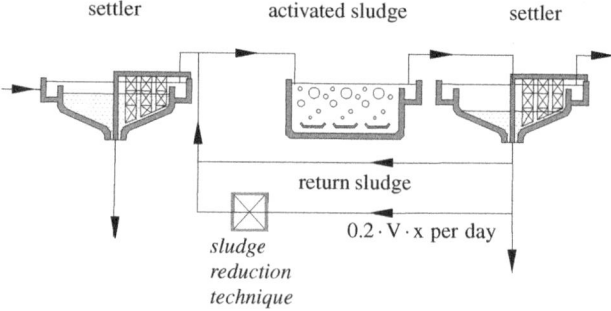

Figure 7.7. Example of treatment frequency for a sludge reduction technique integrated in the wastewater handling units.

7.12 PHYSICAL PROPERTIES OF SLUDGE

The evaluation of the physical properties of sludge is important as this knowledge allows us to predict sludge behaviour when handled and submitted to various utilisation/disposal operations (Spinosa and Wichmann, 2006). The main parameters are well reviewed by Spinosa and Wichmann (2006) and include:

- *capillary suction time* (CST): widely applied as a fast and simple method to evaluate sludge dewaterability by filtration;

- *specific resistance to filtration* (SRF): indicates the suitability of sludge to be dewatered by means of a filtration process; useful to predict the performance of full-scale filtering devices and the optimal dosage of chemicals;
- *compressibility*: complementary to SRF and useful for evaluating the best range of pressures to be adopted for filtration;
- *settleability*: used for calculating the rate of sludge settling and the sludge volume index (SVI);
- *thickenability*: for further concentration of suspended sludge solids in settling under gravity, determined after settling under gentle stirring.

8
Biological treatments

8.1 INTRODUCTION

The microorganisms employed in the biological processes of a WWTP – where the organic matter and nitrogen compounds are assimilated – can be classified on the basis of their energy source, origin of cell carbon and terminal electron acceptor as per Table 8.1 (Low and Chase, 1999a).

The biological processes are based on heterotrophic bacteria and autotrophic (especially nitrifying) bacteria which require an aerobic environment to oxidise organic compounds and ammonia/nitrite respectively. In the absence of dissolved oxygen, heterotrophic bacteria can use nitrate as an electron acceptor in the process known as denitrification. Aerobic, anoxic and anaerobic environments will determine the availability of reducing equivalents thus influencing the metabolic efficiency, as described in more detail in § 8.3. Such conditions can be found in the WWTPs in separate tanks, or in the same tank in different zones, or even in microenvironments formed inside biological flocs.

© 2010 IWA Publishing. *Sludge Reduction Technologies in Wastewater Treatment Plants.* By Paola Foladori, Gianni Andreottola and Giuliano Ziglio. ISBN: 9781789065305. Published by IWA Publishing, London, UK.

Table 8.1. Synthetic classification of microorganisms involved in biological processes in WWTPs.

Group	Origin of cell carbon	Energy source/electron donor	Electron acceptor	
heterotrophic bacteria	organic compounds	organic compounds	aerobic	oxygen
			anoxic	nitrate, sulphate
			anaerobic	organic compounds
autotrophic bacteria	inorganic carbon	ammonia, H_2S, Fe^{2+}	oxygen	

Downstream digestion processes for the stabilisation of excess sludge can operate both anaerobically or aerobically to reduce the volume and the mass of sludge to be dewatered and disposed of. During digestion, when the external organic substrate is completely degraded, cell death and lysis occur and the cell products can be reused as a carbon source in microbial metabolism and a portion of the carbon is liberated as carbon dioxide (and methane in anaerobic digestion) resulting in reduced overall biomass when these conditions are prolonged. In general, the solid reduction during stabilisation processes is not expected to be higher than approximately 40% even after a digestion time of 15–25 days, because sludge comprises an inert fraction including endogenous residue and inert solids which enter with wastewater.

Aerobic sludge stabilisation is the most common process for sludge treatment in small-medium WWTPs applying a digestion time of <25 days. Microorganisms aerobically degrade the residual available organic matter and endogenous metabolism proceeds once exogenous biodegradable organic matter is completely depleted. The end-products are carbon dioxide, water and ammonia; subsequently, ammonia is removed through nitrification. The bacterial biomass is represented by the typical composition $C_5H_7NO_2$ and reacts with oxygen as follows:

$$C_5H_7NO_2 + 5O_2 \rightarrow 5CO_2 + 2H_2O + \underbrace{NH_3}_{\downarrow \atop NO_3} + \text{energy}$$

Aerobic digestion can be considered as an extension of the activated sludge process under endogenous conditions. The VSS reduction which can be obtained in the aerobic digestion process normally carried out at ambient temperature, reaches 30%, even for very long digestion time of up to 50 d. Aerobic digestion under thermophilic conditions allows this percentage to be increased.

A significant reduction of pathogens can only be obtained by operating at long digestion time at ambient temperatures (for example 40 d at 20°C or 60 d at 15°C) otherwise at 55–60°C (for 10 d).

Both conventional (ambient temperature) and thermophilic or autothermal digestion (50–60°C) require a sufficient availability of dissolved oxygen for optimal performance. Usually very low dissolved oxygen concentration is measured in aerobic digesters, especially when nitrification occurs. This fact is related to the low oxygen transfer rate at high solid concentrations and at relatively high temperatures in the case of thermophilic digestion. Furthermore, a high air flow rate is often associated with scum production, which represents one of the main obstacles to satisfy the oxygen requirements (Roš and Zupančič, 2003), which could be overcome by using pure oxygen instead of air.

Anaerobic digestion is the most widely used process to reduce both primary and biological excess sludge. The choice of an anaerobic digester is usually preferred in larger WWTPs with capacity above 20,000–30,000 PE, because the energy recovered by exploiting the methane gas produced becomes economically advantageous in larger plants.

Anaerobic digestion is achieved through three consecutive phases: enzymatic hydrolysis, acidogenesis followed by methanogenesis and it is performed mainly under mesophilic conditions (33–35°C), although some applications under thermophilic conditions (53–55°C) are used. The rate-limiting step is the enzymatic hydrolysis of solids. This is the reason why many researchers have focused their efforts on applying several pre-treatment methods (mechanical, thermal and/or chemical treatments, ozonation, etc...) in order to increase the hydrolysis rate and the sludge degradation.

Conventional mesophilic anaerobic digestion leads to sludge stabilisation by converting about 30–50% of total COD or VSS into biogas in 30 d, which is valuable as a renewable energy source, and by producing stabilised sludge with TSS reduction usually by 30–40%. In particular, the VSS reduction during the treatment of primary + biological sludge in mesophilic anaerobic digestion is around 40%[1], which is higher than the reduction during aerobic digestion. The biogas production is 400–600 L/kgVSS removed and energy recovered is typically 6.5 KWh/Nm^3 of biogas produced.

---

[1] Primary and biological excess sludge behave differently during anaerobic digestion. In particular:
- for primary sludge the VSS reduction reaches 60%;
- for biological excess sludge the VSS reduction reaches 30%.

For example, considering a sludge made up of: *40%* of primary sludge (in terms of mass) and *60%* of secondary sludge, the expected sludge reduction is equal to: *0.40* · 0.60 + *0.60* · 0.30 = 0.42 (42%).

The return flows produced by dewatering after anaerobic digestion and recirculated in the wastewater handling units, are characterised by moderate flow rates, but very high concentrations of COD (1000–12000 mg/L) and ammonia (1000–2000 mgN/L). When the return flows reach the wastewater handling units the COD concentration is not usually a problem, while the ammonia load could reach up to 50% of total nitrogen load in the inlet of the plant (Zupančič and Roš, 2008). With respect to anaerobic digestion, the concentrations of COD and ammonia in return flows after aerobic digestion + dewatering are lower, due to oxidation during the aerobic digestion itself.

8.2 RECENT INSIGHTS ON DEGRADABILITY OF SLUDGE UNDER AEROBIC AND ANAEROBIC CONDITIONS

The approaches to the design/modelling of stabilisation in WWTPs usually consider that the biodegradability of excess sludge is the same under aerobic or anaerobic conditions.

However, in recent years several contributions have highlighted that the biodegradability of sludge is influenced by factors other than those identified in existing models (Jones *et al.*, 2009). In conventional approaches the main phenomenon of sludge digestion is attributed to the decay of heterotrophic biomass. In the research carried out by Novak *et al.* (2003) the authors also suggest the existence of other mechanisms in floc destruction during anaerobic and aerobic conditions, as described in the following sections.

8.2.1 Aerobic conditions

Sludge is made up of a microbial consortium, organic and inorganic matter aggregated in a floc matrix formed by EPS (made up mainly of proteins and carbohydrates), which can be of two types (Novak *et al.*, 2003):

(a) iron-bound biopolymers;
(b) divalent cation-bound biopolymers.

Bruss *et al.* (1993) showed that storage of activated sludge under anaerobic conditions would lead to poorer dewatering and attributed the changes to the reduction of iron and the release of discrete bacteria into solution.

Divalent cations are an integral part of the floc structure which improve floc stability by bridging the negatively charged biopolymers (proteins and polisaccharides) (Novak *et al.*, 2003).

During aerobic digestion a release of calcium (Ca^{2+}) and magnesium (Mg^{2+}) into solution was observed (Novak *et al.*, 2003) in conjunction with VS destruction. The concentration of Ca^{2+} increased by a factor of 5 or more compared to initial value and Mg^{2+} increased by a factor of 3 (Novak *et al.*, 2003). Because of the important role that cations play in the formation of flocs and the binding of both protein and polysaccharides, the increase of cations in solution as digestion proceeds it is expected to be in conjunction with the solubilisation of polysaccharides and proteins, as shown in Figure 8.1. The release of proteins during aerobic digestion is less than that observed during anaerobic digestion, whilst the release of polysaccharides is similar.

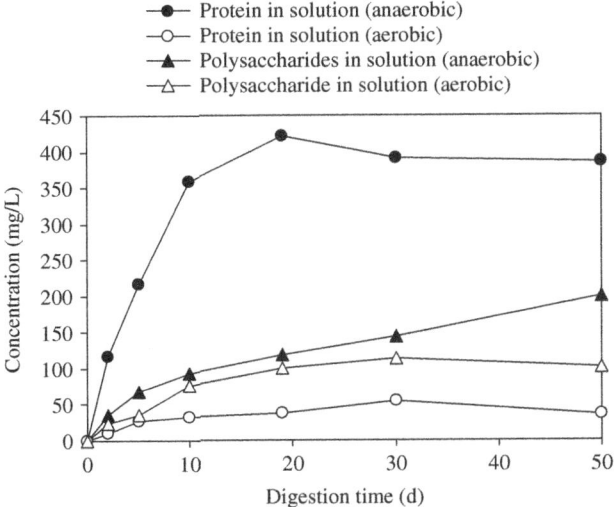

Figure 8.1. Concentration of soluble proteins and polysaccharides as a function of the duration of aerobic digestion and anaerobic digestion at 20°C (*modified from* Novak *et al.*, 2003).

8.2.2 Anaerobic conditions

Under anaerobic conditions, the cations that increased were the monovalent cations, potassium and ammonium, and a large release of proteins was observed, reaching 300–400 mg/L compared to 30–50 mg/L measured during aerobic digestion, as indicated in Figure 8.1 (Novak *et al.*, 2003). However, the release of divalent cations, Ca^{2+} and Mg^{2+} did not change. The high release of soluble proteins during anaerobic digestion was due to the loss of selective binding between protein and Fe(III) under reducing conditions. In other words, when sludge undergoes reducing anaerobic conditions, iron is reduced and solubilised, weakening the bond with proteins. Most of the soluble iron precipitates as FeS. The consequence is the deflocculation of sludge described in more detail in § 8.2.3 (Novak et al., 2003; Nielsen and Keiding, 1998); furthermore, the increased content of soluble proteins can cause deterioration of sludge dewatering properties (Novak et al., 2001; Cetin and Erdincler, 2004). Moreover, the observed increase in ammonium ions under anaerobic conditions is the result of protein degradation.

8.2.3 Disintegration of sludge flocs under anaerobic conditions and in the presence of sulphides

When sludge is stored anaerobically in storage tanks for a few days before dewatering, significant sulphate reduction can be observed, causing the production of sulphides. The presence of sulphides may cause a deterioration and deflocculation of sludge floc structure, as a consequence of the reduction of Fe(III) to Fe(II) and the further formation of FeS according to the following chemical reaction:

$$2\,Fe^{3+} + 3\,S^{2-} \rightarrow 2FeS + S^0$$

Fe(III) has a better flocculation capacity than Fe(II) due to the higher number of positive charges and to the lower solubility of Fe(III), which undergoes hydrolysis followed by precipitation as $Fe(OH)_3$.

The reduction of Fe(III) can occur through a microbial reduction, using Fe(III) as an electron acceptor, or through a chemical reduction by sulphides (e.g. by H_2S). In both cases the reduction of Fe(III) causes a reduction in floc strength leading to the release of particles to bulk water, dissolution of EPS and partial floc disintegration, especially when a shearing force is applied by pumping or intensively stirring (Nielsen and Keiding, 1998). Most Fe(III) is assumed to be organically bound since many EPS have a high affinity for Fe(III) (Rasmussen et al., 1996).

An example of the floc structure and the effect of disaggregation caused by Fe(III) reduction by sulphides is indicated in Figure 8.2. The solubilised compounds are mainly certain EPS components. Filamentous bacteria are dispersed and released in the bulk water. The compounds which are not expected to disintegrate are any organic compounds or bacteria strongly bound to high density inorganic particles, organic fibers or bacteria aggregated in microcolonies (Nielsen and Keiding, 1998).

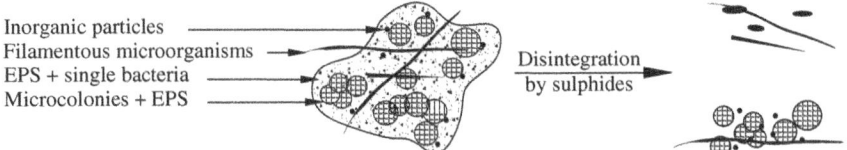

Figure 8.2. Example of floc structure and disaggregation effect caused by Fe(III) reduction by sulphides (*modified from* Nielsen and Keiding, 1998).

This effect of floc disintegration by sulphides could negatively affect dewaterability by reducing filterability.

Iron (Fe) is commonly present in activated sludge from WWTPs, either from influent wastewater or added as a reagent for phosphorus precipitation. Iron reduction, according to the reactions described above, may occur in anaerobic process tanks with biological phosphorous removal or in anaerobic storage tanks prior to dewatering (Rasmussen *et al.*, 1996), or in processes such as OSA (see § 8.4.1) in which an anaerobic side-stream reactor is integrated with the conventional activated sludge. Furthermore, iron reduction could take place in the innermost anaerobic part of flocs even when aerobic bulk conditions exist (Rasmussen *et al.*, 1996).

8.3 THE INFLUENCE OF AEROBIC/ANOXIC/ ANAEROBIC CONDITIONS ON HETEROTROPHIC MAXIMUM GROWTH YIELD

The heterotrophic maximum growth yield, and therefore the metabolic efficiency, depends strongly on the proportion of substrate electrons which are used in the synthesis of new cell mass and the proportion which are ceded to the terminal electron acceptor, such as (see Table 8.1):

- O_2 under aerobic conditions
- NO_3 under anoxic conditions (without oxygen)
- organic matter or CO_2 under anaerobic conditions (without oxygen and nitrate).

These conditions can be achieved purposely in dedicated reactors of a WWTP or can even occur spontaneously in the microenvironment inside biological flocs as a function of the oxygen concentration maintained in the bulk liquid (Figure 8.3).

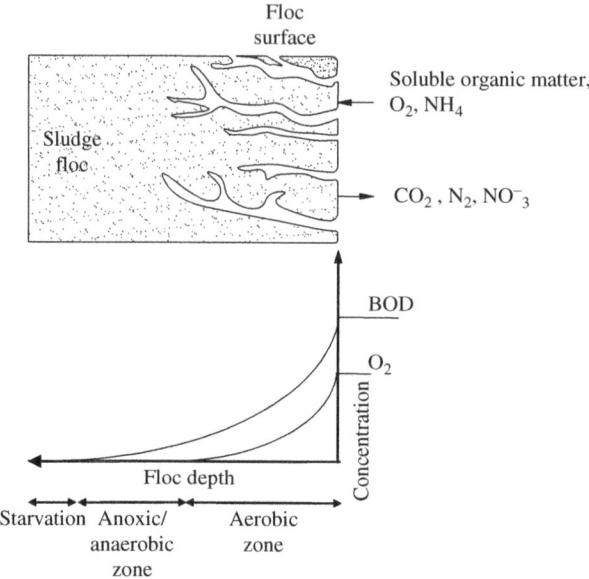

Figure 8.3. Profiles of concentrations of oxygen and organic matter inside a floc (*from* Low and Chase, 1999a).

The amount of energy produced in the oxidation-reduction reactions is reduced progressively when passing from oxygen to nitrate or to carbon dioxide as electron acceptor. Table 8.2 shows the typical values of the maximum growth yield for the more common donors and acceptors of electrons (Tchobanoglous *et al.*, 2003).

As indicated in Table 8.2 the heterotrophic maximum growth yield assumes lower values under anoxic conditions, when compared to aerobic ones. In the absence of oxygen (but with nitrate) the maximum growth yield (0.36 gVSS/gCOD) is reduced by 20% compared to the value in the presence of oxygen. When nitrate serves as electron acceptor, approximately only 2 mol of ATP are formed per pair of electron moles transferred in the electron transport chain to nitrate, compared to the 3 mol of ATP when the transfer is to oxygen (Muller *et al.*, 2003). This difference reduces the energy captured by the organism in the oxidation of substrate when nitrate serves as electron acceptor compared to oxygen, resulting in a reduction of the anoxic growth yield compared to the aerobic value (Muller *et al.*, 2003).

Considering several investigations of the literature, well reviewed by Muller *et al.* (2003), the heterotrophic anoxic growth yield is 78–85% of the aerobic yield.

Table 8.2. Typical values of maximum growth yield for the more common biological reactions in wastewater treatment (Tchobanoglous et al., 2003; Muller et al., 2003; Copp and Dold, 1998).

Growth conditions	Electron donor	Electron acceptor	Maximum growth yield (Y_H)
Aerobic	Organic compounds	Oxygen	0.67 gCOD/gCOD 0.45 gVSS/gCOD
Anoxic	Organic compounds	Nitrate	0.41–0.53 gCOD/gCOD 0.30–0.36 gVSS/gCOD
Anaerobic	Organic compounds	Organic compounds	0.06 gVSS/gCOD
Anaerobic	Acetate	Carbon dioxide	0.05 gVSS/gCOD

The growth yield under anaerobic conditions assumes the minimal value, 0.05–0.06 gVSS/gCOD, which is 12% compared to the aerobic yield.

The value of the maximum growth yield influences the mass of heterotrophic organisms and therefore, from a theoretical point of view, it may affect sludge production.

8.3.1 Process of denitrification + nitrification

On the basis of the values of the aerobic growth yield ($Y_{H,AE}$) and the anoxic growth yield ($Y_{H,AX}$), in an activated sludge process involving nitrification + denitrification, a lower production of sludge is theoretically expected, according to the following calculation:

- in the aerobic phases $Y_{H,AE}$ is 0.45 gVSS/gCOD (see Table 8.2);
- in the anoxic phase $Y_{H,AN}$ is 0.30 gVSS/gCOD, 33% lower than the value under aerobic conditions;
- an approximate duration of the anoxic phase up to 12 h/d (50% of the day) can be considered; during this period the growth yield is effectively 0.30 mgVSS/mgCOD, while in the remaining aerobic period the growth yield is 0.45 mgVSS/mgCOD;
- bacterial biomass in the sludge is typically around 10–30% of VSS in sludge (see § 2.3.3).

Overall, a theoretical reduction of sludge produced (as VSS) of around 5–10% can be estimated. In synthesis, in conventional activated sludge systems, the addition of an anoxic stage – such as pre-denitrification – has a less significant effect on the net sludge production, due to the slightly lower biomass produced under anoxic conditions compared to that produced under aerobic conditions.

An intermittently aerated process fed with synthetic wastewater was applied at lab-scale to evaluate TSS reduction (Jung et al., 2006). The main parameters for controlling intermittent aeration are the length of the aerobic period (T_{AE}) and the anoxic period (T_{AN}) and the operation cycle time ($T_C = T_{AE} + T_{AN}$). Experiments were conducted comparing various alternating aerobic/anoxic periods (T_{AE}/T_{AN}) with continuous aerobic ($T_{AE}/0$) and anaerobic ($0/T_{AN}$) conditions used as controls. In the continuous aerobic configuration (control) the TSS reduction was 61.9%. Using intermittent aeration (different combinations of T_{AE} from 1 to 20 h; T_{AN} from 2 to 20 h) the TSS reduction was between 40 and 70%. The highest sludge reduction of 70% was obtained under 4 h-aerobic/4 h-anoxic conditions, and corresponds to an increase of 13% compared to the continuous aerobic case (Jung et al., 2006).

The authors found that the shorter the operation cycle time was, the more effective the sludge reduction was, as indicated in the diagram in Figure 8.4 where the ratio between the reduction of sludge produced (RSP) under aerobic+anoxic conditions (RSP_{AE+AN}) and under all aerobic conditions (RSP_{AE}) is indicated. Moreover, the sludge reduction was higher when T_{AE} was equal to or longer than T_{AN} (Jung et al., 2006).

These experimental results are in agreement with the theoretical calculation that the reduction in sludge production after the addition of an anoxic phase is marginal. Furthermore, a reduced decay rate of heterotrophic bacteria under anoxic conditions compared to aerobic may compensate for the reduced anoxic growth yield (Siegrist et al., 1999).

8.3.2 Digestion with alternating aerobic/anoxic/anaerobic conditions

With respect to the conventional aerobic digestion – in which the tank is completely and continuously aerated – alternating aerobic, anoxic (and/or anaerobic) phases can be introduced, in order to obtain certain advantages in terms of performance and energy economy (Figure 8.5).

A synthesis of the results referred to in the literature regarding the reduction of VSS and total N in conventional aerobic digestion and in the digestion with alternating aerobic/anoxic phases, is indicated in Table 8.3 and Table 8.4.

Biological treatments 119

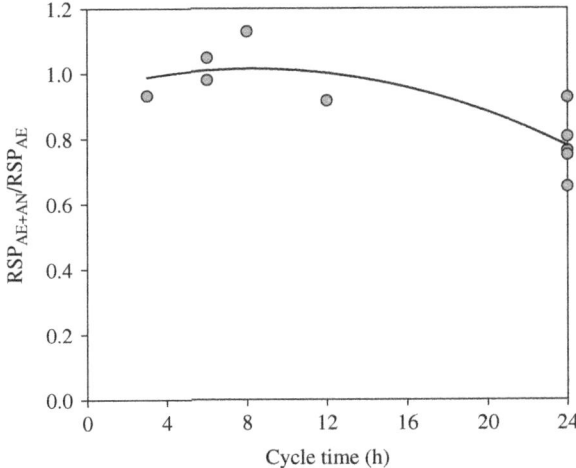

Figure 8.4. Ratio between the reduction of sludge production (RSP) under aerobic + anoxic conditions (RSP_{AE+AN}) and under all aerobic conditions (RSP_{AE}) as a function of cycle time (T_{AE+TAN}) (*data from* Jung *et al.*, 2006).

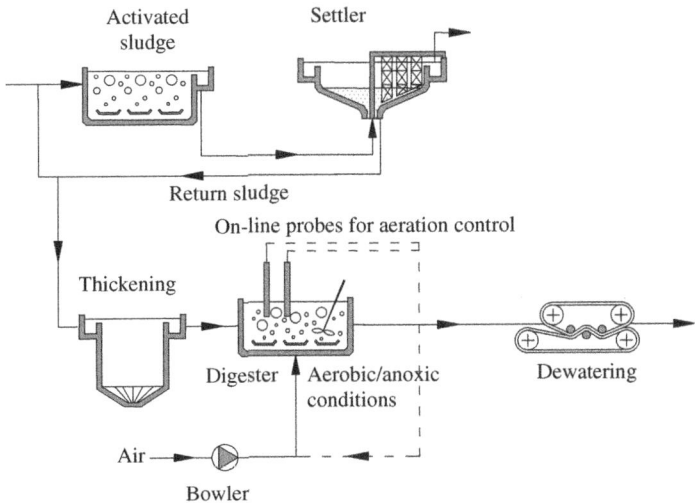

Figure 8.5. Digestion with alternating aerobic/anoxic phases.

Table 8.3. Comparison of VSS reduction in conventional aerobic digestion and in alternating aerobic/anoxic phases.

Parameter	Conventional aerobic digestion	Alternating aerobic/anoxic phases	References
VSS reduction	33%	32–48%	Hashimoto et al. (1982)
	Similar in the two systems		Warner et al. (1985)
	39–46% (similar in the two systems)		Matsua et al. (1988)
	–	40%	Hao and Kim (1990)
	–	26%	Peddie and Mavinic (1990)
	Similar in the two systems		Hao et al. (1991)
	42.7%	42.4–43.7%	Al-Ghusain et al. (2002)

Table 8.4. Removal of total N in conventional aerobic digestion and in alternating aerobic/anoxic phases.

Parameter	Conventional aerobic digestion	Alternating aerobic/anoxic phases	References
Total N reduction	2%	13–27%	Hashimoto et al. (1982)
	–	31–41.5%	Matsua et al. (1988)
	–	25%	Hao and Kim (1990)
	–	32%	Jenkins and Mavinic (1989a,b)

The results obtained by several authors agree on two common aspects:

(1) the VSS reduction in the alternating aerobic/anoxic phases is similar to that obtained in conventional aerobic digestion, as demonstrated by almost all the experimental studies indicated in Table 8.3;
(2) the total N reduction increases significantly due to the introduction of denitrification during the anoxic phase (Table 8.4).

The adoption of aerobic and anoxic phases during digestion – with a variable number and length of anoxic phases for a total of between 8 and 16 h per day – certainly ensures good results in terms of total N removed, with lower nitrogen loads recirculated in the wastewater handling units. This strategy also favours energy saving thanks to the shorter duration of aeration in the digestion tanks. The optimisation of digestion alternating aerobic/anoxic phases can achieved through the control of the process by using ORP and pH probes (Al-Ghusain and Hao, 1995) or directly using NO_3 and NH_4 probes.

The introduction of an additional, strictly anaerobic stage operating at very low ORP values (around −250 mV) allows more efficient conditions for sludge reduction compared to the aerobic/anoxic ones (see § 8.4).

8.4 SIDE-STREAM ANAEROBIC REACTOR (AT AMBIENT TEMPERATURE)

The transition from aerobic to anaerobic conditions favours the introduction of uncoupled metabolism (§ 4.3) with a consequent reduction of the observed biomass yield (Harrison and Maitra, 1969; Harrison and Loveless, 1971).

Chudoba et al. (1991) and Chudoba et al. (1992a) also indicated that when microorganisms undergo anaerobic conditions without exogenous substrate, an important consumption of ATP stored at intracellular level occurs. The anaerobically treated microorganisms are thus subjected to a physiological shock created by a lack of oxygen and food and use ATP as a source of energy (Chudoba et al., 1992a). They are able to rebuild ATP only when again subjected to aerobiosis and supplied with exogenous substrate. Then they rebuild their energy reserves at the expense of growth, causing an overall reduction of sludge production.

The initial development of the Oxic-Settling-Anaerobic (OSA) process (Chudoba et al., 1992a; Chudoba et al., 1992b) demonstrated the significant sludge reduction as described more in detail in § 8.4.1. More recently Novak et al. (2007) found that sequential exposure of activated sludge to aerobic and anaerobic conditions in the Cannibal® system (see § 8.4.2) resulted in a significant sludge reduction compared to a fully aerated process, suggesting that the mechanisms of degradation of solids under aerobic and anaerobic conditions are not the same.

8.4.1 Oxic-Settling-Anaerobic process

The OSA system is a modification of the conventional activated sludge process by inserting an anaerobic tank in the sludge return line between the secondary settler and the aeration tank (Chudoba et al., 1992a; Chudoba et al., 1992b). The recirculated sludge undergoes alternating oxic (in the activated sludge reactor) and anaerobic (in the additional anaerobic reactor) processes. This alternance of *"sludge fasting/feasting"* was proposed to explain the process, according to the scheme of Figure 8.6 (Chen et al., 2001a):

(1) the sludge *fasting* refers to the exposure of the settled sludge to an anaerobic environment where food is insufficient;

(2) the sludge *feasting* means that when the fasted microorganisms return to the activated sludge stage (under oxic environment with sufficient food), they can harvest cell energy, resulting in ATP storage.

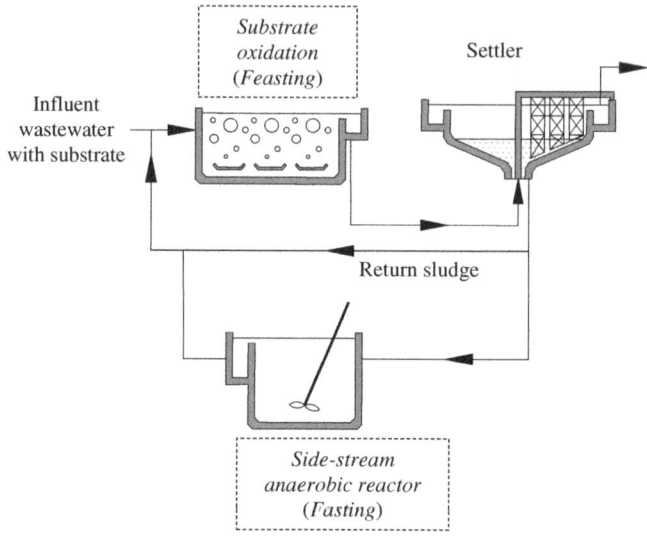

Figure 8.6. OSA process integrated in the activated sludge stage.

However, the mechanisms that allow the sludge reduction with the OSA system are not yet fully understood. On the basis of earlier investigations the process could be based on metabolic uncoupling (see § 4.3). A decrease of ATP content in sludge occurs when the sludge remains in the anaerobic reactor, without external substrate and at low ORP levels. When the sludge returns to the activated sludge stage optimal conditions for the formation of new ATP (oxygen and availability of substrate) are restored. This cyclic alternation of ATP content in sludge uncouples catabolism and anabolism, which causes a decrease in biomass yield favouring sludge reduction (Harrison and Maitra, 1969; Harrison and Loveless, 1971; Chudoba *et al.*, 1991; Chudoba *et al.*, 1992a).

Another possible explanation, introduced more recently, is that the rate of cell death is accelerated when sludge undergoes low ORP values in the anaerobic reactor (Chen *et al.*, 2003; Saby *et al.*, 2003).

In the anaerobic reactor an increase of soluble COD was observed, which passes from concentrations of 20–40 mgCOD/L in the sludge extracted from the secondary settler, to 40–90 mgCOD/L in the discharge from the anaerobic reactor[(2)] (Saby et al., 2003). The solubilised COD could be related to cellular death. According to this hypothesis the anaerobic reactor behaves like a cell lysis reactor and when the sludge returns to the aerobic activated sludge stage, the soluble COD supports cryptic growth, which leads to an overall sludge reduction.

Considering this explanation, the OSA process could be based on the mechanism of cell lysis-cryptic growth instead of the mechanisms of uncoupled metabolism. This hypothesis is not fully demonstrated and some doubts remain. According to Siegrist et al. (1999), the fact that the decay rate of heterotrophic and nitrifying bacteria is lower under anoxic or anaerobic conditions with respect to aerobic ones, also contributes to the uncertainty about the explanation of the mechanisms involved in the OSA process.

The anaerobic reactor has to operate at a high biomass concentration and with a sufficiently long retention time, without any wastewater feeding and maintaining a low redox potential (ORP).

The anaerobic reactor used in the OSA process thus differs significantly from aerobic + anoxic stages used in the wastewater handling units to obtain nitrification + denitrification, due to:

(a) no source of external substrate is present in the anaerobic reactor, because only the sludge feed is provided;
(b) solid concentration in anaerobic reactor is higher (approximately double) compared to that typical of activated sludge, which further guarantees the achievement of anaerobic conditions;
(c) the absence of nitrate or its rapid consumption if a moderate concentration is present.

The retention time in the anaerobic reactor should be sufficiently long. In the work of Saby et al. (2003) the retention time in the aerobic stage and in the anaerobic reactor was 6 and 10.4 h respectively; in this way the retention time under anaerobic conditions was 60% of the retention time in the entire OSA system (activated sludge + settler + anaerobic reactor).

The first study applying the OSA process to evaluate the reduction of excess sludge was carried out by Chudoba et al. (1992a). At lab-scale and using synthetic wastewater, the sludge production of a conventional activated sludge

---

[(2)] the concentration of soluble COD at the outlet from the anaerobic reactor depends from the ORP level maintained in the anaerobic reactor (soluble COD increases when ORP decreases).

system was compared to that of the OSA process. In the case of the OSA process, the anaerobic reactor was maintained at ORP of −250 mV. The observed sludge yield in the activated sludge system was 0.37 kgTSS/kgCOD$_{removed}$, while in the OSA process was 0.22 kgTSS/kgCOD$_{removed}$ (reduction of 40%).

An important parameter to be controlled in the anaerobic reactor is the oxidation-reduction potential, ORP (Chen et al., 2001b; Saby et al., 2003). Maintaining ORP at values around −250 mV prevents the prevalence of anaerobic conditions under which the COD removal ability may be reduced when the sludge returns from the anaerobic tank to the activated sludge tank (Saby et al., 2003).

The production of sludge in the OSA system (fed with synthetic wastewater) operating with an additional reactor maintained at ORP of +100 mV, −100 mV and −250 mV is indicated in Figure 8.7, compared to a conventional activated sludge system in which the sludge production was considered as 100%. The reduction of sludge in the OSA process was enhanced by lower values of ORP. For example, at ORP of −250 mV, the excess sludge was reduced by 50% compared to conventional activated sludge system.

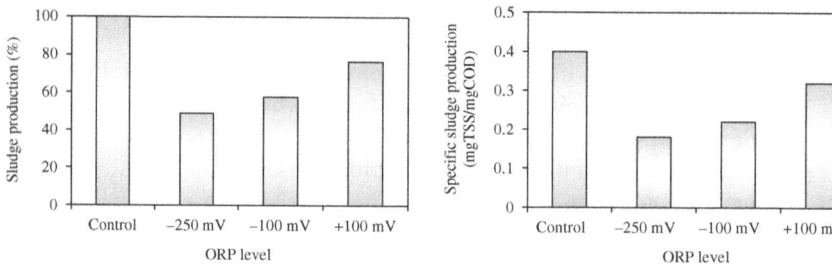

Figure 8.7. Variation of sludge production (as dry mass) obtained with an OSA process as a function of various ORP levels in the side-stream reactor (*from* Saby et al., 2003).

The observed sludge yield in the OSA system was 0.18–0.32 kgTSS/kgCOD$_{removed}$ compared to 0.40 kgTSS/kgCOD$_{removed}$ in the conventional activated sludge process (Figure 8.7).

The presence of the anaerobic reactor also affects the release of phosphorus and in particular, a higher increase of PO$_4$ concentration in the treated effluent from the wastewater handling units was observed for lower ORP values.

Using an OSA process, better settleability was observed (Chen et al., 2001b; Saby et al., 2003).

Microbiological analyses applied to investigate bacterial activity in sludge by using fluorescent staining with DAPI and CTC[3], demonstrated that the exposure of bacteria to anaerobic environments significantly reduces their activity (reduction of 50% at −250 mV) compared the conventional activated sludge process.

In the OSA process a significant presence of sulphate-reducing bacteria was observed and they were more active at ORP of around −250 mV.

In full-scale applications of the OSA process, optimal values of ORP should be set in the range between −200 and −250 mV, in order to avoid the formation of hydrogen sulphide and further problems when sludge returns to the activated sludge stages.

From an economic point of view, the upgrading of an existing WWTP inserting an OSA process, only requires the addition of a mixed anaerobic reactor. After some weeks of operation, ORP should independently reach low levels of around −250 mV (only at lab-scale a supply of N_2 gas may be needed to reduce ORP).

8.4.2 Cannibal® system

The process for sludge reduction known as Cannibal® system (Siemens Water Technologies Corp.) consists of both physical and biological processes and is based on a side-stream anaerobic digester (operating at ambient temperature), treating a part of the sludge return flow and integrated in the wastewater handling units. The flow scheme of the process, shown in Figure 8.8, is schematically composed of:

- extraction of a part of return sludge (about 50% of return flow rate);
- screening with very fine drum screens (i.e. at 250 μm): the separated material – characterised by 30–40% of dry solids and 90% of VS – is claimed to be inert organic material, primarily made up of hair and cellulose fibres. This physical process alone separates 20–30% of TS (Johnson *et al.*, 2008);
- an intermediate tank;
- treatment of a fraction of sludge in hydrocyclones to separate inert solids (grit);
- return of most of the sludge flow to activated sludge stage;

---

[3] DAPI and CTC are two fluorescent dyes used for the identification of bacteria cells under epifluorescence microscopy. In particular, CTC is utilised for discriminating active cells in mixed populations.

- introduction of the remaining part in the anaerobic reactor, also known as the interchange reactor;
- return of the sludge treated anaerobically to activated sludge stage.

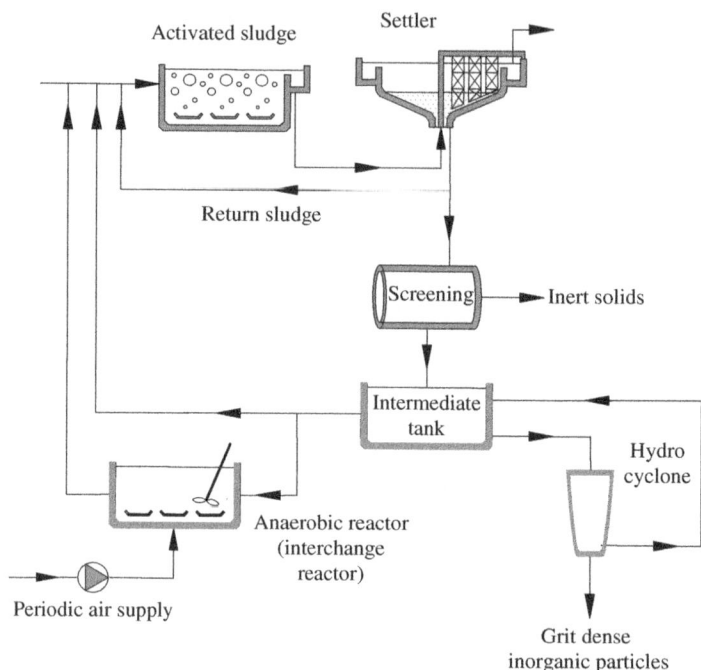

Figure 8.8. Flow scheme of the Cannibal® system integrated in the wastewater handling units.

Current applications do not include primary settling prior to secondary treatment (Johnson et al., 2008).

The most important stage in the Cannibal® system is the anaerobic reactor (interchange reactor) which operates with SRT of 8–15 d (typically 10 days). The HRT of the reactor differs in general from SRT due to a settling period coupled with surnatant extraction (using a SBR type decanter).

The anaerobic reactor is operated in a sequencing batch reactor mode with intermittent periods of aeration and quiescent holding; an example is shown in Figure 8.9 (Johnson et al., 2008).

In order to work near the boundary between anoxic and anaerobic conditions, the anaerobic reactor in equipped with ORP control and intermittent bubble aeration is supplied to produce nitrate (Johnson et al., 2008). In between aeration phases, the sludge is allowed to settle before decanting and subsequently mixed when solid discharge is desired.

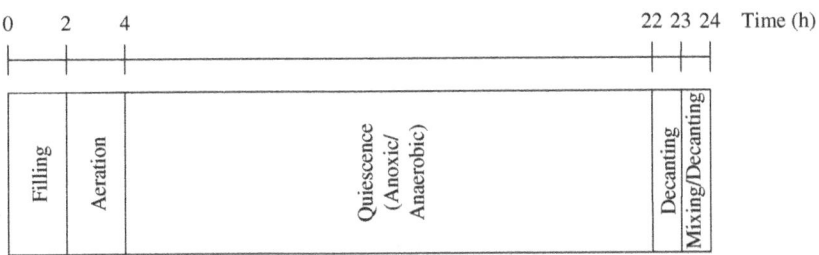

Figure 8.9. Sequence of phases in the Cannibal® system.

Reports from field operations indicate that the Cannibal® system reduces 40–70% of sludge mass (considering both biological mineralisation and physical separation), but only a few studies have been conducted up to now to fully understand and explain the basic mechanisms causing sludge reduction. Two recent works, giving some explanations on the mechanisms that account for the mass loss and the exact quantification of the amount of solids reduction, have been presented by Novak et al. (2006; 2007).

In the investigation reported by Novak et al. (2006; 2007) the Cannibal process was applied at lab-scale using synthetic wastewater.

The most important operational parameters were:

(1) the retention time in the anaerobic reactor which depends on the ratio between the volume and the influent sludge flow rate; HRT higher than 7–8 days was advised (Novak et al., 2006, Johnson et al., 2008);
(2) the interchange rate, which is the rate of solids passed through the Cannibal reactor, expressed as a percentage per day of the biomass in the activated sludge stage. An increase of the interchange rate from 4% to 7% resulted in a significant decrease in sludge production in the Cannibal process (Novak et al., 2007).

The generation of approximately 60% less solids in the Cannibal process than in the control was found by Novak et al. (2007), originating an observed sludge yield of 0.11 gVSS/gCOD (at the higher interchange rate of 7%), compared to 0.28 in the control process (conventional activated sludge + aerobic digestion). Under steady-state conditions the Cannibal process operated at an SRT of approximately 80 days, compared to 30 days of the control.

These results are confirmed by the finding referred to by Datta et al. (2009), evaluating the sludge reduction in an SBR system integrated with a side-stream anaerobic reactor. In this system the recycled biomass was subjected to feasting and fasting at high SRT (overall SRT of 100 d). The observed sludge yield in the SBR + side stream anaerobic reactor (fed with synthetic wastewater) was estimated to be 0.07 gTSS/gCOD, representing a net 63% sludge reduction compared to the control SBR. This sludge reduction can be sustained along with 90% phosphorus removal and complete nitrification.

In the study carried out by Goel and Noguerra (2006) at lab-scale SBRs, the Cannibal process achieved a 62% TSS reduction compared to a system without a side-stream anaerobic bioreactor. The authors highlighted that an appropriate comparison should be made between the Cannibal process and a control line equipped with conventional (usually mesophilic) anaerobic digestion. In this case the Cannibal system is estimated to achieve a 16–37% reduction of TSS with respect to a control system equipped with anaerobic digestion (Goel and Noguerra, 2006).

In the studies by Novak et al. (2007), Datta et al. (2009) and Goel and Noguerra (2006), the activated sludge systems were fed with synthetic wastewater. It must be noted that real wastewater has a significant presence of inert particulate COD (refractory and accumulating in the sludge) which will contribute up to 30–40% of total solids, limiting the overall sludge reduction.

In their work, Novak et al. (2007) evaluated the Cannibal process comparing two lab-scale systems, one of which had SRT of approximately twice the other. They demonstrated that the sludge reduction is not the result of the high SRT, but, on the contrary, the high SRT is the result of the lower sludge production in the Cannibal process (Novak et al., 2007). Therefore care must be taken when using the actual SRT (see § 7.8) as design parameter for activated sludge processes.

With regards to the explanation of the mechanisms involved in sludge reduction in the Cannibal process, Novak et al. (2007) evaluated field data collected at the Byron WWTP (Illinois), which operated a Cannibal process. The measurement of protein and polysaccharides across the system demonstrated clearly that after anaerobic treatment:

- soluble proteins are released, and these proteins are readily degraded in a subsequent aerobic environment;
- polysaccharides also increase, but less than protein.

These findings suggest that the mechanism for solids reduction by the Cannibal process is the release of organic matter in the anaerobic reactor, and this material is then biodegraded under aerobic conditions (Novak et al., 2007).

The Cannibal process seems to contribute to phosphorus removal, presumably because the anaerobic environment in the side-stream anaerobic bioreactor favours the growth of fermenting bacteria and biological phosphorus removal. However, the fate of P in the system remains enigmatic, because the measurement of total P accounted for only 67% of the influent P load in the study by Goel and Noguerra, (2006), but the reason for this is not yet understood.

8.5 THERMOPHILIC ANAEROBIC DIGESTION

If the anaerobic digestion is carried out under thermophilic conditions, instead of mesophilic, some advantages can be obtained:

- acceleration of biochemical reactions,
- lower solids retention time: conventional mesophilic digesters are normally designed for a retention time of 20–30 days, while in the case of thermophilic digestion the retention time can be reduced to around 10–12 days (Nielsen and Petersen, 2000),
- higher biogas production rate,
- considerable pathogen inactivation,
- higher resistance to foaming (Palatsi *et al.*, 2009).

These advantages of thermophilic anaerobic digestion result in more compact reactors for treating the same amount of sludge. The VSS reduction during thermophilic digestion reaches 40–45%, corresponding to a TS reduction of 30%, similar to conventional anaerobic digestion but obtained in a shorter digestion time.

Despite the major energy needs, thermophilic digestion is a good option if heat recovery from the effluent sludge is applied. Furthermore, the biogas produced in the digester is normally used for both power and heat production. In order to optimise these benefits it is important that the energy consumption for heating the sludge is kept as low as possible, by means of (Nielsen and Petersen, 2000):

- increasing of the solids content in the fed sludge; to obtain solid content higher than 3% TS, mechanical thickening is generally necessary;
- recovering part of the heat from the digested sludge by means of sludge/sludge heat exchangers; in full-scale WWTPs the typical heat recovery is 50–70%.

In Figure 8.10 a typical flow diagram for thermophilic anaerobic digestion with heat recovery is shown.

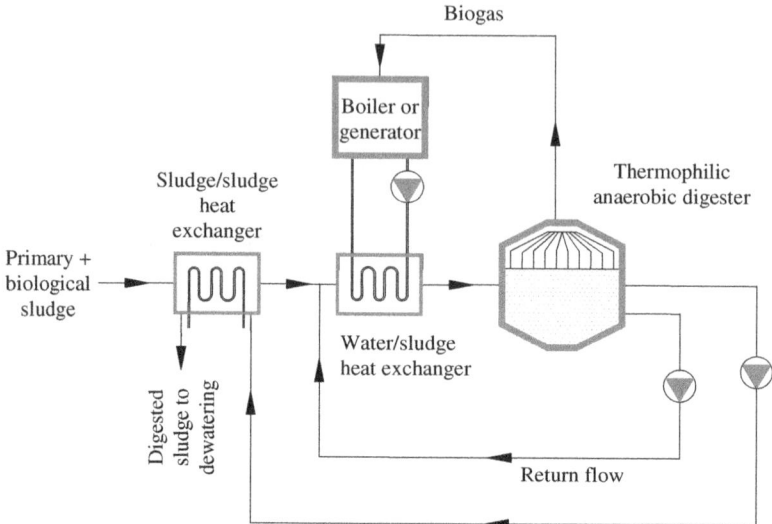

Figure 8.10. Flow diagram for thermophilic anaerobic digestion with heat recovery (*modified from* Nielsen and Petersen, 2000).

At the Prague Central WWTP at the end of the '90s, the temperature increase from the mesophilic to thermophilic level in the full-scale digesters led to better use of the existing facilities and a higher degradation efficiency associated with higher biogas production and improvement of the energy balance of the process. Due to a careful and slow increase of the operating temperature, the thermophilic process was very stable (Zábranská *et al.*, 2000b). Biogas was used in cogeneration units for electricity and heat energy production: 2.2 kWh of electric energy and 3.4 kWh of heat energy was obtained per m^3 of biogas (approximately 10 kWh per 1 Nm^3 CH$_4$). The higher energy consumption for the thermophilic temperature was also influenced by the insulation properties of tanks and air temperature. With air temperature below −5°C, 4300 m^3 of biogas per day and per digester (V = 4823 m^3) are required, fully covered by the heat energy from increased biogas production (Zábranská *et al.*, 2000b).

Methanogenic activity increases passing from mesophilic to thermophilic conditions, but the specific growth rate of thermophilic microorganisms is significantly lower than mesophilic ones, because of the greater maintenance energy required by thermophilic organisms.

Another advantage of thermophilic digestion is related to the pathogen inactivation induced by the high temperatures.

One drawback of thermophilic digestion is odour formation: in fact, the significant presence of ammonia and VFA causes persistent unpleasant odours in the sludge which is produced by thermophilic digestion rather than mesophilic.

8.6 THERMOPHILIC AEROBIC REACTOR

The transformations of sludge occurring during a thermophilic aerobic treatment are the result of a thermal action (see § 11.2) and an enzymatic (biological) attack carried out under aerobic or microaerobic conditions, described in this section.

The thermophilic aerobic treatment was applied for sludge reduction integrated in the wastewater handling units as described in § 8.6.1 and called S-TE PROCESS® (Sakai *et al.*, 2000; Shiota *et al.*, 2002).

The thermophilic aerobic treatment can also be integrated in the sludge handling units as a pre-treatment before anaerobic digestion; this application is described in § 8.6.2.

Thermophilic bacteria

Yan *et al.* (2008) demonstrated that after 1 h of thermal treatment at 60°C nearly 98% of the mesophilic bacteria in sludge died, leaving 2% of thermophilic bacteria, which were able to secret protease and grow.

Some strains of thermophilic bacteria – belonging to *Bacillus* spp. – and in particular the strain *Bacillus stearothermophilus* SPT2-1 (gram-positive) – are capable of solubilising organic sludge efficiently by secreting hydrolytic enzymes, such as heat stable extracellular proteases, which help to degrade the microbial cells by hydrolysing the cell wall compounds (Hasegawa *et al.*, 2000; Song and Hu, 2006).

The bacterium type SPT2-1 is rod-shaped, able to form endospores, and it belongs to the group of facultative aerobes or microaerophiles, although it is not able to grow in anaerobic environments. It grows in a range of pH between 5.0 and 8.5 and at an optimal temperature of 55–70°C (Hasegawa *et al.*, 2000). Other strains which are also capable of hydrolysing sludge were isolated by Song and Hu (2006): all of them were gram-positive and able to form endospores. Although strains capable of lysing bacterial cells such *E. coli* have been isolated, little is known about the mechanisms involved in the lysis process.

In general, the process of enzymatic hydrolysis carried out by bacterial strains such as *Bacillus* spp. is associated with a thermal treatment, which also renders the cells more susceptible to enzyme attack.

The influence of temperature in the thermophilic range

The influence of temperature in the range 50–80°C on sludge solubilisation is indicated in Figure 8.11, which refers to the results obtained by Hasegawa et al. (2000) at lab-scale. An approximate value of 30% of VSS was solubilised due to the thermal effect (Figure 8.11). A further increase in VSS solubilisation was observed in the range 55–70°C, due to thermophilic bacteria whose hydrolytic activity is highest at these temperatures (Hasegawa et al., 2000). Some organic matter solubilised by thermophilic bacteria is also assimilated by them.

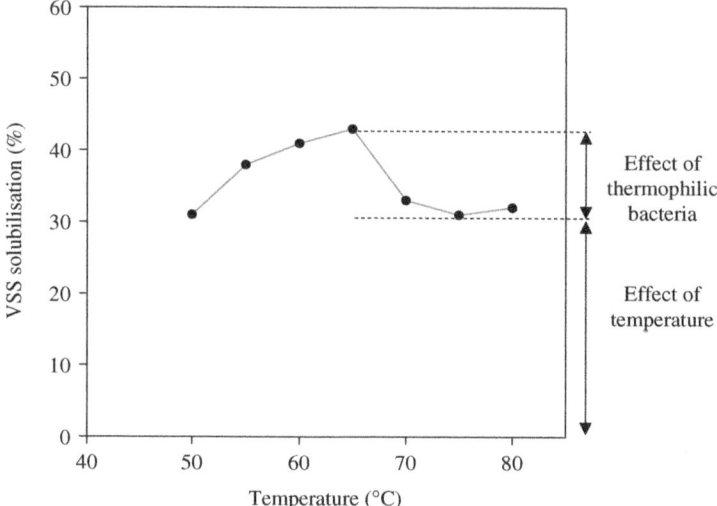

Figure 8.11. Influence of temperature (in the thermophilic range) on sludge solubilisation (*from* Hasegawa et al., 2000).

In synthesis we can conclude that:

(1) at T < 70°C sludge hydrolysis is due to thermal action + thermophilic bacteria, which contribute to oxidise a part of the solubilised compounds;
(2) at T > 70°C the solubilisation is only due to the thermal action, whilst the biological one is negligible.

The thermostable proteases produced by *Bacillus* spp. shows at 75°C the optimum temperature for maximum extracellular endoproteolytic activity (Kim et al., 2002). However, this temperature is not ideal for the thermophilic aerobic process, which should be around 60°C for pratical operations.

The influence of divalent ions

The enzymatic hydrolysis of proteins is considered to be the rate limiting step during sludge digestion. The enhancement of the activity of the enzyme protease excreted by *Bacillus stearothermophilus* was investigated by Kim et al. (2002) and a low concentrations of Fe^{2+} and Zn^{2+} and a relatively high concentration of Ca^{2+} enhance the efficiency of the thermophilic aerobic digestion of industrial sludge, obtaining reductions in TSS, DOC and proteins.

The influence of aerobic and microaerobic conditions

The influence of the dissolved oxygen (DO) concentration on the activity of thermophilic aerobic bacteria and sludge solubilisation was investigated by Hasegawa et al. (2000) at lab-scale. Figure 8.12 shows the effect of aeration intensity on sludge solubilisation and VFA accumulation during thermophilic aerobic treatment. The values of aeration intensity indicated in Figure 8.12 are expressed as $v_a/v_r/min^{(4)}$, establishing the following conditions (Hasegawa et al., 2000):

(1) aerobic conditions at 0.4 $v_a/v_r/min$ (DO = 1–3 mgO_2/L);
(2) microaerobic conditions at 0.08 $v_a/v_r/min$ (DO = 0–0.1 mgO_2/L).

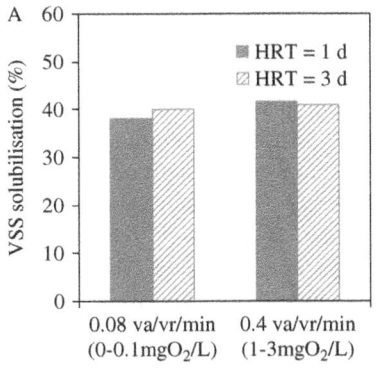
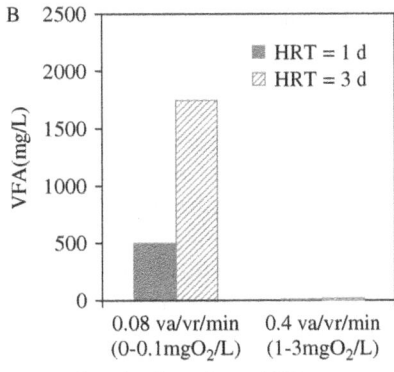

Figure 8.12. Influence of aeration intensity on (A) sludge solubilisation and (B) VFA production (*from* Hasegawa et al., 2000).

---

[4] $v_a/v_r/min$ indicates the volume of air supplied (v_a) per unit of volume of the reactor (v_r) and per minute (min).

Under aerobic conditions (1–3 mgO$_2$/L) the VSS solubilisation is about 40% or higher after HRT of just one day. Organic matter removal was 16% (due to oxidation) and no accumulation of VFA was observed (Figure 8.12A and B).

Under microaerobic conditions (0–0.1 mgO$_2$/L) a comparable VSS solubilisation (about 40%) was found, but a large amount of VFA was accumulated, which can not be removed due to the lack of oxygen (Figure 8.12A and B) (Hasegawa *et al.*, 2000).

It must be observed that in the thermophilic aerobic process, there is an increasing difficulty in the transfer of oxygen in the bulk liquid, due to its low solubility at high temperatures. Higher concentrations of DO at T > 50°C can only be obtained by supplying pure oxygen.

The influence of contact time

In the investigation by Kim *et al.* (2002), the TSS concentration decreased rapidly (20% of TSS reduction) in the first 12 h of sludge contact in the thermophilic aerobic reactor, due to the solubilisation of solid particles caused by temperature increase (see the effects of thermal treatment at temperatures around 65°C in § 11.2). The reduction of TSS is slower after the first 12 h and up to 80 h of treatment (35% of TSS reduction) and is mainly due to the lytic activity.

8.6.1 Integration in the wastewater handling units (S-TE PROCESS®)

The thermophilic aerobic treatment can be applied for sludge reduction integrated in the wastewater handling units. In this configuration, when thermophilic aerobic bacteria are subjected to optimal conditions for their growth, such as in the thermophilic aerobic reactor, they perform enzymatic hydrolysis at expenses of other microbial cells. When they return to the activated sludge stages, their activity stops and some will form spores, which will be recirculated in the thermophilic reactor with the return sludge (Figure 8.13).

The scheme of the S-TE PROCESS® (S-TE = Solubilisation by thermophilic enzyme) proposed by Kobelco Eco-Solutions Co. (Japan) integrated in the wastewater handling units is indicated in Figure 8.14 (Sakai *et al.*, 2000; Shiota *et al.*, 2002). The process is based on the cell lysis occurring in a thermophilic aerobic reactor treating a part of the return sludge. The sludge is pre-heated in a heat-exchanger and enters the thermophilic aerobic reactor, where it is solubilised both by the thermal effect and by the enzymatic action of thermophilic bacteria. The lysated sludge passes through the heat-exchanger again to heat the inlet sludge and is then recirculated to the activated sludge stages where cryptic growth occurs at the expense of the lysate (using the mechanism of cell lysis-cryptic growth, see § 4.2).

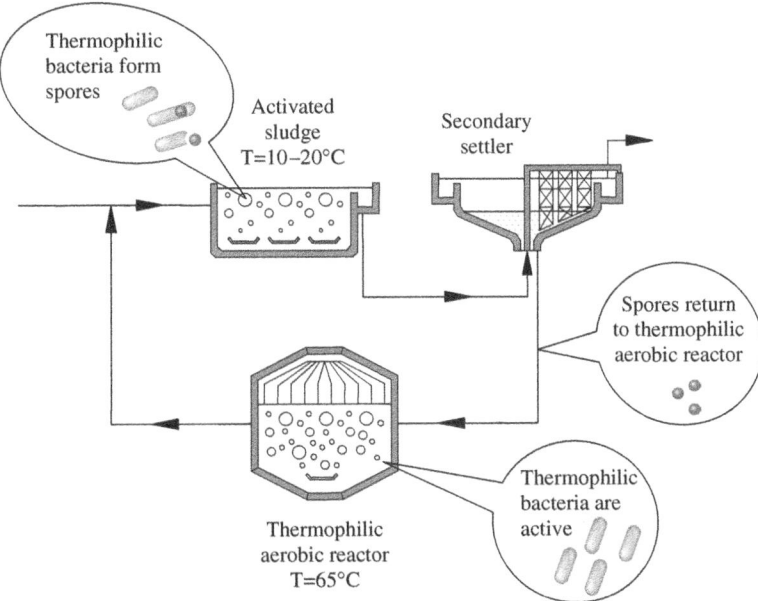

Figure 8.13. Thermophilic aerobic bacteria pass from the thermophilic aerobic reactor to the activated sludge stage, where they become inactive and form spores which return again to the thermophilic aerobic reactor.

The S-TE process was applied at full-scale in a small WWTP in Japan (*Sawatari Water Quality Control Center*, Sakai *et al.*, 2000; Shiota *et al.*, 2002) with the following characteristics:

– average flow rate of influent wastewater of 250 m^3/d;
– activated sludge (volume of 400 m^3);
– thermophilic aerobic reactor (volume of 5 m^3), completely mixed, aerated and operating at 65°C with HRT from 1 to 3 d.

Before entering the thermophilic reactor, the sludge is mechanically thickened in order to save energy when heating the sludge thus avoiding energy wastage in the liquid phase.

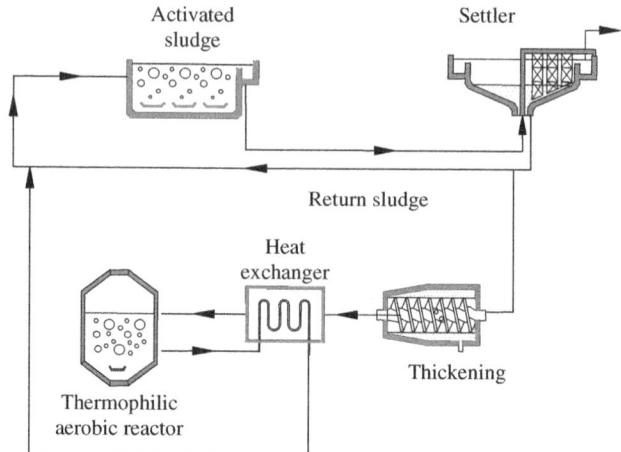

Figure 8.14. Scheme of thermophilic aerobic reactor integrated in the wastewater handling units.

The VSS solubilisation in the termophilic reactor – affected mainly by the retention time – was 30–40% of fed VSS. The solubilisation of sludge caused an increase of BOD and COD in the lysate, which is degraded in the activated sludge stage. The specific sludge production – expressed as kgTS per unit of influent wastewater volume – was 0.028 kgTS/m^3 in the system where the S-TE process was integrated, compared to 0.11 kgTS/m^3 in the same plant but before the installation of the S-TE process (sludge reduction of 75%) (Shiota et al., 2002). In the case of the treatment of municipal wastewater, even after the installation of an S-TE process, a certain percentage of waste sludge always remains and it is not possible to achieve zero production, because of the presence of non-biodegradable organic matter and inorganic inert solids.

The concentrations of TSS, BOD and COD in the effluent from the wastewater handling units remained optimal, while the P removal decreased because of the reduced use of P for bacteria synthesis. The VSS/TSS ratio of sludge passed from 0.85–0.90 before the introduction of the S-TE process, to 0.70–0.75 with the S-TE process in operation (Shiota et al., 2002).

The costs of the S-TE process are mainly due to the building of the reactor, heating at thermophilic temperatures, which is in any case modest due to the compact volume of the reactor (due to the short retention time).

As far as we are aware, to date there are some sites applying this process in Japan and other countries. This technology is supplied by Kobelco (Japan) and was also licensed to Ondeo-Degrémont (France) in 1999 (under the name Biolysis®E).

8.6.2 Integration in the sludge handling units. Dual digestion

In the two-stage scheme where the thermophilic aerobic reactor – operating with a short retention time, of 1–2 days – is applied as a pre-treatment before a conventional mesophilic anaerobic digestion (Figure 8.15), some advantages can be achieved:

- pathogen inactivation,
- great flexibility among the final options for disposal/re-use of sludge,
- enhancement of solid reduction,
- increase of biogas production.

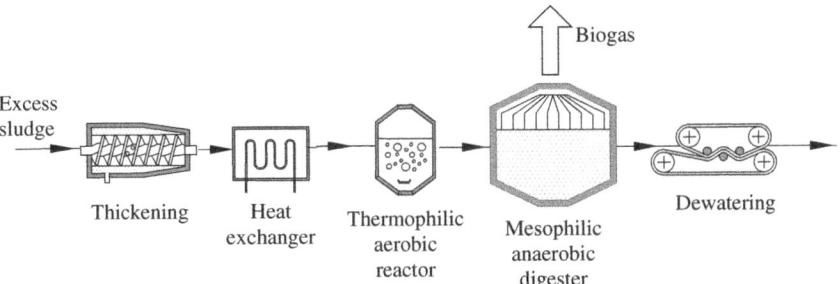

Figure 8.15. Thermophilic aerobic reactor + mesophilic anaerobic digestion (dual digestion).

The solubilisation of organic matter in the sludge occurs through the thermal effect and the hydrolysis caused by thermophilic aerobic bacteria.

Considering the results from the VSS solubilisation and the VFA production in the thermophilic aerobic reactor, depending on the aerobic or microaerobic conditions (see pag. 133), it can be observed that the maximum release of VFA – which are extremely useful to improve biogas production in the subsequent mesophilic anaerobic digestion – can be obtained by maintaining the sludge under microaerobic conditions. The microaerobic conditions allow other advantages to be achieved, such as:

- VSS solubilisation ranging from 10–20% to 40% after just 1 d of HRT;
- lower energetic consumption for aeration;

– lower concentration of residual oxygen in the sludge to be fed in the mesophilic anaerobic digester.

The biogas production after the thermophilic aerobic pre-treatment was investigated by Hasegawa *et al.* (2000) and the results are indicated in Figure 8.16 and compared to the biogas production without pre-treatment (control).

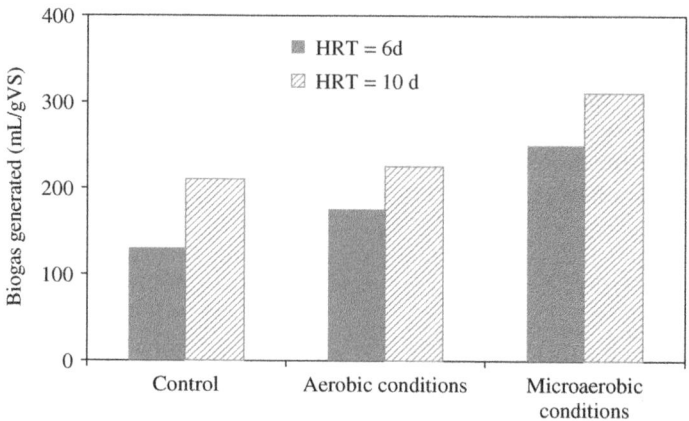

Figure 8.16. Production of biogas during mesophilic anaerobic digestion after pre-treatment in the thermophilic aerobic reactor at different aeration intensities compared to the control (without pre-treatment) (*from* Hasegawa *et al.*, 2000).

A moderate increase of biogas production was observed in the sludge pre-treated in the thermophilic reactor under aerobic conditions, whilst a 1.5-fold biogas increase was generated in the sludge pre-treated under microaerobic conditions (Hasegawa *et al.*, 2002).

8.6.3 Integration in the sludge handling units. Autothermal thermophilic aerobic digestion

Thermophilic aerobic digestion of sludge is exothermic and thus it can be autothermal with appropriate heat retention and heat exchange.

The operating principle of the autothermal thermophilic aerobic digestion – introduced some 30 years ago as a means of retrofitting conventional digesters – is based on the conservation of the heat energy released during the aerobic biodegradation of sludge with the aim of maintaining thermophilic temperatures (55–70°C).

Biological treatments

The main advantages of this technology are the following:

- removal of organic matter of around 50%;
- short retention time (SRT = 6–12 d) compared to other configurations of aerobic or anaerobic digetsers, resulting in smaller volumes and space saving;
- fast pathogen inactivation, due to the high operating temperature, contributing to the pasteurisation of sludge.

The most widely used configuration is characterised by approximately 6 d retention time, in one reactor or two in series, with daily feeding (Figure 8.17). In the two-stage configuration, the sludge undergoes the aerobic process at 35–50°C in the first stage and at 50–70°C in the second stage.

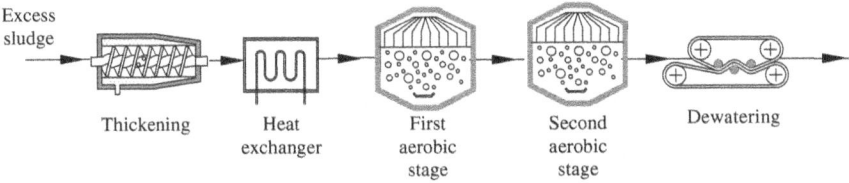

Figure 8.17. Autothermal thermophilic aerobic digestion.

The hydrolysis and degradation kinetics of particulate matter in the autothermal thermophilic aerobic digestion were investigated by Csikor *et al.* (2002). Cell lysis is accelerated by heat stable exoenzymes, excreted by thermophilic bacteria and by a physico-chemical effect caused by a strong heat shock when the sludge enter a hot environment at 50–60°C.

A comprehensive model for autothermal thermophilic aerobic digestion was developed by Gómez *et al.* (2007), aimed at predicting the dynamics of temperature, liquid and gas compounds on the basis of formulations of biochemical reactions, physico-chemical transformations and thermal energy balances. Kovács *et al.* (2007) proposed an adaptation of the Activated Sludge Model No. 1 (ASM1), to include also thermophilic bacteria and their kinetics.

With regards to the aeration of the digesters, optimisation is needed in order to avoid (Zambrano *et al.*, 2009):

- over-aeration, which causes an increase in costs, a cooling effect on the sludge with the consequent risk of pasteurisation temperatures not being reached;

- under-aeration, which limits the efficiency for stabilisation, reduces heat generation and promotes anaerobic conditions associated with potential undesired odours in the outlet off-gas.

Recently, a generation of autothermal thermophilic aerobic digestion, based on automatic control of the aeration system through more sophisticated devices has been proposed, in which the flow-rate of the injected gas stream is automatically controlled. Conventional industrial sensors are often not appropriate for withstanding the aggressive operational environments such as in this technology, characterised by the corrosive nature of the digesting sludge, high temperatures and high TSS concentrations (Zambrano et al., 2009). Among the sensors available, Zambrano et al. (2009) proposed red-ox potential (ORP) and temperature sensors for control operations during autothermal thermophilic aerobic digestion. These authors modelised a new control strategy based on ORP for regulating external aeration in a non-continuously fed reactor; the control allows the complete depletion of biodegradable organic substrate fed into the digester (occurrence of a bend-point in the temperature graph, called "bCOD knee") to be detected automatically. The lack of substrate to be oxidised makes external aeration unnecessary, allowing overall aeration costs to be reduced (Zambrano et al., 2009). As far as we are aware, experimental verification of the simulation results has not yet been reported.

Several full-scale investigations have evaluated the efficiency of autothermal thermophilic aerobic digestion for sludge disinfection: these studies have demonstrated that most indicator organisms are already significantly reduced in the first aerobic reactor, while enterococci are the most resistant group of bacteria (Zábranská et al., 2003).

8.7 ENZYMATIC HYDROLYSIS WITH ADDED ENZYMES

The use of enzymes for the hydrolysis of specific components (e.g. cellulose) in wastewater and municipal solid waste or for the enhancement of sludge dewaterability was widely studied in the '80s and '90s. Various types of enzymes were identified for wastewater and sludge treatment purposes (Karam and Nicell, 1997). These enzymes can derive from influent wastewater or can be produced by the sludge itself as extracellular enzymes. The latter can be located on cell walls, or dissolved in the bulk liquid, although most extracellular hydrolytic enzymes are immobilised by adsorption in the EPS matrix (Parmar et al., 2001; Vavilin et al., 1996).

The EPS, which play a key role in binding floc components together, are mostly composed of carbohydrates, proteins, lipids, nucleic acids and humic substances. The main forces involved in these binding functions are van der Waals, hydrophobic, and electrostatic interactions, which are mediated by cationic polymer bridging (Palmer *et al.*, 2001). Changes in EPS composition occur, for example, during anaerobic storage of sludge (see § 8.2.2, § 8.2.3).

Hydrolytic enzymes break down EPS through multi-step processes summarised as follows (Ayol *et al.*, 2008; Wawrzynczyk *et al.*, 2008):

(1) initially the hydrolytic enzymes adsorb to the sludge-substrate and attack the polymeric substances forming enzyme-substrate complexes;
(2) small polymers loosely bound to the surface are hydrolysed;
(3) cell lysis transfers cell content into the medium;
(4) solubilisation of sludge solids occurs;
(5) the end products are biodegraded by microbial metabolism;
(6) the more compact part of sludge flocs is hydrolysed at a lower rate, because enzymes diffuse with difficulties inside the floc matrix.

It is considered that the main mechanisms which occur during enzymatic treatment are sludge solubilisation, cell lysis and cryptic growth.

Due to the high presence of proteins, carbohydrates and lipids in the composition of excess sludge, the addition of protease, lipase, cellulase, emicellulase and amylase could be advisable.

Sludge solubilisation by enzymes is more intense the nearer the temperature is to the 50°C, considered optimal for hydrolitic enzyme activity. For this reason, the enzymatic treatment is best applied with anaerobic digestion in mesophilic or thermophilic conditions.

The use of commercial enzymes for the hydrolysis of organic matter has often been proposed for the improvement of sludge biodegradation and reduction, the enhancement of biogas production when used as a pre-treatmet prior to anaerobic digestion (inter alia Parmar *et al.*, 2001; la Cour Jansen *et al.*, 2004), or to enhance organic waste degradation (Lagerkvist and Chen, 1993).

However, many of these enzymatic products are patented and their exact composition is generally confidential; in some cases these mixtures contain other stimulatory nutrients as well as enzymes.

The mechanisms for floc disintegration by externally added enzymes during aerobic and anaerobic bioprocessing have recently been described by Ayol *et al.* (2008). In lab-scale tests, different hydrolytic enzymes (0.1%) were applied to an anaerobic reactor (granular anaerobic sludge at 37°C) and an aerobic reactor (suspended sludge at 28°C). The effects of enzyme addition were: improvement of floc disintegration, significant reductions in EPS in both aerobic and anaerobic reactors, while better filterability and higher biogas production for the anaerobic reactor (Ayol et al., 2008).

Wawrzynczyk et al. (2008) investigated the mechanism of enzyme adsorption on sludge flocs and the role of cations such as Ca^{2+}, Mg^{2+} or Fe^{3+} on the enzymatic solubilisation of sludge. The removal of these cations leads to the disruption of flocs and to the consequent release of proteins, carbohydrates and humic substances (Dey et al., 2006). This suggests that the treatment of sludge with cation-binding agents would cause a disintegration of sludge, resulting in an increase of the specific surface area available for enzymatic hydrolysis (Dimock and Morgenroth, 2006). For example, cation-binding agents were used for the extraction of EPS from sludge (Park and Novak, 2007; Wawrzynczyk et al., 2007).

Some types of cation-binding agents are ethylenediaminetetraacetic acid (EDTA), citric acid and tripoliphosphate sodium (STPP) (Wawrzynczyk et al., 2008). Reduced sulphur compounds (H_2S, HS^-, S^{2-}) are also a cation-binding agent able to favour the reduction of floc size and enhance the enzymatic activity during methanogenesys (Watson et al., 2004).

Sludge solubilisation by enzymes (lipase, cellulase, α-amilase, endo-xylanase, dextranase, protease: 68.5 mg of enzymes per 1 g TS of sludge) was improved by the addition of these cation-binding agents and citric acid was the most effective (Wawrzynczyk et al., 2008). However, the dosages used are very high and authors did not recommend them for practical applications at large/industrial scale.

In a lab-scale investigation, Parmar et al. (2001) evaluated the effects of a combination of commercial protease, lipase, cellulase and hemicellulase on the reduction of anaerobically digested sludge (3.1% solids). Tests were carried out with a dosage of pure enzymes of 0.1% w/w and by incubation at 40°C (tests are representative of conditions found in mesophilic digesters). The TSS reduction after enzymatic treatment with contact time of 48 h, was 22.6% compared to 6.1% observed in untreated sludge (Parmar et al., 2001). The greatest percentage reduction among organic components was for protein and proteases were the most effective in reducing sludge solids and in improving settling. The removal of protein is thought to weaken the floc, which explains the role of proteases in solid settling (Parmar et al., 2001).

Lab-scale experiments were also carried out by la Cour Jansen et al. (2004) to evaluate enzyme treatment on the basis of solubilisation of particulate matter and methane production. After the addition of enzyme dosages of 1% w/w (6 different enzymes with 0.06 % w/w of each enzyme, final concentration per 1% w/w of the sludge TS) in a mixture of primary + secondary sludge, the percentage of solubilised particulate COD was 35–38%. When used as pretreatment before anaerobic digestion, methane production increased by about 20% compared with untreated sludge. Dewatering properties were also improved, leading to a reduction of total sludge volume, but no information was given on sludge mass reduction.

Enzymatic treatment + anaerobic digestion was applied at full-scale for a 6-month period, by adding two glycosidic enzymes (Recktenwald et al., 2008). The authors previously verified that glycosidic enzymes gave better sludge solubilisation than proteases and lipases at lab-scale. The dosage of glycosidic enzymes was 2.5 kg of each enzyme solution per tonne of TS fed to the digester. The dosage point was the heat exchanger system (Figure 8.18), where the sludge was heated to 55°C, in the optimal temperature range for these enzymes (45–60°C). The installation of a dosage system for the enzymes represents the only investment cost.

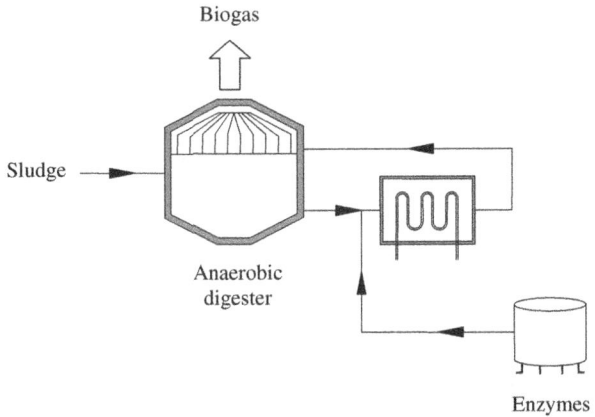

Figure 8.18. Scheme of enzyme addition in the anaerobic mesophilic digestion (*according to* Recktenwald et al., 2008).

The enzyme-treated sludge showed an increase in gas production of 10 to 20% compared to the reference digester, because hydrolytic enzymes improved the hydrolysis step prior to acidogenesis. In the enzyme-treated sludge the dry content after dewatering reached 31–33% (reference level of 27–30%) and the polymer dosage for dewatering decreased by 10–20%. There was no great impact on the quality of the water phase recirculated in the wastewater handling units (Recktenwald et al., 2008).

Folic acid (vitamin B) has been reported to contribute to reducing sludge yields and an industrial application was reported to in a U.S. boardmill, but no more information has been found by us (Akerboom *et al.*, 1994, referred to by Mahmood and Elliott, 2006).

Enzyme mixtures have been proposed for sludge reduction, but the actual cost and the optimal dosages are the most important factors for their use at full-scale. To date, there have been several experiences using pure enzymes, but promising results are often related to very high enzyme dosages. For this reason the effective viability of the application of extracellular enzymes at full-scale has not yet been thoroughly investigated.

8.8 ADDITION OF CHEMICAL METABOLIC UNCOUPLERS

Types of metabolic uncouplers

The process of uncoupled metabolism has been studied since the '90s for the reduction of sludge production (Chen *et al.*, 2000; Low and Chase 1998; Low *et al.*, 2000; Mayhew and Stephenson, 1998; Okey and Stensel, 1993).

Many chemicals of varying types and mode of action have been reported to cause the metabolic uncoupling in bacterial and eukaryotic cells alike. The antibiotic group of chemicals contain several uncouplers and inhibitors (Mayhew and Stephenson, 1997).

One of the first chemical metabolic uncouplers studied since the middle of the 20th C. was 2,4-dinitrophenol (DNP). Others uncouplers of early interest were chlorinated and nitrated phenols: 4-chloro-2-nitrophenol (CNP), 2,4-dichlorophenol (DCP), 2,4,5-trichlorophenol (TCP), ortho-chlorophenol (oCP), para-nitrophenol (pNP), meta-chlorophenol (mCP), meta-nitrophenol (mNP), pentachlorophenol (pCP), *o*-nitro-*p*-chlorophenol, 2,4,6-tribromophenol, 2,6-dibromo-4-nitrophenol.

Also 3,3',4',5-tetrachlorosalicylanilide (TCS), a component in the formulation of soaps and shampoos, can reduce the transfer resistance of cell membranes to proton, thus upsetting the balance of protons within the cell, resulting in excessive energy use to recover this balance (Ye and Li, 2005). As a consequence, energy used in anabolism is reduced and the cell growth rate decreases.

Metabolic uncouplers include various types of molecules but many are lipophilic weak acids and protonophores. They diffuse relatively freely through the phospholipid bilayer and the transport rate is proportional to the concentration gradient across the membrane. Once inside the membrane, the deprotonation of the phenolic hydroxyl dissipate the proton gradient which is a driving force for ATP production.

The presence of chemical uncouplers in the culture medium causes energy spilling; this phenomenon results in an effective pathway for bacteria to protect themselves against abnormal living conditions, but at the cost of consuming non-growth energy (Liu, 2000).

As a result, higher concentrations of metabolic uncouplers favour a higher energy dissociation, a reduction of maximum growth yield and thus sludge reduction is expected.

Reduction of observed sludge yield (Y_{obs})

Studies on pure microbial cultures have demonstrated effective reduction of the observed growth yield in the presence of these metabolic uncouplers and the feasibility of achieving sludge reduction. In a chemostat with *Pseudomonas putida*, the dosage of 100 mg/L pNP reduced the efficiency of biomass production by up to 62% with a simultaneous increase in the specific substrate uptake rate (Low and Chase, 1998).

Efficiency in the reduction of sludge production depends on the type of compound added and on the dosage. Some results are described below and summarised in Table 8.5:

(a) Strand *et al.* (1999) compared the effects of 12 chemical metabolic uncouplers on Y_{obs} in activated sludge. The most efficient compounds were TCP and CNP. When 10 mg/L TCP was applied, the reduction of Y_{obs} was 50%. For lower dosages and for longer periods, acclimatisation phenomena can occur which lead to the progressive bacteria ability to degrade the compounds with resulting loss of efficiency;

(b) activated sludge treated with DNP (4 mg/L) demonstrated a lower Y_{obs} (0.30) than the control (0.42) as referred by Mayhew and Stephenson (1998); the BOD removal capabilities of both control and DNP treated activated sludge were comparable;

(c) at lab scale, testing activated sludge, Low *et al.* (2000) obtained a reduction of growth yield of 30% after addition of 100 mg/L pNP. By adding pNP the efficiency of substrate removal decreased by 25%, settleability worsened causing the loss of solids in the effluent, protozoa were reduced and a proliferation of filamentous bacteria was observed;

(d) at the dosage of 100 mg/L pNP in the activated sludge stage the reduction of excess sludge was:
- 16% of TSS (18% of VSS) in the case of wastewater handling units fed with raw wastewater;
- 27% of TSS (30% of VSS) in the case of wastewater handling units fed with settled wastewater (Chase *et al.*, 2007);

(e) at 20 mg/L pCP, Y_{obs} decreased by 58% and the reduction in COD removal was 8.9%, as compared to the control without the addition of pCP (Yang et al., 2003);

(f) at the dosage of 20 mg/L mCP, a Y_{obs} reduction of 86.9% was achieved, while the COD removal efficiency was reduced by only 13.2% (Yang et al., 2003); mCP was found to be more effective in reducing sludge production compared to pCP;

(g) at an mNP concentration of 20 mg/L, Y_{obs} decreased by 65.5% and a reduction of 13.2% in COD removal efficiency was observed (Yang et al., 2003);

(h) at an o-nitrophenol (oNP) concentration of 20 mg/L, sludge production was reduced by 86.1% and COD removal efficiency decreased by 26%; therefore o-nitrophenol appears to be a stronger metabolic uncoupler in terms of sludge reduction compared to mNP (Yang et al., 2003);

(i) in activated sludge Y_{obs} was reduced by around 40–46% (in batch activated sludge cultures) by the addition of 0.8–1.0 mg/L TCS (Chen et al., 2001b; Chen et al., 2002). Similar findings are referred by Ye and Li (2005) which indicated a reduction of Y_{obs} of approximately 30% at a dosage of 1 mgTCS/gTSS (around 3 mg/L), without adversely affecting the substrate removal capability and sludge settleability. One drawback was the increase of the effluent nitrogen concentration after TCS addition: in fact, the mean effluent NH_4-N concentrations in the system with TCS addition were almost twice those of the control system and the mean effluent TN levels in the two systems were 47.71 and 58.20 mg/L.

Yang et al. (2003) found that dissociation of energy metabolism is proportionally related to the concentration of metabolic uncouplers. It is important to consider that the addition of metabolic uncouplers does not block electron transport along the respiratory chain to oxygen, and therefore the efficiency of substrate removal often remains good. For example, TCS is able to induce the decrease of energy level in the cells, without changes to cell storage, cell viability, cell division and cell size (Chen et al., 2000).

Other authors also refer that substrate removal efficiencies were not affected by the presence of TCS (Ye and Li, 2005), as also reported for TCP (Strand et al., 1999), DNP (Mayhew and Stephenson, 1998); in the case of pNP, the decrease of substrate removal efficiency was attributed to a surmountable effect arising from the design of the apparatus (Low et al., 2000).

Table 8.5. Comparison of the reduction of sludge yield and of the carbonaceous substrate removal for various metabolic uncouplers.

Type of uncoupler	Dosage of uncoupler (mg/L)	Type of ww or sludge*	Reduction of Y_{obs}	Reduction of substrate removal	Reference
DNP	4	r	28%	negligible	Mayhew and Stephenson (1998)
mNP	20	s	65.5%	13.2%	Yang et al. (2003)
oNP	20	s	86.1%	26%	Yang et al. (2003)
pNP	100	s	30%	25%	Low et al. (2000)
pNP	100	r	16–27%	–	Chase et al. (2007)
mCP	20	s	86.9%	13.2%	Yang et al. (2003)
pCP	20	s	58%	8.9%	Yang et al. (2003)
TCP	10	s	50%	negligible	Strand et al. (1999)
TCS	0.8	s	40% (batch cultures) 69% (continuous cultures)	20%	Chen et al. (2000) Chen et al. (2001b) Chen et al. (2002)
TCS	1.0	s	46% (batch cultures)	negligible	Chen et al. (2002)
TCS	≈3.0	s	30%	2.8%	Ye and Li (2005)

*s = synthetic; r = real; ww = wastewater.

The acid constant (Ka) of a metabolic uncoupler significantly affects its effectiveness in sludge reduction and therefore a metabolic uncoupler with a lower pKa value has a higher potential for reducing sludge production (Yang et al., 2003).

In batch cultures of activated sludge containing chemical metabolic uncouplers, Y_{obs} decreased upon increasing the ratio of initial uncoupler concentration (U) to initial biomass concentration, X_0; this ratio (U/X_0) better reflects the real strength which a metabolic uncoupler exerts on a biomass unit, rather than using uncoupler concentration alone (Liu, 2000).

In conclusion, the above results demonstrate that the dosage of metabolic uncouplers can be efficient in sludge reduction, but little is known about the effect over long periods in which acclimatisation could play a role, or about the optimal conditions for the process. Furthermore, the dosage of these compounds may cause negative side-effects (Wei *et al.*, 2003a):

- possible reduction of the COD and BOD removal efficiency by using certain uncouplers (not in all cases);
- increase of oxygen consumption;
- possible worsening of sludge properties, such as settleability and dewaterability;
- uncouplers are not very cost-effective (Ye and Li, 2005; Ginestet, 2007c).

The choice of a type of metabolic uncoupler and the optimal dosage should be based on a compromise between the desired sludge reduction and system performance (Yang *et al.*, 2003).

Although many compounds applied as uncouplers are xenobiotics and therefore incompatible with the discharge of treated effluents in surface water bodies, others compounds such as TCS are non-toxic.

However, to date, this application has been limited to pilot-scale or lab-scale and uncouplers are often difficult to use at full-scale unless cheaper uncouplers are introduced.

8.9 PREDATION BY PROTOZOA AND METAZOA

Predators, naturally occuring in wastewater treatment processes, are mainly protozoa and metazoa. In activated sludge protozoa represent 5% of the dry weight of sludge (about 50,000 protozoa per mL) and 70% of protozoa is made up of ciliates (Ratsak *et al.*, 1996). Metazoa are mainly Rotifera, Nematoda and Anellida. It is well known that protozoa and metazoa play an important role in activated sludge processes, thanks to their grazing effect resulting in a clearer effluent. The presence of predators limits the growth of bacteria in suspension, favours the growth of floc-forming bacteria which are able to grow inside the flocs which are therefore more protected against predation.

Protozoa have often been monitored in activated sludge in order to obtain information about the performance and efficiency of the biological wastewater treatment processes. Some examples of protozoa found commonly in activated sludge are indicated in Figure 8.19.

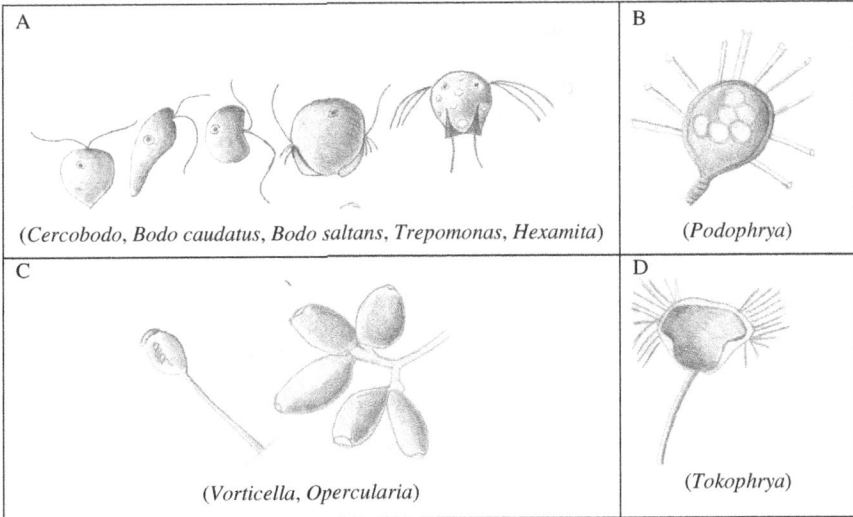

Figure 8.19. Examples of organisms belonging to (A) Flagellata; (B,C,D) Ciliata.

Activated sludge contains many metazoa and worms, in particular, are usually more abundant in WWTPs with a specific organic load below 0.1 $kgBOD_5$ $kgTSS^{-1}$ d^{-1}.

The number of worms varied both seasonally and between the aeration tanks (Wei *et al.*, 2003a). In the oxidation tanks or in the digesters, high densities of these worms appear when the "worm blooms" occur and this phenomenon is usually accompanied by a sludge reduction.

The reduction of sludge by microfauna predation is caused by the loss of energy in the food chain and by the change of part of the sludge to liquid or gaseous compounds (Liang *et al.*, 2006b). When sludge in solid form is ingested and consumed by microfauna, a sludge fraction is synthesized into the bodies of the microfauna. Then a fraction is decomposed by catabolic excretion releasing inorganic and organic carbon in the bulk liquid. As a consequence of respiration, inorganic CO_2 is released as gas, enhancing the reduction of sludge mass.

150   Sludge Reduction Technologies in Wastewater Treatment Plants

Studies with lab cultures of bacteria and protozoa have demonstrated that protozoa grazing on bacteria cause a considerable biomass reduction (Ratsak *et al.*, 1994). However, in activated sludge processes the density of protozoa is generally high, but their main role is to work as filter feeders only grazing dispersed bacteria so their contribution to sludge reduction is not always significant (Huang *et al.*, 2007).

The performance of metazoa in sludge reduction has attracted more attention compared to protozoa. In fact, the role of metazoa, especially aquatic worms, which are the largest organisms observed in activated sludge, could be more important in sludge reduction than protozoa due to their larger size which allows them to graze sludge flocs directly (Liang *et al.*, 2006a).

8.9.1 Types of predators

Different aquatic organisms that are frequently found in WWTPs have been proposed for use in reducing excess sludge belonging to Nematoda, Rotifera and Anellida phylum (Figure 8.20).

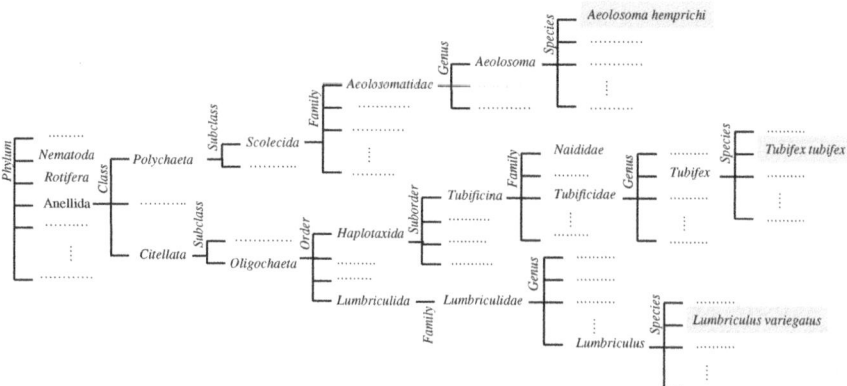

Figure 8.20. Nematoda, Rotifera and Anellida phylum and phylogenetic tree of the Anellida (Integrated Taxonomic Information System, ITIS).

Examples of organisms belonging to Nematoda and Rotifera are shown in Figure 8.21.

Among metazoa used for sludge reduction, the rotifers *Bdelloid* spp. (Lapinski and Tunnacliffe, 2003), which are ubiquitous in wastewater and sludge, are around 300 µm long and are capable of consuming several times their body weight of organic matter per day. The name "rotifer" is due to the corona of cilia located around the mouth and used to produce a vortex of water, that draws particles suspended in water into the rotifer's mouth. The size range of particles in sludge consumed by rotifers is 0.2–10 μm due to their ability not only to feed on bacteria suspended in the bulk liquid, but also on the sludge particles themselves. Lapinski and Tunnacliffe (2003) highlighted that a limited availability of particles in the size range ingestible by rotifers could limit rotifer growth. The authors suggested carrying out a partial disintegration of sludge particles prior to a rotifer grazing stage; combining limited sludge disintegration with rotifer predation might be an efficient way of reducing sludge production in WWTPs.

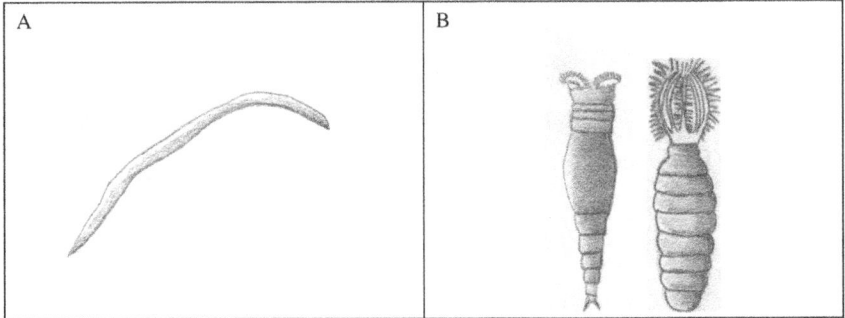

Figure 8.21. Examples of organisms of (A) Nematoda; (B) Rotifera.

The common groups of Anellida are:

– *Aeolosomatidae* and *Naididae*: free-swimming worms, the most common worms appearing in activated sludge processes; they propagate by division into an anterior and a somewhat smaller posterior part (Wei *et al.*, 2003b);
– *Tubificidae*: sessile worms, which do not normally occur in activated sludge suspensions but are more frequent in sludge sediments on the bottom rather than in suspension; their reproduction takes place sexually in most cases (Wei *et al.*, 2003b).

Aeolosoma hemprichi (belonging to Aeolosomatidae, Figure 8.22) is about 0.05 mm in width and 0.8 mm in length and it was been investigated many times for sludge treatment (Liang *et al.*, 2006a; Liang *et al.*, 2006b).

Due to its larger size (length up to 100 mm and width of less than 1 mm), *Tubifex tubifex* (sessile worm belonging to Tubificidae, Figure 8.23) is the most widely used in studies on sludge reduction with predators (Rensink and Rulkens, 1997; Ratsak and Verkuijlen, 2006; Huang *et al.*, 2007), because it is extremely tolerant of pollution in sludge and is able to maintain growth, even at very low oxygen levels (Rensink & Rulkens, 1997).

152   Sludge Reduction Technologies in Wastewater Treatment Plants

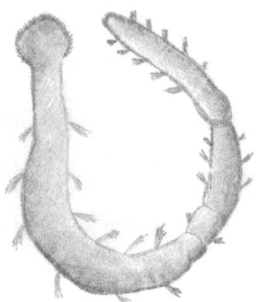

Figure 8.22. Aeolosoma hemprichi.

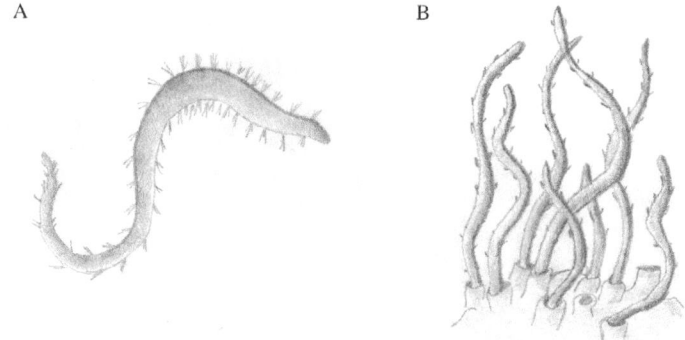

Figure 8.23. (A) Tubificidae; (B) Tubifex tubifex.

The aquatic worm species *Lumbriculus variegatus* (*Lumbriculidae*, Figure 8.24), 2–7 cm long when grown on activated sludge, has also been used for sludge reduction by transforming it into worm faeces (Elissen *et al.*, 2006).

During the digestion process, sludge properties change significantly when the sludge passes through the gut of the worms. Various theories have been developed, considering that heterotrophic bacteria in sludge may be killed during the passage through the gut or may survive the gut passage and their growth may be stimulated.

8.9.2 Process configuration and sludge reduction

The occurrence of predators in activated sludge or in attached biomass systems strongly depends on the operational conditions in the WWTP, such as applied organic load, oxygen concentration, sludge age, and other unknown factors. Among these, sludge age is very important because it has to be long enough with respect to the growth rate of predators, in order to avoid their washout. For example, free-swimming species of aquatic worms only occur in (ultra) low loaded systems with enough oxygen available for growth and reproduction.

Figure 8.24. Lumbriculus variegatus.

Despite the efforts to control the growth and reproduction of predators in the biological systems, the conclusion is that it is very difficult to manage predators directly within activated sludge. Recently, extensive monitoring of worm densities and operational conditions at four full scale WWTPs in the Netherlands confirmed this conclusion (Elissen *et al.*, 2008).

Only a few scientific works have been referred to in the literature on the effect of predators for excess sludge reduction in WWTPs: some research is aimed to evaluate lab-scale tests inoculated with some kind of predators. Because of the limited information available, the process is still not fully understood and sometimes results can be contradictory and optimal conditions for guaranteeing the effective full-scale application are difficult to predict.

Furthermore, the different results and sometimes contradictory opinions about the efficacy of sludge reduction by predation are often caused by lack of effective methods for predicting the sludge reduction caused by microfauna (Liang *et al.*, 2006b).

Initially, in the mid '90s, a predation process based on a two-stage reactor system (a modification of conventional activated sludge process) was proposed, but recently research is focusing on new alternatives for additional specialised predator-reactors kept separate from activated sludge stages, as described in the following sections.

An additional specialised predation-reactor for treating excess sludge after production can be integrated with:

(1) the wastewater handling units, recirculating the sludge after predation to the activated sludge stage,
(2) the sludge handling units, where the sludge is discharged after predation and sent on to other treatment units or disposal.

Other investigations of oligochaetes used for sludge reduction in MBRs are described in § 8.11.

Two-stage reactor system (chemostat integrated in activate sludge process)

The presence of predators grazing dispersed bacteria favours the growth of floc-forming bacteria, and thus most of the bacterial biomass formed in conventional processes is protected against predators. One way to overcome this selection pressure is to design a two-stage system (Figure 8.25) as proposed by Ratsak et al. (1994) and Lee and Welander (1996a; 1996b) with the following objectives:

(1) the first aerobic reactor with the aim of favouring bacterial growth: this stage is a chemostat without biomass retention. In this reactor SRT is equal to HRT. This short retention time allows the growth of fast-growing dispersed bacteria which consume the readily biodegradable organic matter in the wastewater, preventing the formation of bacterial aggregates and the growth of predators grazing on bacteria. However retention time must be long enough to avoid the washout of dispersed bacteria (Ghyoot and Verstraete, 1999);
(2) the second reactor is optimized for the growth of predators (ciliates and metazoa) grazing the bacteria; this stage can be designed as a biofilm or a suspended biomass reactor, with a long SRT to favour the growth of predators.

Salvado and Puigagut (2007) proposed to feed the chemostat with an influent flow rate of 70% using an average contact time of 4 h.

A two-stage system, in which the second stage was a biofilm system, was applied for the treatment of pulp and paper industry wastewater, obtaining Y_{obs} of 0.01–0.23 gTSS/gCOD$_{removed}$, with respect to the value of 0.2–0.4 gTSS/gCOD$_{removed}$ obtained for the treatment of similar wastewaters in a conventional activated sludge process (Lee and Welander, 1996a).

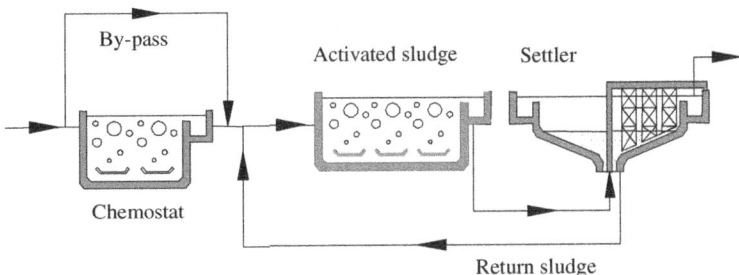

Figure 8.25. Two-stage reactor system (chemostat integrated in activated sludge process).

The implementation of this process at the Norske Skog Folla CTMP Mill in Norway was the first installation at full-scale (cited by Mahmood and Elliott, 2006). The plant treated 5000–6000 m^3/d and the configuration of the WWTP was a 1300 m^3 selector followed by a 15,700 m^3 aeration basin and a secondary clarifier. Y_{obs} decreased from above 0.15 kgTSS/kgCOD to 0.03 kg TSS/kgCOD, while the oxygen demand increased by 30% compared to the conventional activated sludge process (cited by Mahmood and Elliott, 2006).

It has to be taken into account that municipal wastewater contains low concentrations of readily biodegradable organic matter compared to industrial wastewater and therefore the effectiveness of the predation process in a municipal WWTP may be reduced (Lapinski and Tunnacliffe, 2003).

Wei et al. (2003a) suggested that for a two-stage process, retention time in the first aerobic reactor should be very long, but this would greatly increase the volume and its operation costs. Thus it is generally not feasible to apply the two-stage process in practice for municipal wastewater.

As for application in practice, some authors suggest that the two-stage process may have some limits: (1) grazing of protozoa and metazoa on nitrifying bacteria, which could result in a critical decrease of nitrifying capacity; (2) excessive growth of filaments and flagellates in activated sludge stages may result in bulking events (Ghyoot and Verstraete, 1999). This predation process may be cost-effective due to limited energy requirements, but from an economic point of view the main drawback is related to the increased area required for the application of low-loaded reactors where sludge age is sufficient for the growth of predators.

Predation-reactor integrated in the wastewater handling units

In this configuration soluble substances produced in the additional specialised predation-reactor could be used by the subsequent activated sludge process, enhancing the rate of material cycle and using the potential mineralisation ability of activated sludge as much as possible (Huang et al., 2007).

Huang et al. (2007) investigated the sludge reduction in an activated sludge reactor combined with a recycled sludge reduction reactor (RSRR) where *T. tubifex* was inoculated (Figure 8.26). In the reactor, the worms were mixed with the sludge and air was supplied for the needs of the worms and the sludge itself. The sludge reduction capacity of *T. tubifex* in RSRR was 0.65–1.08 gVSS L^{-1} d^{-1} (expressed per unit of volume of the predation-reactor, equivalent to 0.18–0.81 gVSS $gT.tubifex^{-1}$ d^{-1}). Operational conditions were 2.5 $gT.tubifex$ L^{-1} and sludge recycled ratio of 1. With the breaking up of sludge flocs, COD, NH_4-N and phosphorus were released, but the removal efficiency of COD and NH_4-N in activated sludge did not change significantly. An increase of phosphorus in the effluent and hence a slight decrease in P removal was observed compared with the control reactor.

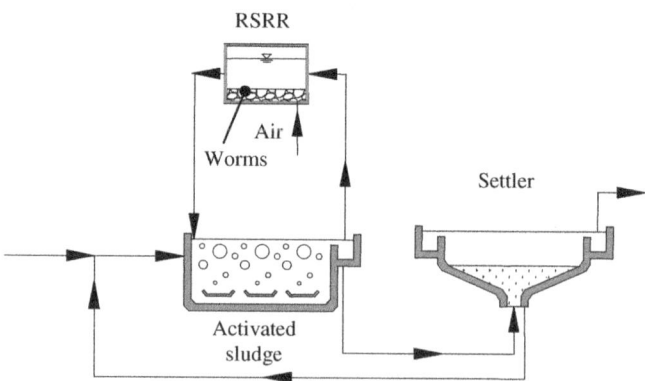

Figure 8.26. Recycled sludge reduction reactor (RSRR) inoculated with *T. tubifex* in the wastewater handling units.

In scaling up the system to full-scale applications, one drawback is the very large space required for the predation-reactor, which needs to be designed in a flat shape (Huang et al., 2007).

Guo et al. (2007) proposed a plug-flow worm-reactor (Figure 8.27) integrated in the wastewater handling units or in the sludge handling units (Guo et al., 2007). The worm-reactor was inoculated with Tubificidae (a mixture of *Branchnria Sowerbyi* and *Limnodrilns* sp.), mixed and aerated (DO at 0.5–3.0 mg/L). The sludge was reduced by both Tubificidae predation and endogenous metabolism simultaneously. The sludge reduction in the worm reactor was lower when integrated in the wastewater handling units (10%) rather than integrated in the sludge handling units (46.4%) (Guo et al., 2007). The contribution to sludge reduction by endogenous metabolism was less than 7% in both cases. No significant difference occurred in viscosity, specific resistance, and floc size distribution when the sludge was treated with Tubificidae (Guo et al., 2007).

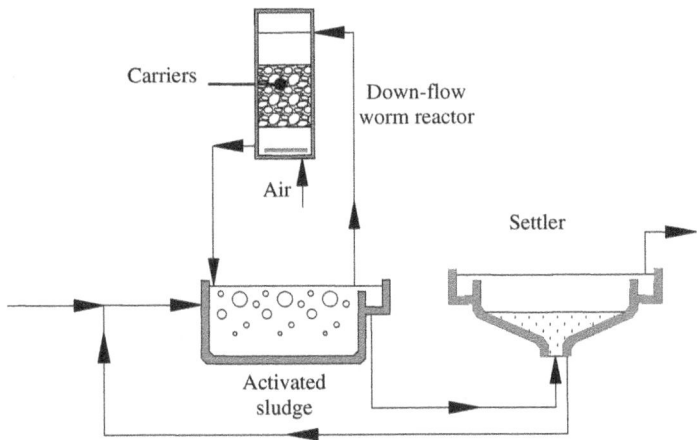

Figure 8.27. Plug-flow worm reactor integrated in the wastewater handling units.

Predation-reactor integrated in the sludge handling units

Wei and Liu (2005) developed a combined worm-reactor (at lab-scale) for oligochaete growth which consisted of three sections (see Figure 8.28): (1) a section for free swimming worm growth, (2) a section for sessile worm growth, filled with solid medium and (3) a section for sludge settling.

The worm-reactor was aerated and fed with the excess sludge discharged from an activated sludge system. *T. tubifex* was inoculated. In case of worm wash-out, part of the sludge separated in section (3) was returned to section (2).

In the preliminary results an average sludge reduction of 48–59% was achieved, much higher than in the control reactor with the same configuration but not inoculated with worms (Wei and Liu, 2005; Wei and Liu, 2006).

A modest improvement of sludge settling characteristics caused by worms was observed (Wei and Liu, 2005).

The concentrations of NO_3, NO_2 and PO_4 in both the influent sludge and the effluent sludge were similar (Wei and Liu, 2006). On the other hand, high NH_4 or NO_2 concentrations should be avoided because they inhibit worm growth.

158   Sludge Reduction Technologies in Wastewater Treatment Plants

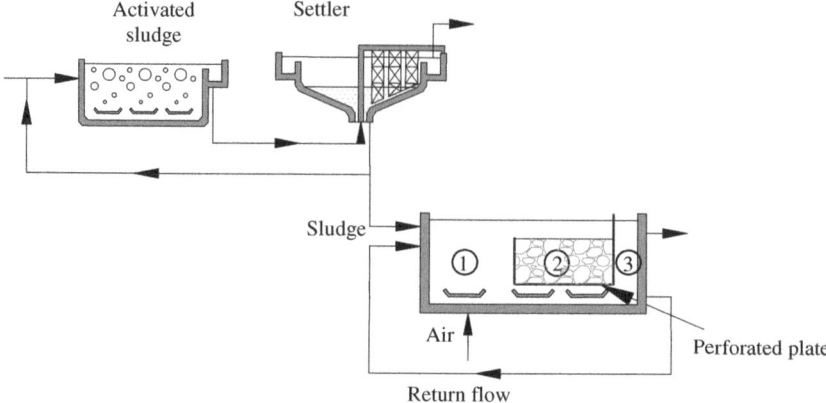

Figure 8.28. Worm-reactor integrated in the sludge handling units proposed by Wei and Liu (2005). For number explanation see the text.

Elissen *et al.* (2006) investigated at lab scale the capacity of an aquatic worm species (*Lumbriculus variegatus*) for sludge reduction. A new reactor concept was proposed (Figure 8.29), in which the separation of the worms from the sludge was obtained by immobilisation of *L. variegatus* in a carrier material.

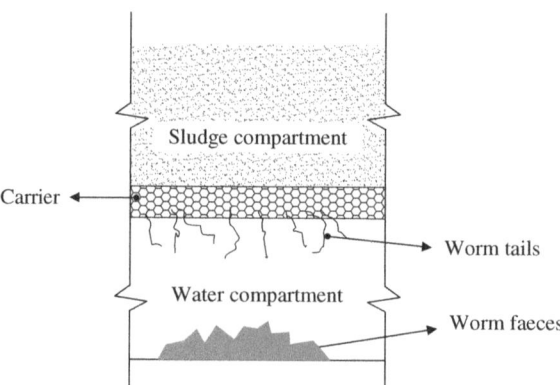

Figure 8.29. Scheme of the reactor concept for sludge predation by *Lumbriculus variegatus*, proposed by Elissen *et al.* (2006).

The worms graze the mass contained in the sludge compartment on one side of the carrier material, whilst their tails protrude through the carrier so that they defecate on the other side in the water compartment where the worms also find dissolved oxygen. Thus oxygen can be supplied to the worms without the sludge consuming a considerable part of it (Hendricks *et al.*, 2009). Furthermore, one purpose of this reactor is to recover the worms to obtain a valuable protein-rich product from the waste sludge.

In initial experiments, a TSS reduction of 75%, compact worm faeces with a high settleability (SVI half that of the waste sludge) was observed (Elissen *et al.*, 2006). Later experiments, with the same sludge and under similar conditions (at the worm density of 2–3 kg wet weight of worms per m^2), showed a TSS reduction of 36%–77% depending on the carrier area and the DO (Hendrickx *et al.*, 2009a). At low DO (<2.5 mg/L) the TSS reduction was high (77%), due to the lower faeces production, but long term operation at low DO may affect the survival of the worms.

Ammonia is also released at a rate of 20 mgN/gTSS digested by the worms. When its concentration rises excessively the water in the compartment has to be returned to the wastewater handling units for nitrogen removal: however this recirculation leads to an additional nitrogen load of less than 5% (Hendrickx *et al.*, 2009a).

A modification of the system presented in Figure 8.29 was introduced by Hendrickx *et al.* (2009b) in order to realise a continuously operating reactor with *L. variegatus*. The aim was to immobilise worms on two types of carrier material with a mesh of 300 μm and 350 μm. Spontaneous worm growth and avoidance of loss of worms – crucial for long-term continuous operation – was improved when using a mesh of 350 μm. In the continuously fed system, the worms consumed 40–67% of the VSS of sludge fed to the worm reactor (of this percentage, around 70–80% was digested and transformed into faeces and around 20–30% was mineralised, Figure 8.30). Another 5–17% of VSS reduction was due to natural sludge breakdown. The average natural sludge breakdown rate was estimated at 0.016–0.026 gVSS gVSS^{-1} d^{-1} (Hendrickx *et al.*, 2009b).

As a result, the net TSS reduction of the sludge consumed by the worms was 16–26%. The average sludge consumption rate was estimated at 39–92 mg TSS/g$L.variegatus^{-1}$ d^{-1}, higher when the organic fraction of sludge was lower (Hendrickx *et al.*, 2009b).

The application of this process is still at lab-scale and the need for large areas of the carrier material may affect the economic feasibility of this reactor concept.

8.9.3 Pros and cons of microbial predation

Although the advantages of microbial predation may lead to substantial and cost-effective sludge reduction, some drawbacks remain. In fact, at the moment knowledge of the process is not completely clear and therefore the application of microfauna for sludge reduction at full-scale is still not completely manageable and predictable, mainly due to:

- unstable predator growth: e.g. alternating blooms and absence of predators (Wei and Liu, 2006, Wei *et al.*, 2009);
- unknown factors that trigger the initiation of blooms;
- high variation of organism density;
- partially unknown factors affecting growth.

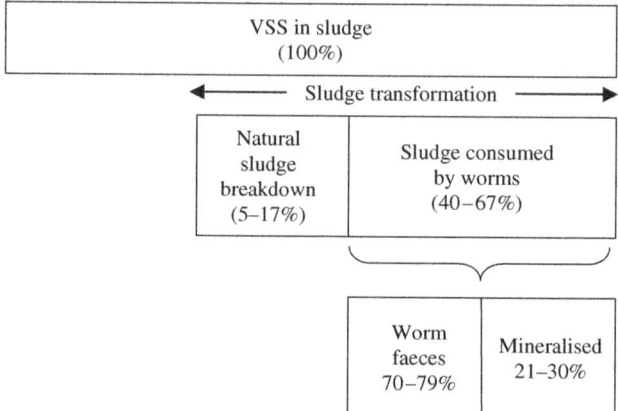

Figure 8.30. Scheme of sludge transformation in a worm-reactor.

The process may be technically feasible only if the predators are able to grow and reproduce (either sexually and/or asexually) on the sludge (Ratsak and Verkuijlen, 2006).

The process is less manageable in configurations where predators grow alongside bacteria in the same bioreactor, because processes such as activated sludge are designed for bacterial growth rather than for predator growth and this system may not be ideal for their stable growth and their retention (Wei *et al.*, 2009).

Because of these limitations, even in spite of substantial research efforts in the last two decades, to date, as far as we are aware, rare systems with sludge consumption by microfauna has been put into practice, restricted to industrial wastewater (Buys *et al.*, 2008; Elissen *et al.*, 2006; Mahmood and Elliott, 2006).

In the case of predators, such as aquatic worms, used to digest sludge contaminated with heavy metals, the organic part is digested but the heavy metals or other potential pollutants (e.g. micropollutants) may remain in the sludge, and thus contact with the human food chain is inadvisable (Elissen et al., 2010). Possible re-use of the worm biomass – when the worm faeces and the worms can be separated – has been well reviewed by Elissen et al. (2010) and may be addition to non-food animal feed, fertilizer or energy; only worm biomass grown on clean sludge could have a broader application potential, e.g. as fish feed (Elissen et al., 2010; Ratsak and Verkuijlen, 2006).

8.10 EXTENDED AERATION PROCESSES

When an activated sludge system operates at a sufficiently long sludge age and low F/M ratio, such as in extended aeration systems, the excess sludge production is generally reduced (Wang et al., 2006a). This fact is usually attributed to the lower biomass yield (which depends on SRT, see § 7.9), and to concepts such as maintenance energy requirements, endogenous respiration, cell decay and grazing by predators. The observed biomass yield is in fact inversely dependent on the SRT and endogenous respiration in steady state activated sludge processes. This relationship also provides a theoretical basis in biological process design to control total sludge production by adjusting the SRT during the wastewater biological treatment. For example, sludge production may be reduced by 60% when SRT is increased from 2 to 18 days (Liu and Tay, 2001).

Extended aeration systems require long aeration times and low applied organic loads, around 0.04–0.1 $kgBOD_5$ $kgTSS^{-1}$ d^{-1} compared to 0.08–0.15 $kgBOD_5$ $kgTSS^{-1}$ d^{-1} in conventional activated sludge processes. Theoretical zero sludge production is not achievable, although, in practice, a significantly lower quantity of excess sludge is produced compared to conventional activated sludge processes.

In the upgrading of a municipal WWTP, in which the conventional activated sludge process was converted to extended aeration to reduce sludge production, a 30% decrease was observed, with a resulting saving of U.S.$785,000 p.a. for lower landfill disposal use. Other benefits were dewatering improvement (dry content from 21% to 27%), lower odour potential and better sludge stability. One drawback was the increased cost for aeration (U.S.$51,000 p.a.) (Mahmood and Elliott, 2006).

Other drawbacks of processes with long sludge ages – applied to reduce sludge production – may be the formation of pinpoint flocs characterised by poor settleability (Bisogni and Lawrence, 1971) and the excessive costs for aerating larger oxidation tanks.

8.11 MEMBRANE BIOLOGICAL REACTORS (MBR)

In the MBR processes (*Membrane Biological Reactors*) the separation of sludge and effluent takes place in a highly efficient membrane module rather than in a conventional gravity settler. The main advantages of MBR compared to conventional activated sludge are: (i) high efficiency in TSS removal, (ii) application of high sludge concentrations in the reactor, (iii) reduced sludge production, (iv) small footprint. Drawbacks of MBRs are membrane fouling, higher energy consumption compared to conventional activated sludge and, to date, high investment costs.

The sludge production in the MBRs is usually expected to be lower than in conventional activated sludge processes, as a consequence of the low F/M ratio and long SRT, due to the higher sludge concentration which can be maintained in a MBR system, from 7–12 gTSS/L up to 15–20 gTSS/L. A long SRT or a low F/M ratio do not favour cell growth causing increased energy maintenance requirements. MBR processes have been chosen in the past for their expected reduction of excess sludge production, especially when applied at organic loads lower than 0.1 kgCOD kgSSV^{-1} d^{-1}.

Some experiences investigated the possibility of managing MBRs with complete sludge retention (i.e. almost without sludge discharge), aimed to achieve zero net growth (Laera *et al.*, 2005; Pollice *et al.*, 2004, 2008). The feasibility of a process with complete sludge retention was demonstrated and high removal efficiency and very limited sludge production was confirmed. Bhatta *et al.* (2004) investigated the sludge production in a lab-scale MBR system fed with synthetic wastewater in the case of very long SRT, up to 500 d. Operating with an organic load of 0.5 kgBOD$_5$ m^3 d^{-1} the Y$_{obs}$ decreased from 0.16 to 0.09 kgTSS/kgBOD$_5$, passing from SRT of 90 d to 500 d. This value is very low compared to Y$_{obs}$ expected for conventional activated sludge systems, around 0.57–0.69 kgTSS/kgBOD$_5$ (see § Table 2.3).

Results reported in the literature are not always consistent with respect to the accumulation of inert solids in the biomass under complete sludge retention (Laera *et al.*, 2005), and this important aspect is not always taken into account, especially when working with sludge fed with synthetic wastewater.

Laera *et al.* (2005) – in a long-term experimental work on a MBR fed with real municipal wastewater and with complete sludge retention – measured a very low observed sludge yield (around 0.02 gSSV/gCOD) at organic loads of 0.07 kgCOD kgSSV^{-1} d^{-1}, and evaluated zero sludge production at organic load of about 0.02 kgCOD kgSSV^{-1} d^{-1}. Under these conditions the maximum growth rate was slightly higher the decay rate, confirming the difficulties for microbial populations to grow, spending energy for maintenance requirements.

However, the reduction of sludge production is limited due to the potential adverse effects of high TSS concentrations on membrane expected in practice, such as membrane fouling, increased membrane cleaning requirements, oxygen transfer limitations, increased sludge viscosity, worsening in sludge filterability, and reduction of biological activity (*inter alia* Visvanathan *et al.*, 2000). In practice, the saving in sludge disposal costs should be higher than the increased operational costs and the aeration costs (Yoon *et al.*, 2004b).

Pollice *et al.* (2008), in a 4-year research at lab-scale, demonstrated that an MBR operated well at sludge age higher than 40 days without significant drawbacks in terms of biological activity and cleaning needs, but complete sludge retention appeared less feasible with regard to some operational parameters.

A long SRT in MBRs is also very suitable for favouring the abundant growth of bacteria predators such as protozoa and metazoa. The action of these predators is to enhance mineralisation and further contribute to reduce sludge production. However, the effect of predators on the sludge reduction in MBRs is controversial. Luxmy *et al.* (2001) found that the presence of metazoa in the sludge of an MBR, in concentrations around 1,000–2,000 per mL, was not so effective in sludge reduction. Though metazoa was not found to be so effective, it might have potential for fouling control, because a lot of metazoa was found to be attached to the membrane of the MBR, reducing the formation of cake layer and playing a significant role in removing accumulated sludge from the membrane surface (Luxmy *et al.*, 2001).

The sludge reduction induced by oligochaetes for an extended period was compared in an MBR (F/M = 0.13 kgCOD kgVSS^{-1} d^{-1}) and in a conventional activated sludge process (F/M = 0.26 kgCOD kgVSS^{-1} d^{-1}) by Wei *et al.* (2003b). The worm population introduced in the MBR was unstable, and worm growth and disappearance alternated in the MBR, resulting in a negligible sludge reduction.

8.11.1 MBR + physical, chemical treatments

Since the sludge reduction by increasing SRT may not necessarily be advantageous because of the potentially adverse effects of high TSS, some authors proposed to introduce additional disintegration techniques. The aim is to increase the decay rate of biomass maintaining a relatively low TSS concentration in MBR. Some of the treatments proposed for the integration with MBRs are synthetically described following.

- *MBR + alkaline treatment and ozonation*: a synergistic effect can be obtained coupling alkaline treatment (with 22.3 meq/L NaOH, pH = 11, 3h) with ozonation (0.02 gO$_3$/gTSS); alkaline treatment serves not only as a sludge solubilizing agent reducing the ozonation cost, but also as a buffering reagent preventing pH drops by the ozone treatment (ozone consumes a considerable amount of alkali, see § 13.9.4) and by the nitrification process in the MBR (Oh *et al.*, 2007). The COD solubilisation after alkaline treatment was 14% and increased to 20% after the subsequent ozonation (level comparable to 0.05 gO$_3$/gSS, but obtained with the relatively cheaper alkaline reagent). The Y_{obs} in the control line (MBR without chemical treatments) was less than 0.1 gTSS/gCOD, while approximately zero sludge production was observed in the MBR with the chemical treatments. No significant deterioration of the effluent quality and membrane performances was observed (Oh *et al.*, 2007).
- *MBR + ultrasonic disintegration*: this combined process was applied with the purpose of zero excess sludge production in the MBR fed with synthetic wastewater and with an average organic load of 0.12 kgBOD$_5$ kgTSS^{-1} d^{-1} (Yoon *et al.*, 2004a). Sludge extracted from the aeration tank of the MBR (10 gTSS/L) was disintegrated with ultrasonic equipment (20 kHz) and then recirculated into the MBR. An additional MBR (without ultrasonic disintegration) was used as a control. The ultrasonic treatment was applied once per day at stress frequency around 0.12 (see § 7.11) and with an applied E_s (defined in § 10.3) of 216,000 kJ/kgTSS. This level of E_s is very high and reach the maximum energy level applied in the literature, as shown in § 10.5, Figure 10.6 and Figure 10.7. A slight increase in effluent COD concentration, worsening in sludge filterability and foaming was observed. Zero excess sludge production was achieved in this system fed with synthetic wastewater, but in the case of the real wastewater the presence of inert solids (insoluble inorganic particles such as silts constitutes by silicon, aluminum and iron) may lead to an inevitable accumulation in the sludge and sludge reduction may become lower.

8.12 GRANULAR SLUDGE

The application of granular sludge systems, which are based on a self-immobilisation of microorganisms in granules, is recently of great interest for the treatment of municipal and industrial wastewater. The characteristics of this process, when compared with conventional activated sludge systems, are the following:

- very high biomass concentration which can be maintained in the reactor, from 15–20 up to 60 kgTSS/m^3, compared to the typical TSS concentration of activated sludge of around 3–4 kgTSS/m^3;
- very high organic load which can be treated, up to 10 kgCOD m^{-3} d^{-1};
- very low sludge production.

With respect to activated sludge – which require large areas for oxidation tanks and final settlers – the granular sludge presents a smaller foot-print.

The aerobic granular sludge present a strong microbial structure and generally good settleability. Up to now the mechanisms which describe the formation of granules is not fully understood, but is seems that the process is favoured in SBR systems, in the case of discontinuous feeding of biodegradable substrate and depends on hydrodynamic conditions.

The observed sludge yield in granular sludge reaches only 0.07–0.15 kgTSS/ kgCOD$_{removed}$ compared to conventional activated sludge systems which presents a typical value of 0.27–0.35 kg TSS/kgCOD (see § 2.4). The low sludge production observed in these systems is due to the endogenous metabolism and the high maintenance requirements occurring when the biomass grows in very compact and dense granules.

Ramadori *et al.* (2006) proposed a *Sequencing Batch Biofilter Granular Reactor* (SBBGR) filled with aerobic granules characterised by high biomass concentration (up to 60 g/L) for the treatment of municipal wastewater aimed to obtain high carbon removal and nitrification (Figure 8.31). This system is a submerged upflow filter equipped with an external recirculation flow to obtain a homogeneous distribution of substrate and oxygen through the bed.

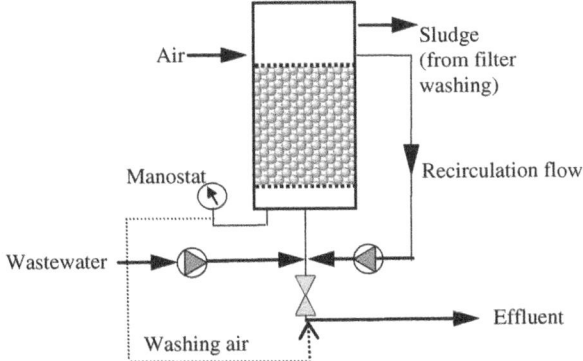

Figure 8.31. Scheme of a Sequencing Batch Biofilter Granular Reactor (SBBGR) filled with aerobic granules (*modified from* Ramadori *et al.*, 2006).

Treating municipal wastewater, the maximum organic load compatible with significant nitrification efficiency was 5.7 kgCOD m^{-3} d^{-1}. Denitrification occurred in this continuously aerated reactor, as demonstrated by the low nitrate concentration in treated effluents, always lower than 5 mgN/L. As well explained by the authors, denitrification probably occurs in the internal part of the granules, not reached by oxygen but where carbon sources are present as internal storage products or as hydrolysis products of particulate organic matter present in the feed (Ramadori et al., 2006).

In this plant the sludge production was very low, a magnitude order lower than that commonly reported for conventional systems. This result was confirmed by a mass balance considering oxygen and COD consumption, which demonstrated that 95% of COD was removed by oxidation. Under these conditions, the energy available for anabolism was very low, and thus the bacterial growth resulted strongly limited.

9
Mechanical disintegration

9.1 INTRODUCTION

The mechanical disintegration processes originate in the biotechnological field in order to achieve instant cell rupture and the resulting immediate release of intracellular compounds.

The application of mechanical disintegration techniques to the treatment of surplus sludge produced in WWTPs has been proposed since the mid '90s with the aim to disrupt sludge flocs, disintegrate bacteria cells, improve sludge biodegradability and reduce sludge production. In particular, the application of a mechanical treatment induces the following changes to the sludge properties (Müller *et al.*, 2004):

- *damage of microorganisms:* damaged microorganisms undergo a rapid lysis and the loss of intracellular compounds followed by hydrolysis. This phenomenon favours the sludge reduction due to the mechanism of

© 2010 IWA Publishing. *Sludge Reduction Technologies in Wastewater Treatment Plants.* By Paola Foladori, Gianni Andreottola and Giuliano Ziglio. ISBN: 9781789065305. Published by IWA Publishing, London, UK.

cell lysis-cryptic growth (see § 4.2). Larger microorganisms are damaged more easily; gram-positive bacteria are more resistant due to the strength of the cell walls;
- *floc size reduction*: the effect is an increase in accessibility among bacteria, substrates and enzymes;
- *sludge solubilisation*: the consequence of disintegration is the solubilisation of organic compounds released in the bulk liquid;
- *improvement/worsening of settling and dewatering*: if there is a high level of disintegration the sludge dewatering efficiency is enhanced, achieving greater dry content compared to untreated sludge. In the presence of a low level of disintegration, settling and dewatering efficiency decreases due to a partial disaggregation of flocs. The treatment of sludge with high filamentous bacteria content (cases of bulking) produces an improvement in settleability due to the break up of bridges between filaments.
- *foaming reduction*: in some cases, foaming reduction in anaerobic digesters was observed by introducing a mechanical pre-treatment;
- *increased flocculant demand*: reducing particle size and increasing the specific surface could lead to a greater electrical charge on particle surfaces; this lead to a greater demand for chemicals to neutralise the charges during conditioning of sludge. As a result a greater quantity of flocculants are needed for sludge dewatering;
- *viscosity reduction*: disintegration causes a reduction of sludge viscosity which facilitates mixing and pumping operations.

Biological or thermo-chemical reactions are not generally observed during the mechanical disintegration.

9.2 TYPES OF EQUIPMENT FOR MECHANICAL DISINTEGRATION

Mechanical disintegration can be achieved using various types of equipment. The disintegration efficiency and the specific energy required depend on both the equipment used and the nature of the sludge feed, such as particle size, solid content, etc...

The choice of a suitable mechanical disintegration process has to take into account both efficiency and investment, management and energy costs. Another important aspect is related to the wear of the equipment used.

Up to now the main techniques proposed for sludge disintegration, evaluated at lab-scale or already in use at full-scale are:

- lysis-thickening centrifuge (§ 9.4);
- stirred ball mills (§ 9.5);
- high pressure homogenisers (§ 9.6);
- high pressure jet and collision (§ 9.7);
- rotor-stator disintegration system (§ 9.8);
- ultrasonic disintegration (Chapter 10).

The lysis-thickening centrifuge, stirred ball mills and high pressure systems (homogenisers or jets) operate with continuous flow, while the ultrasonic disintegration presents configurations with both continuous flow and batch mode.

Mechanical disintegration can be integrated in the wastewater handling units or in the sludge handling units, according to Figure 9.1. All the configurations shown in the figure can be built, but it is more economical to operate with thickened sludge, in order to reduce the energy requirement for disintegration.

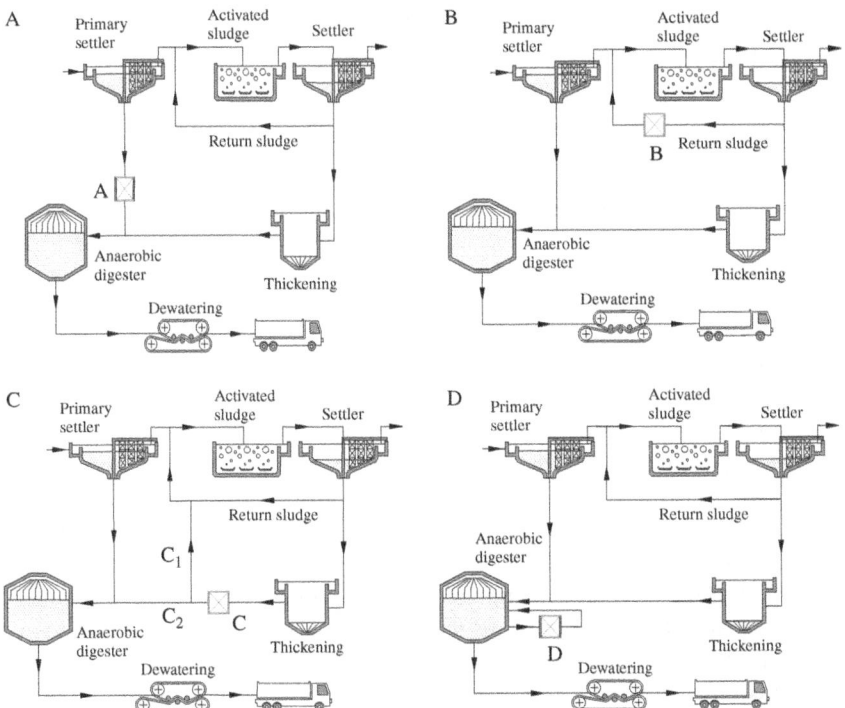

Figure 9.1. Alternatives for the integration of mechanical disintegration in the wastewater handling units (B, C_1) and in the sludge handling units (A, C_2, D). The symbol ⊠ indicates the mechanical treatment equipment.

Configuration A is rarely applied because the primary sludge is rich in biodegradable solids and therefore the benefit of the treatment is not significant.

In the case of *configurations B and C1* (integration in the wastewater handling units) the COD solubilised during the mechanical treatment can be used as an additional carbon source in the pre-denitrification stage.

Configurations C2 and D show the integration in the sludge handling units and the anaerobic digestion and help to increase biogas production.

The economic viability of some technologies has been already demonstrated, especially when integrated with anaerobic digestion for improving biogas production (Zábranská *et al.*, 2006). However, these technologies are economically sustainable when sludge disposal is particularly expensive (see Chapter 3).

9.3 ENERGY LEVELS REQUIRED FOR SLUDGE DISINTEGRATION

For all the mechanical disintegration systems described in detail in the following sections, these general observations can be made:

(1) at low energy levels the rapid disaggregation of flocs is observed, with a reduction of particle size;
(2) at higher energy levels the rupture of microorganisms is also observed: damage to cell walls and the release of intracellular compounds.

The energy associated with the bridges between microorganisms is low compared to the energy required for disintegrating the cell walls of bacteria. As a consequence the application of low energy levels causes the rapid separation of microorganisms and biopolymers and the disaggregation of biological flocs. Only higher energy levels permanently damage the cellular structure (Müller, 2000a). Moreover, the rupture of cell walls by mechanical action occurs at different energy levels, according to microorganism type and cell wall composition. Generally, the damage to larger microorganisms is firstly obtained, while the smaller bacterial cells are more resistant and damaged only by very high energy levels (Müller, 2000a).

The specific energy (E_s) expressed as kJ per unit of TSS (or TS) represents the most widely used parameter for describing results of mechanical disintegration. In particular, E_s is calculated as follows:

Mechanical disintegration

$$E_s = \frac{P \cdot t}{V \cdot x \cdot 1000} \quad (kJ/kgTS)$$

where:
- P = applied power (W);
- t = duration of the treatment (s);
- V = treated volume (m³);
- x = sludge concentration (kgTS/m³).

9.4 LYSIS-THICKENING CENTRIFUGE

This is a thickening centrifuge equipped with additional rotating cutting tools for sludge disintegration. The special impact gear is incorporated in the thickening centrifuge and is located at the end of the machine where the thickened sludge leaves it (Figure 9.2).

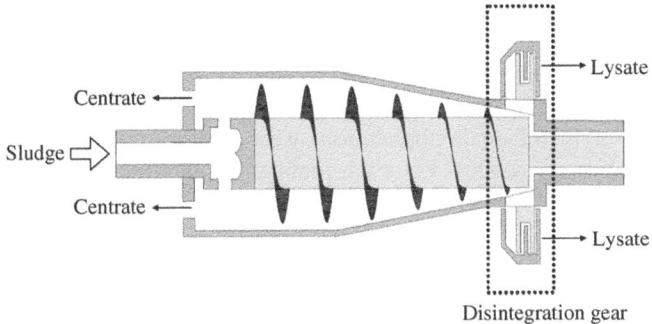

Figure 9.2. Layout of a lysis-thickening centrifuge.

This equipment has the capacity to treat large volumes of sludge with moderate energy consumption (Zábranská et al., 2000a; Zábranská et al. 2006).

By using this equipment thickening and partial disintegration of sludge can be obtained simultaneously. The degree of disintegration (DD_{COD} as defined in § 7.3.1) is approximately 8–18% (Müller, 2000a; Müller et al., 2004; Zábranská et al., 2006; see Figure 9.12).

Considering a soluble COD concentration in the thickened sludge before disintegration less than 1% of total COD (concentrations of 70–120 mg/L), the disintegration treatment caused an increase of soluble COD to 5% of total COD (Dohányos et al., 2000).

One of the first full-scale lysis-thickening centrifuges was implemented on a mod. BSC-4-2 (*Baker Hughes*) with an inlet flow rate of 100–110 m^3/h, equipped with different types of impact gear, installed at the *Central Prague* WWTP (1,200,000 PE). The inlet sludge flow rate was about 4,600 m^3/d (7 gTS/L) and the production of thickened sludge was 650 m^3/d (50 gTS/L) (Dohányos *et al.*, 2004).

The use of this lysis-thickening centrifuge produced a significant increase in sludge biodegradability, with an increase of biogas production during anaerobic digestion. In fact, disintegration cause an increase of biogas production when treating biological excess sludge alone of $+31.8\%$ on average, while the increase treating primary + biological sludge was of $+13.6\%$ on average (Dohányos *et al.*, 1997). The mix of primary and biological sludge generally leads to a lower increase in biogas production because the biodegradability of primary sludge is already high before disintegration and therefore the additional benefit is negligible. Other results indicate that disintegration increased biogas production of $+11.5$–31.3% (Dohányos *et al.*, 2000) and $+15$–26% (Zábranská *et al.* 2006).

The Prague WWTP was able to double its biogas production over the period 1993–2005, from about 7 million m^3/year to more than 16 million m^3/year, by enhancing sludge biodegradability via lysis-thickening centrifuge and increasing digestion capacity by rising the operational temperature to the thermophilic range (Jenicek, 2007; Zábranská *et al.*, 2000b; see § 8.5).

Three different full-scale applications were long term monitored as reported by Zábranská *et al.* (2006). For each plant the following principal parameters were measured (Table 9.1):

(1) the degree of disintegration, DD_{COD};
(2) the specific biogas production in anaerobic digestion, expressed as Nm^3/kgVS in the inlet;
(3) the VS removal during anaerobic digestion.

Liberec plant (Czech Republic) – The plant had a capacity of 100,000 PE and was equipped with three anaerobic mesophilic digesters, each with a volume of 4400 m^3 and with HRT of about 40 days. The sludge disintegration device was installed in the centrifuge (mod. BSC 3054 SDC-1) with an inlet sludge flow rate of 39 m^3/h and a rotating speed of 3,140 rpm. The results obtained are summarised in Table 9.1.

DD_{COD} was in the range 9–17.5% depending on centrifuge parameters (thickening efficiency), the quality of the inlet sludge and the wear level of rotating parts[1] The biogas production increased from an average value of 0.335 Nm^3/kgVS to an average value of 0.422 Nm^3/kgVS, giving an increase of

---

[1] in this experience the rotating parts of the disintegration device were replaced after 1.5 years of use.

26%. The annual biogas production in this plant rose from 837,828 to 1,055,413 Nm^3, giving an increase of 217,585 Nm^3 (Zábranská et al., 2006).

Table 9.1. Comparison of the performance of mechanical pre-treatment before anaerobic digestion in 3 full-scale plants (from Zábranská et al., 2006).

Parameter	Liberec plant	Fürsten-Feldbruck plant	Aachen-Soers plant
Degree of disintegration, DD_{COD}	9–17.5%	8.5–10.7%	–
Specific biogas production before mechanical pre-treatment installation (Nm^3/kgSV)	0.335	0.462	0.326
Specific biogas production after mechanical pre-treatment installation (Nm^3/kgSV)	0.422	0.529	0.402
Increase in specific biogas production	26%	14.5%	23.3%
VS reduction before mechanical pre-treatment installation	24–42%	58.5%	–
VS reduction after mechanical pre-treatment installation	45–63%	62%	–

Another advantage of the mechanical sludge disintegration using the lysis-thickening centrifuge relates to the significant reduction of sludge viscosity. The pumping limit for thickened sludge is generally held to be around 6% TS. Thickening to a higher degree could cause clogging problems in the head piping, especially when pumping with conveying pumps over longer distances. Disintegration allows sludge to be thickened up to 9–11% TS, while maintaining pumpability. A sludge with a higher degree of thickening leads to smaller anaerobic digesters (or a longer retention time) and therefore helps to reduce energy needed for heating (Zábranská et al., 2006).

Fürstenfeldbruck plant (Germany) – The plant (capacity of 70,000 PE) consisted of two anaerobic mesophilic digesters, each with a volume of 1,800 m^3 and an HRT of 35 d. The thickening centrifuge was type BSC 3-01, with an inlet flow rate of 12 m^3/h and a rotating speed of 2,250 rpm. The results obtained from the monitoring are summarised in Table 9.1 (Zábranská et al., 2006).

DD_{COD} was in the range of 8.5–10.7%. The specific biogas production was 0.462 Nm^3/kgVS before the installation of the disintegration device and was 0.529 Nm^3/kgVS after installation, giving an increase of 14.5%. The VS reduction in anaerobic digestion passed from 58.5% to 62% after installation of the disintegration device (Zábranská et al., 2006).

Aachen-Soers plant (Germany) – The plant (with a capacity of 650,000 PE) consisted of 4 anaerobic mesophilic digestors with a total volume of 20,000 m^3. Two thickening centrifuges already installed in the plant – each with a flow rate of 200 m^3/h – were equipped with the additional disintegration device. The specific biogas production varied from 0.326 m^3/kgVS to 0.402 m^3/kgVS (Zábranská et al., 2006). This increase in biogas production allowed energy production to be enhanced by 880,000 kWh p.a., equivalent to a cost saving of approximately 60,000 € p.a.[2] (Zábranská et al., 2006).

9.5 STIRRED BALL MILLS

A stirred ball mill consists of a cylindrical grinding chamber (vertical or horizontal) with a volume of up to 1 m^3, equipped with a rotating central crankshaft. The chamber is almost completely filled with grinding spheres; a rotor fitted with blades of variable shape forces the beads into a rotational movement. The spheres are made of steel or ceramic materials and are forced to collide with each other. The disintegration of sludge is caused by shearing and pressure forces between the spheres. For a continuous operation the spheres are held back by a sieve while the suspension flows through the grinding chamber (Winter, 2002).

A simplified configuration of a stirred ball mill working continuously is shown in Figure 9.3.

Sludge and microbial cells are disrupted in the contact zone of the spheres by compaction or shearing action and by energy transfer from balls to cells, although an exact understanding of the mechanism is not available (Middelberg, 1995).

Temperature rises during milling due to the large amount of energy input transferred with the high rotational speeds, while some heat dissipation is present. Power input ($P = \omega^3 \phi^5$) is related to the agitator speed (ω) and the diameter of the agitator (ϕ). Power input thus increases with the rotational speed and as the mill is scaled up (Middelberg, 1995).

The main operating parameters of stirred ball mills are:

– size of grinding spheres
– sphere material
– filling ratio of spheres in the grinding chamber
– grinding time

---

[2] Energy cost for industrial uses was 0.07 €/kWh.

- rotation speed of the rotor
- sludge concentration.

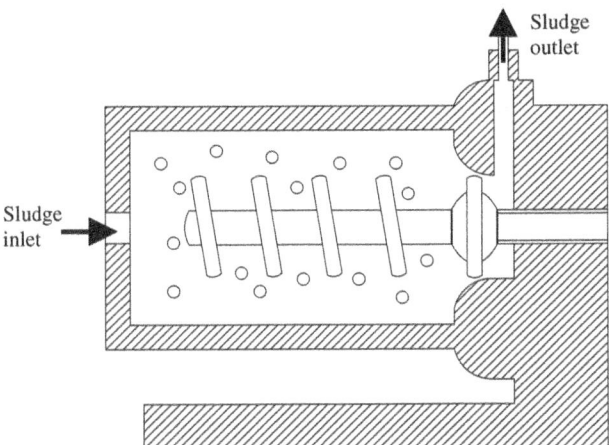

Figure 9.3. A simplified configuration of a stirred ball mill.

Sphere size – Several studies have observed that efficiency improves when the sphere size decreases. Spheres of 0.10–0.15 mm were used to obtain the disaggregation of biological flocs and also the disintegration of bacterial cells. But in full-scale applications, spheres of a greater size are more suitable (0.4–0.6 mm) because it is easier to separate them from the sludge at the sieve (Geciova *et al.*, 2002).

Sphere material – Glass and zirconium balls were used by Baier and Schmidheiny (1997), but only a modest difference was observed, with the glass spheres performing better. The effect of rotational speed and sphere size overshadowed the influence of sphere material. Thus the sphere material does not significantly affect disintegration efficiency, while it plays an important role in wear resistance.

Filling ratio – An increase of the filling ratio of spheres in the grinding chamber allows the degree of disintegration to be increased due to the greater number of collisions among spheres; at the same time, temperature also increases and wear of mechanical parts occurs. The optimum filling ratio is considered 80–85%, because the disadvantages related to a filling ratio of 90% do not outweigh the small increase in disintegration efficiency (Geciova *et al.*, 2002).

Grinding time – The contact time of sludge in stirred ball mills is few minutes.

Rotation speed – When the rotation speed increases (approximately 2,000–4,000 rpm, circumferential speed >10 m/s), sludge solubilisation increases, but not in a linear way (Baier and Schmidheiny, 1997). A speed of >10 m/s is recommended for the disintegration of bacterial cells (Geciova et al., 2002).

Sludge concentration – Thickened sludge having higher solid concentration are less suitable to ball milling than sludge with lower concentrations (Baier and Schmidheiny, 1997).

Examples of equipment referred in the literature are:

- Dyno Mill (WAB Company, Switzerland), water cooled horizontal milling chamber at lab-scale; revolution speed of 2,000–4,200 rpm; glass and zirconium spheres, diameters 0.2–2 mm (Baier and Schmidheiny, 1997; Jung et al., 2001);
- Netzsch type LME 4, at lab-scale (used by Kopp et al., 1997; Müller, 2000a);
- Netzsch Feinmahltechnik GmbH, Type LME 50 K, at full-scale, equipped with a 37 kW motor; circumferential speed of 15 m/s; grinding chamber volume of 52.8 L, filled to 85% with spheres of varying diameters (0.15–0.80 mm) and materials (Winter, 2002);
- a Draiswerke stirred ball mill, Type Cosmo 25, at full-scale, equipped with a 30 kW motor; grinding chamber volume of about 8.5 L; circumferential speed of 22 m/s.

The spheres are separated from the sludge flow by a sieve, or alternatively by centrifugal force in the Draiswerke stirred ball mill.

The early studies proposing this technology date back to the mid '90s (*inter alia*, Kopp et al., 1997; Baier and Schmidheiny, 1997). More recent studies are referred to by Lehne et al. (2001), Müller (2001), Winter (2002) and Müller et al. (2004).

Baier and Schmidheiny (1997) tested ball mills at lab-scale using two sphere materials (glass and zirconium), two sphere sizes (fine and coarse) and various rotation speeds. The ball mill treatment was applied at a temperature of 60°C and this might have contributed to increase sludge solubilisation rather than the milling effect alone. However, the authors indicated that the combination of high shear forces (ball mill) and low-medium viscosity (elevated temperature) was to some extent effective in sludge solubilisation. With respect to an initial soluble COD of 1.0–5.5% of total COD in the untreated excess activated sludge, COD solubilisation reached an average of 9.7% using coarse balls (1–1.5 mm) and 14% using fine balls (0.2–0.25 mm).

An approximate calculation of energy consumption showed a net energy requirement for milling of 1.0–1.25 kW/m³ of sludge treated per day, but this value was not verified at full-scale (Baier and Schmidheiny, 1997).

Kopp *et al.* (1997) described the results of a lab-scale application of a stirred ball mill for sludge pre-treatment before anaerobic digestion. Figure 9.4 shows the percentage of VSS reduction observed during the anaerobic digestion of sludge disintegrated by the stirred ball mill (reaching a degree of disintegration DD_{O2} of 25%). It can be observed that, for short retention time in the digester, the VSS reduction is higher in the presence of pre-treated sludge, while after 14 days of digestion the VSS reduction is similar for pre-treated and untreated sludge (Kopp *et al.*, 1997).

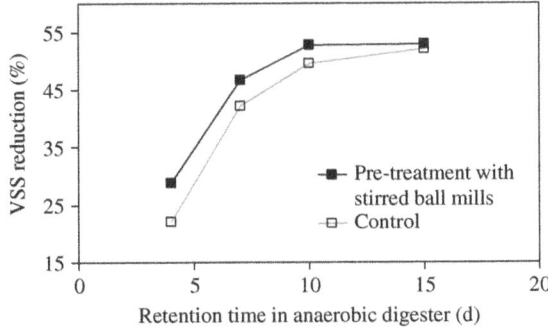

Figure 9.4. Percentage of VSS reduction in anaerobic digestion for untreated sludge (control) and sludge pre-treated using stirred ball mills with DD_{O2} of 25% (*modified from* Kopp *et al.*, 1997).

Winter (2002) tested two types of stirred ball mills at full-scale (Netzsch and Draiswerke) as pre-treatment before anaerobic digestion. The degree of disintegration (DD_{O2}) as a function of the applied E_s is shown in Figure 9.5 (E_s expressed in kJ/kgTS).

Pre-treatment allows biogas production to be increased by 21%.

The sludge treated by stirred ball mills has improved dewatering efficiency, but requires a higher dosage of chemical reagents to neutralise the surface electrical charges related to the increased area of sludge particles (Kopp *et al.*, 1997).

As a consequence of the pre-treatment with stirred ball mills before anaerobic digestion, the TKN concentration in the recycle flows increases by 20–30% compared to the configuration without pre-treatment (Kopp *et al.*, 1997). The sludge pre-treatment does not significantly affect the PO_4-P concentration in recycle flows, which remains quite constant (Kopp *et al.*, 1997).

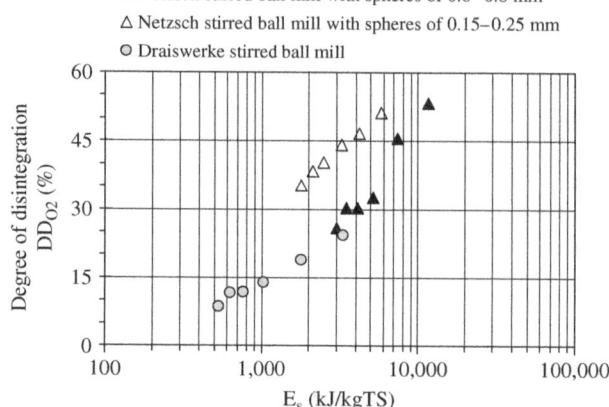

Figure 9.5. Degree of disintegration (DD_{O2}) as a function of E_s for two types of stirred ball mills (*from* Winter, 2002).

9.6 HIGH PRESSURE HOMOGENISER

This is a technology imported from the food, cosmetic and pharmaceutical industries, where the homogenisation of products is common. The traditional use in pharmaceutical industries for recovering enzymes demonstrated that this mechanical treatment conserves the integrity of enzymes after cell disintegration and any chemical modification of the organic matrix is not significant (mainly mechanical modification occurs) (Camacho *et al.*, 2002b).

In the field of sludge reduction, this process was first proposed in the mid '90s (Kunz and Wagner (1994); Müller (2000a, 2000b); Lehne *et al.* (2001)).

In its wastewater treatment configuration, a high pressure homogeniser consists of:

(1) a high pressure pump which compresses sludge up to pressures of several hundred bar;
(2) an adjustable homogenisation valve where sludge decompression to atmospheric pressure takes place (Figure 9.6).

Passing through the homogenisation valve the sludge speed undergoes an increase of up to fifty times (up to 300 m/s), due to the intense restriction. The speed increase causes a rapid drop of pressure to below vapor pressure (cavitation) and induces collisions among sludge particles. All these reactions cause the disaggregation of flocs and the rupture of cells (Strünkmann *et al.*,

2006). The temperature increases by 2°C for each 100 bar applied due to adiabatic compression (Geciova et al., 2002).

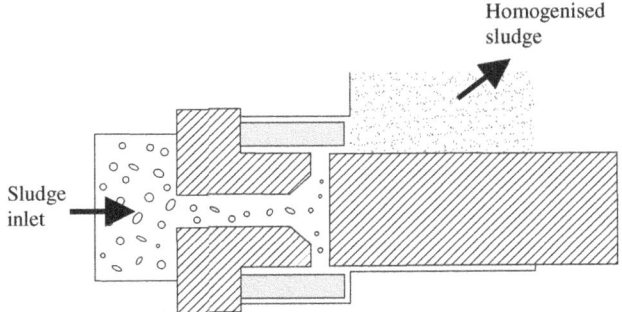

Figure 9.6. Section of a homogenisation valve.

The main operational parameters are:

- applied pressure;
- number of passages through the homogenisation valve;
- manufacturing specifications of the valve;
- sludge temperature.

Engelhart et al. (2000) used a lab-scale high pressure homogeniser with various pressures and the treatment efficiency was evaluated on the basis of the degree of disintegration (DD_{COD}) (Figure 9.7). DD_{COD} rises for increasing values of E_s in a linear way.

Camacho et al. (2002a) tested high pressure homogenisation in pilot plants by applying pressures of 300, 500 and 700 bar and from 1 to 10 passages. The released COD increased proportionally to the pressure up to 500 bar, although limited improvement was observed for 5–10 passages. After the maximum number of passages (10) the released COD levelled off reaching a maximum limit of around 60% (at pressure >500 bar) (Camacho et al., 2002b).

In the pilot plants sludge production was evaluated comparing a line used as control with another line equipped with the homogenisation treatment (SF = 0.2 d^{-1}). The sludge reduction was:

- negligible or moderate for pressures below 300 bar;
- more significant for pressures from 300 to 700 bar, obtaining a ratio of 0.29 $gTSS_{produced}/gCOD_{removed}$ (compared to 0.36 in the control line) and a sludge reduction of 20% (Camacho et al., 2002a).

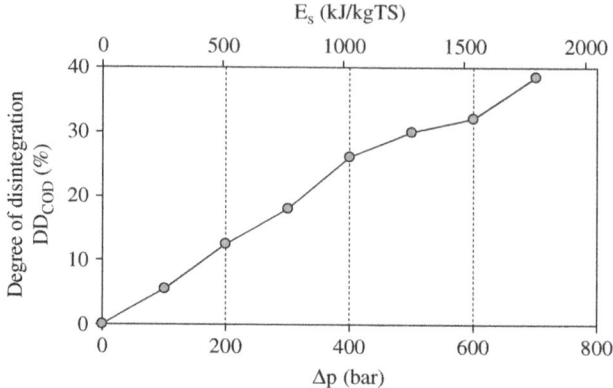

Figure 9.7. Degree of disintegration (DD_{COD}) as a function of applied pressure (Δp) and E_s (*from* Engelhart et al., 2000).

The treatment produces an increase of protein and carbohydrate concentration in the bulk liquid which levels off for pressures above approximately 200 bar. Varying pressure and applied energy, the solubilised protein and carbohydrate maintain an approximately constant ratio, indicating that disintegration affects these compounds similarly.

The action mechanisms of high pressure homogenisation depend on the applied energy level (pressure); in particular:

(1) a low energy level causes modification and physical destruction of the network of sludge. It probably affects cell binding without inducing cell rupture. In fact, high pressure homogenisation reduces the aggregate size to 0.5–3 μm, which is similar to the size of bacterial cells in activated sludge;
(2) a high energy level causes high shear stress and the progressive disintegration of cells.

We can hypothesise that the supplied energy is firstly used to disrupt non-covalent forces between cells (embedded in flocs), and secondly (beyond 2 passages according to Camacho et al., 2002b), to disrupt the cell walls (covalent and non-covalent bonds).

Therefore, to reduce sludge production it would be more advantageous to cause cell disintegration, in order to promote the release of intracellular compounds and their biodegradation, but this may be excessively onerous. Bacterial cells resist high pressures, even up to 1000 bar (Pagan and Mackey, 2000; Hayakawa et al., 1998).

Mechanical disintegration

Many authors recognised that during mechanical treatments such as high-pressure homogenisation, the filamentous structure of activated sludge was destroyed, helping to control bulking in biological reactors and digesters. However, Barjenbruch et al. (2000) applied high-pressure at 600 bar and observed that only limited reduction of foaming was achieved.

Full-scale high pressure homogenisers have been applied in Germany since 2003, used as a pre-treatment before anaerobic mesophilic digestion. These systems have been used at moderate pressures to treat both thickened/un-thickened sludge and both primary and secondary sludge with the aim of increasing biogas production. In the most common configuration, the sludge is taken out of a recirculation loop in the digester, then homogenised and returned to the digester together with the fed sludge.

Onyeche (2004) indicates a layout for the integration of high pressure homogenisation with anaerobic digestion (Figure 9.8). A part of the digested sludge is thickened up to 18–40 gTSS/L utilising, for example, a centrifuge. The thickened sludge first enters a macerator – to roughly reduce the particle size and to avoid clogging the homogenisation valve – and is then pumped into the high pressure homogeniser. In this plant the pressure is set at a moderate 150 bar in order to reduce energy costs. The disintegrated sludge is finally recirculated into the anaerobic digester.

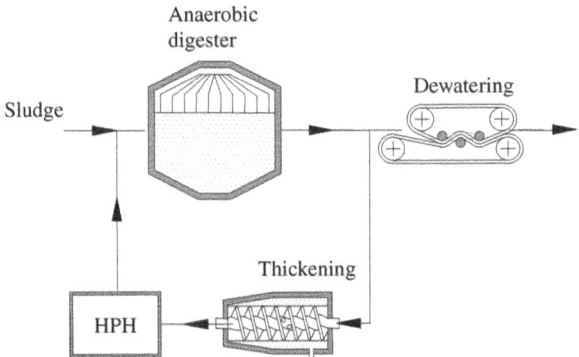

Figure 9.8. Layout of high pressure homogenisation (HPH) integrated in the sludge handling units according to Onyeche (2004).

The results obtained by Onyeche (2004) at full-scale indicate a reduction of sludge production of 24%, with an increase of biogas production of 25%. The centrifuge installed before the homogeniser caused a significant energy consumption. The optimisation of the configuration require the proposal of

182 Sludge Reduction Technologies in Wastewater Treatment Plants

alternatives for sludge thickening with respect to centrifugal treatment, for example choosing fine sieving or other low energy thickeners (Onyeche, 2004).

Working with a high pressure homogeniser up to 1000 bar used as pre-treatement before anaerobic digestion, Barjenbruch et al. (2000) observed an increase of about 20% in the COD and TKN loads in the sludge liquor recirculated into the wastewater handling units.

When the high pressure homogenisation is integrated in the wastewater handling units an improvement of SVI is observed and the nitrification rate does not worsen (Camacho et al., 2002a).

As seen above for stirred ball mills, also in the case of high pressure homogenisation integrated in the sludge handling units, sludge disintegration causes an improvement of dewatering efficiency but a higher dosage of chemical reagents is required (Kopp et al., 1997).

9.7 HIGH PRESSURE JET AND COLLISION SYSTEM

In experiments at lab-scale (Choi et al., 1997; Hwang et al., 1997) and at pilot-scale (Nah et al., 2000) a system based on high pressure jet and collision treatment has been proposed as a pre-treatment before anaerobic digestion. In this system, the sludge is pressurised by a pressure pump and then jetted against a collision surface after passing through a pressure gauge, a T valve and a nozzle (Figure 9.9). In this way, the sludge passes from 5–50 bar pressure to ambient pressure (1 bar) instantaneously and is jetted towards the collision plate at a theoretical velocity about 30–100 m/s.

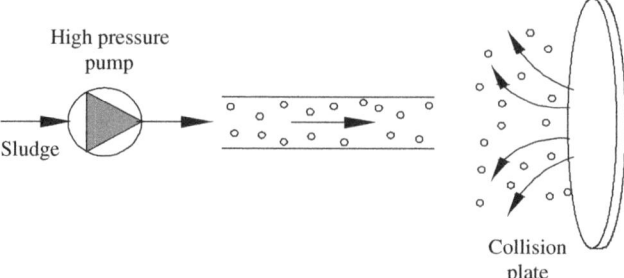

Figure 9.9. Scheme of high pressure jet and collision system (*modified from* Nah et al., 2000).

Sludge solubilisation was demonstrated by the release of soluble proteins which increased for increasing applied pressures, but levelled off over approximately 30 bar and after 2 passages in the system (Choi et al., 1997).

Nah et al. (2000) confirmed these findings, observing an increase of soluble COD of 3.4% of $TSS_{treated}$. Particle size also decreased with increasing pressures, indicating a potential improvement of digestability. An increase in protein, ammonia and total phosphorus (+20%) compared to before pre-treatment was also observed. The supernatant after pre-treatment was more turbid, due to non settleable colloidal and dissolved solids.

When used as pre-treatment before anaerobic digestion (at lab-scale, HRT of 2–26 d) the VS reduction passed from 2–35% using untreated sludge, to 13–50% using sludge treated once (1 passage) at 30 bar (Choi et al., 1997).

At lab-scale it was necessary to sieve the sludge in order to separate coarse solids which may cause nozzle clogging.

Technical aspects of this mechanical treatment are focused on the high-pressure pump (30 bar), which affects the set-up, the operating cost and the durability of the pre-treatment equipment.

One advantage is the potential reduction of digester volume, resulting from the reduced retention time in the digester, from 13 to 6 d without adversely impacting on VS reduction which remained at 30% (Nah et al., 2000).

9.8 ROTOR-STATOR DISINTEGRATION SYSTEMS

The deflaker disintegration system is a technology designed for processing the pulp in paper industries. In the case of sludge disintegration, the sludge enters the machine between a high speed rotating blade (rotor) and a static terminal (stator) with holes. The sludge is pumped into the machine and is driven outwards through the holes in the stator (Figure 9.10). As a consequence of the rotating movement of the blade at high rpm (about 3,000 rpm), sludge aggregates are disintegrated into smaller particles and turbulence, cavitation and shearing occurs.

The equipment used for sludge disintegration in the experience referred to by Kampas et al. (2007) was the 1000 Pilao DTD Spider Deflaker equipped with a 30 kW motor. In this machine the distance between stator and rotor was 0.6–0.9 mm and the rotation speed was 3,000–3,600 rpm.

The disintegration process was obtained by using thickened sludge (4–7% total solids) at retention times of between 2 and 15 minutes. The disintegration causes floc size reduction up to 10 μm. This phenomenon even occurs at $E_s < 2500$ kJ/kgTS (Figure 9.11). For increasing levels of E_s the particle size does not decrease further, but the COD solubilisation increases gradually (Figure 9.11).

The mechanisms occurring in sludge disintegration are the following:

(1) floc disaggregation and solubilisation of a part of COD was observed at $E_s < 2,500$ kJ/kgTS;

(2) further COD solubilisation was observed at E_s in the range 2,500–9,000 kJ/kgTS, due to floc disintegration;
(3) cell rupture and lysis occurs only at $E_s > 9{,}000$ kJ/kgTS.

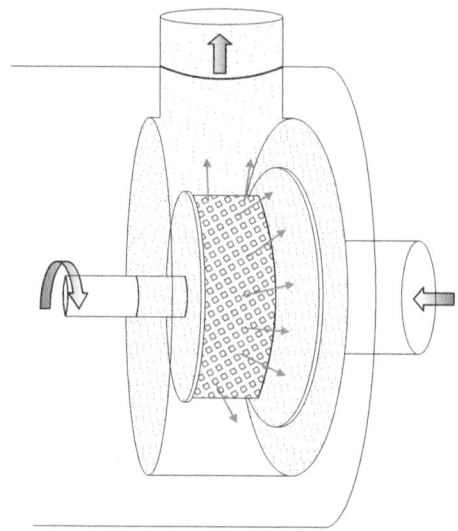

Figure 9.10. A simplified configuration of a deflaker disintegration system.

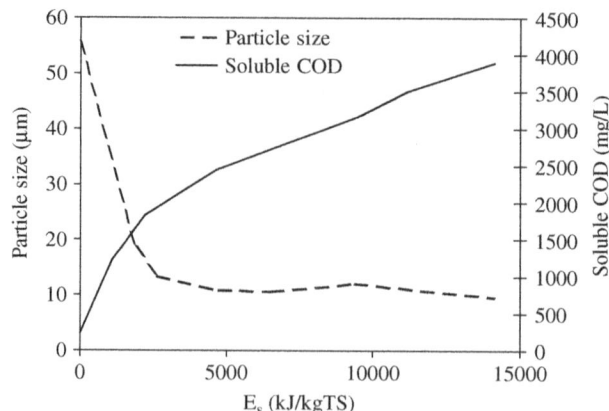

Figure 9.11. Variation of particle size and soluble COD concentration as a function of E_s (*modified from* Kampas *et al.*, 2007).

9.9 COMPARISON OF MECHANICAL DISINTEGRATION TECHNIQUES

The efficiency of mechanical disintegration techniques can be compared by using the respective DD_{COD} values or DD_{O2} values as a function of E_s. Several comparisons are referred to by the *Technical University of Braunschweig* (Germany) and included in the graph of Figure 9.12, in which the DD_{COD} of the following disintegration techniques is compared:

- lysis-thickening centrifuge
- high pressure homogeniser
- stirred ball mill
- rotor-stator disintegration system
- ultrasonic disintegration.

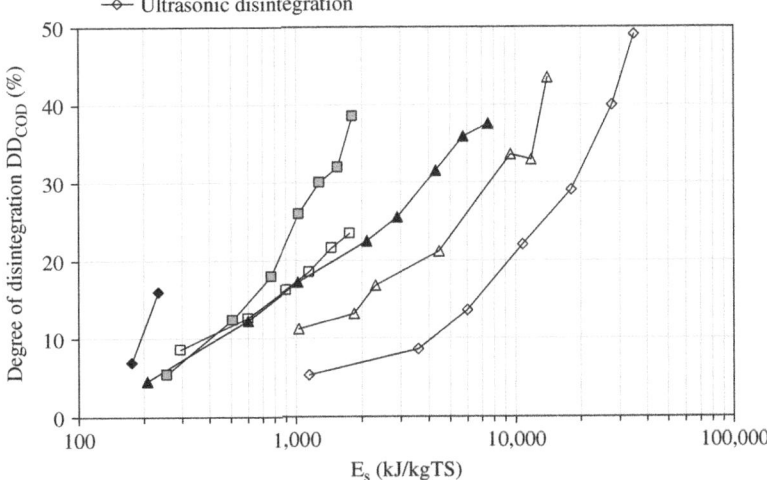

Figure 9.12. Degree of disintegration (DD_{COD}) as a function of E_s for some mechanical disintegration techniques. Where not indicated, reference is Müller (2000a).

186 Sludge Reduction Technologies in Wastewater Treatment Plants

These treatments were applied to surplus sludge with solid concentration between 31 and 44 gTS/kg wet sludge (Müller, 2000a), corresponding to a dry content of about 3–4%. Only the rotor-stator disintegration system was tested with dry content of 6.2–7.2% (Kampas *et al.*, 2007). The ultrasonic disintegration is compared here for completeness, while the results and the performance of this technique are discussed in Chapter 10.

At most the lysis-thickening centrifuge only reaches a DD_{COD} of about 18% (Müller, 2000a), counterbalanced by the lowest energy consumption.

The rotor-stator disintegration system is as efficient as high pressure homogeniser in sludge disintegration, but it consumes much more energy.

The ultrasonic treatment is the only one able to reach DD_{COD} of almost 50%, but it requires the most energy.

An alternative approach to evaluate the efficiency of sludge disintegration consists of using the DD_{O2} parameter (according to the definition given in § 7.3.2), based on the measurement of oxygen consumption by the sludge after treatment. In (Figure 9.13) the high pressure homogeniser (pressure = 100–900 bar), stirred ball mills ($\varnothing_{spheres}$ = 0.35 mm, v_c = 6 m/s, treatment time = 1–60 minutes) and ultrasonic disintegration are compared (Kopp *et al.*, 1997).

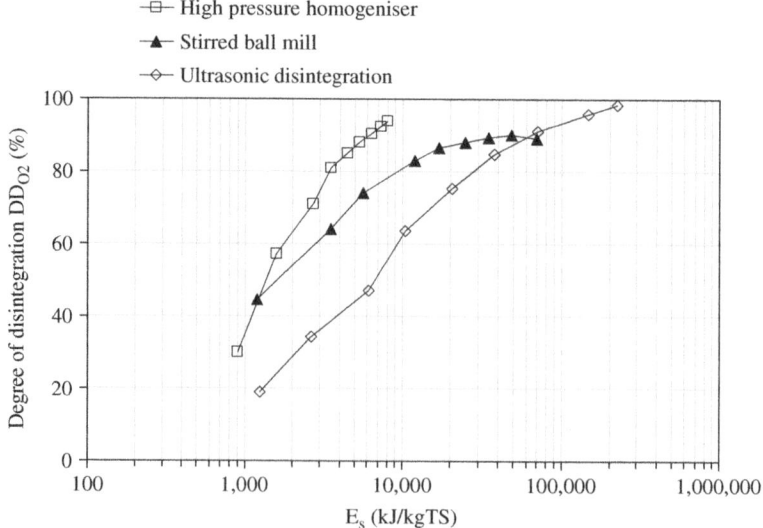

Figure 9.13. Degree of disintegration (DD_{O2}) as a function of E_s for certain mechanical disintegration techniques (*from* Kopp *et al.*, 1997).

Mechanical disintegration

Disintegration using high pressure homogenisers represents the most economical technique, because it allows higher levels of DD_{O2} (about 95%) to be reached, applying energy of 8,500 kJ/kgTS. To reach the same DD_{O2} value, the energy required by ultrasonic disintegration is significantly higher (100,000 kJ/kgTS). Using stirred ball mills, good results (DD_{O2} near 90% over 20,000 kJ/kgTS) are obtained with a long disintegration time (up to 60 minutes), small spheres and a high rotation speed (Kopp et al., 1997).

The comparison of DD_{COD} (but also DD_{O2}) gives an estimated efficiency for each technique in sludge reduction. The evaluation of the increase in VSS reduction in anaerobic digestion after the pre-treatment, carried out using various mechanical disintegration techniques, was reported by Müller et al. (1998; 2004) compared to ultrasonic disintegration (Neis et al., 2000; Nickel and Neis, 2007) and ozonation (Figure 9.14). This example can clarify the calculation: applying ultrasonic disintegration, DD_{COD} is 20%, while the VSS reduction in the anaerobic digester integrated with sonication is 42.4%, compared to 32.3% in the control line without sonication; the increase in VSS reduction results 42.4%-32.3% = 10.1%.

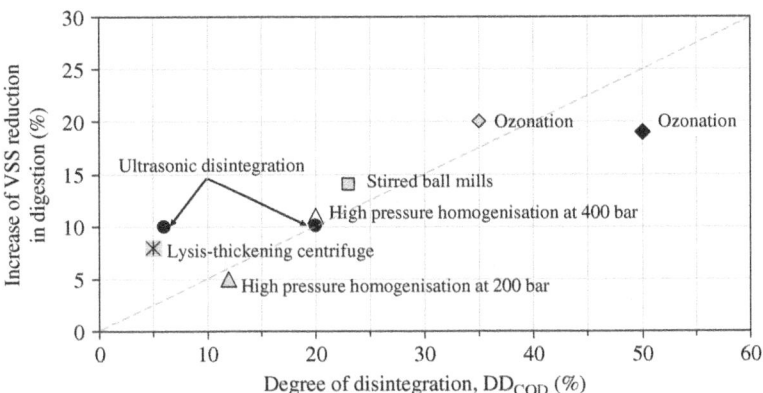

Figure 9.14. Increase in VSS reduction during digestion (HRT 10 d) of pre-treated sludge as a function of the disintegration degree (data from Müller et al., 1998; Müller et al., 2004; Neis et al., 2000; Nickel and Neis, 2007).

It can be seen that the increase of VSS reduction correlates well to DD_{COD}, demonstrating that an increase in the value of DD_{COD} effectively represents a potential enhancement of the sludge reduction expected after the application of a mechanical or a chemical (ozonation) treatment.

In the comparison of the performance of the various mechanical disintegration techniques, as well as treatment efficiency and energy consumption, other factors have to be taken into account when choosing the most suitable technology:

- investment cost;
- availability of the technology for full-scale applications;
- operating problems such as clogging of equipment at restriction points due to fibrous content of sludge in high pressure homogenisers (homogenising valves or similar) or in the spheres separation sieves in stirred ball mills[3];
- machine component wear: not much is known on the subject, since many of the proposed techniques are new or recently installed at full-scale, with little service history. For example:
  - in stirred ball mills, wear of the disintegration spheres occurs,
  - in high pressure homogenisers, erosion of valves and wear of seals in the high pressure pump occur,
  - in ultrasonic disintegration, the sonotrodes are subject to erosion,
  - in the lysis-thickening centrifuge, the cutting blades wear. These phenomena affect the management/maintenance costs of sludge disintegration.

One positive aspect of the mechanical treatments is that the equipment can be time programmed to enable the design of a strategy in which the programmed disintegration produces a lysate subsequently fed in the activated sludge stages only during low-load periods, exploiting the typical day/night, high/low load cycle and obtaining energy saving during periods of low-energy demand.

---

[3] The recent stirred ball mills are based on the centrifugal separation of spheres, in order to avoid the clogging problems of sieving.

10
Ultrasonic disintegration

*Ultrasonic disintegration is a
"no touch, no moving
mechanical parts" technique
(Winter, 2002)*

10.1 INTRODUCTION

Ultrasonic frequencies vary between 20 kHz and 10 MHz. The disintegration effect on sludge is mainly due to ultrasonic cavitation (acoustic cavitation). Cavitation takes place when the local pressure in the bulk liquid becomes lower than the vapor pressure inducing the formation of little bubbles (Tiehm *et al.*, 1997). These bubbles grow rapidly and when they reach critical dimensions they collapse violently, producing intense local heat (up to 5000 K), high pressure at the gas-liquid interface (several hundred or thousand bar), turbulence and intense shearing forces. As a consequence of these extreme conditions, radical species such as OH·, HOO· and H· and hydrogen peroxide are formed, which react with the substances in water, which is an important mechanism in sonochemistry.

© 2010 IWA Publishing. *Sludge Reduction Technologies in Wastewater Treatment Plants.* By Paola Foladori, Gianni Andreottola and Giuliano Ziglio. ISBN: 9781789065305. Published by IWA Publishing, London, UK.

Ultrasonic treatment provides a combination of two effects:

(a) a mechanical action produced by cavitation, especially at low frequencies starting around 20 kHz,
(b) a sonochemical action by radicals, favoured by frequencies higher than 200 kHz.

To produce cavitation – which is more effective in sludge disintegration – the ultrasound power level must be above a certain threshold. The cavitation threshold of pure water is very high but can be reduced significantly by the presence of impurities (Zhang et al., 2008). The predominance of cavitation and mechanical shear forces with respect to the radical effect (OH·) increases progressively with ultrasonic intensity, as demontrated by Wang et al. (2005) by adding $NaHCO_3$ in the bulk liquid to mask the oxidizing effect of OH·.

The ultrasonic treatment of sludge is influenced by three main factors:

(1) applied energy: increasing the level of energy applied increases the sludge disintegration effect;
(2) ultrasound frequency: several experiments have demonstrated that low frequencies are more effective for sludge disintegration;
(3) sludge properties (for example solid content).

Ultrasonic disintegration has been proposed for sludge reduction since the mid '90s. Investigations of the mechanisms involved in ultrasonic disintegration were first carried out by Chiu et al. (1997), Tiehm et al. (1997), Tiehm et al. (2001), Clark and Nujjoo (2000), Neis et al. (2000). The process can be integrated in the wastewater handling units or in the sludge handling units (Figure 10.1). In the first case the treatment is applied to a part of the return flow, while in the second case it is applied as a pre-treatment before anaerobic digestion, obtaining benefits in biogas production.

As in the mechanical treatments described in Chapter 9, ultrasonic disintegration enhances biological flocs disaggregation, organic matter solubilisation, bacteria cell damage, increasing sludge biodegradability. The process is completed with the subsequent biodegradation of lysates in the activated sludge reactors (when the treatment is integrated in the wastewater handling units) or in digestors (when it is integrated in the sludge handling units), by means of the mechanism of cell lysis-cryptic growth, which causes an overall reduction of sludge mass.

Several lab-scale or pilot-scale applications of ultrasonic disintegration have been reported in the literature and several full-scale plants already exist. Some full-scale experiences developed in the UK, Sweden, the USA and Australia have been referred to by Hogan et al. (2004).

Ultrasonic disintegration

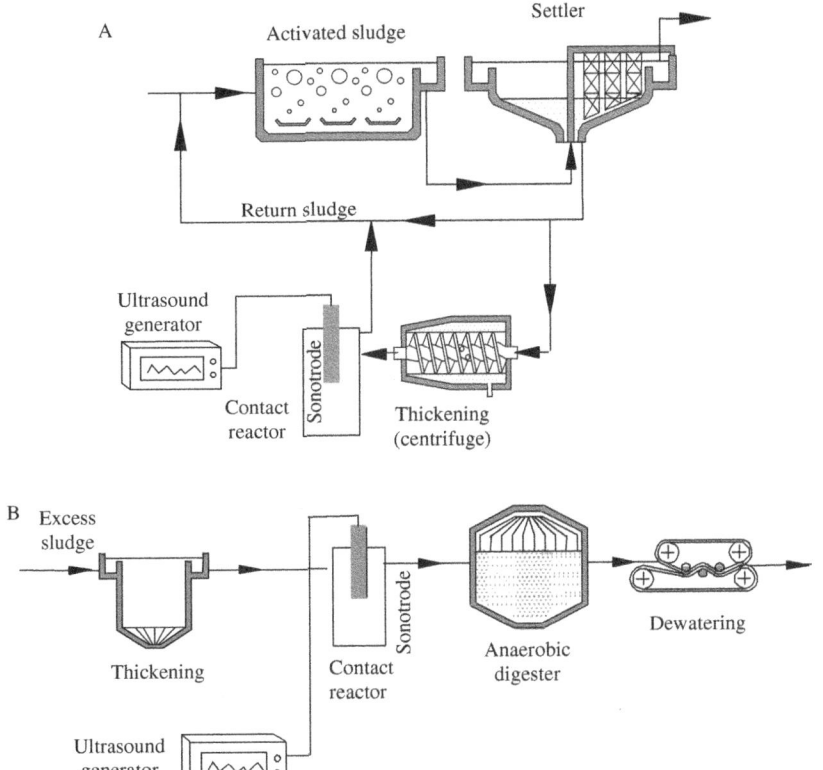

Figure 10.1. Ultrasonic disintegration integrated in the (A) wastewater handling units and (B) sludge handling units.

10.2 CONFIGURATIONS AND EQUIPMENT FOR ULTRASONIC DISINTEGRATION

Ultrasonic disintegration can be achieved with different devices:

- bath type
- probe type (sonotrode)
- flat type.

Some of these devices are used exclusively in laboratories – such as the bath type – while the probe type (called also horn type) is widely applied in full-scale plant applications (Figure 10.2).

192   Sludge Reduction Technologies in Wastewater Treatment Plants

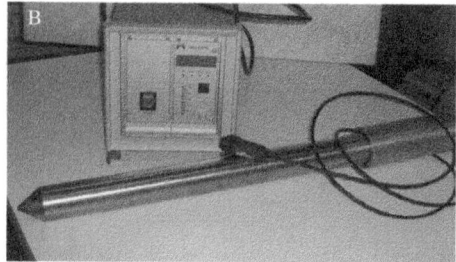

Figure 10.2. Example of (A) a lab-scale sonotrode, (B) full-scale sonotrode.

Since the most important mechanism of ultrasonic disintegration is ultrasonic cavitation, responsible for floc disaggregation and the rupture of bacterial cells, it is advantageous to apply ultrasounds at low frequencies and at high energy levels.

The configurations proposed are batch systems, made up of a simple reactor equipped with the probes inserted from above, or continuous flow systems (Figure 10.3). In both cases the reactors can be equipped with more than one sonotrode. Ultrasounds are produced by an external generator.

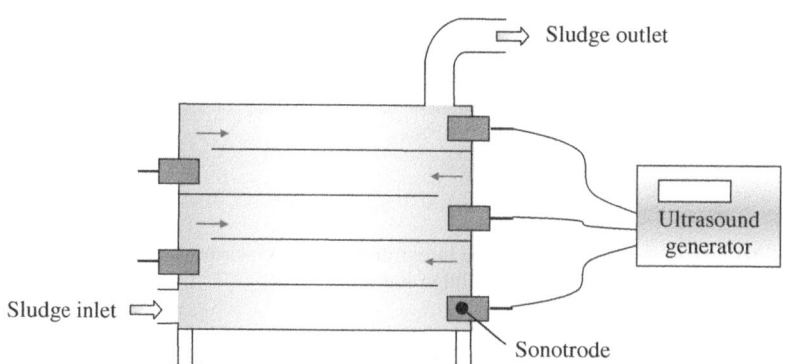

Figure 10.3. Configuration of the full-scale ultrasound reactor operating continuously (*modified from* Nickel and Neis, 2007).

In the case of the full-scale continuous flow reactor proposed by Ultrawaves GmbH (Hamburg, Germany) and by Hamburg University of Technology (Figure 10.3), the plug-flow reactor has a total volume of 29 L and it is equipped with five 20 kHz sonotrodes. Each sonotrode is supplied by a 2 kW generator. The intensity can be adjusted in a range from 25 to 50 W/cm^2 (Nickel and Neis, 2007). The sludge is pumped upwards from the bottom of the reactor – in order to prevent any accumulation of gas bubbles produced by degassing of the sludge bulk liquid – and proceeds against the propagation of the cavitation field. A minimum contact time of about 30 s (typical 90 s) is adopted. The basis principle of this sytem is to enhance transient cavitation which favours mechanical shear forces, only produced in a high energy zone near the sonotrode.

The results of sludge solubilisation (in term of DD_{COD}) are indicated in Figure 10.7, where this full-scale reactor is compared with batch sonication tests carried out using the same type of sonotrode, but immersed in a fully mixed vessel (Nickel and Neis, 2007).

The batch and continuous configurations have already been applied at full-scale in several WWTPs.

Mathematical models to simulate the sound field in ultrasonic reactors have been proposed in the literature, especially applied in the case of pure water or with diluted solutes, with the aim of predicting the creation of the cavitation bubble field. In the case of sludge sonication, modelling is much more difficult because sludge is a complex three phase system composed of solids, a water phase and gases. Therefore, the ultrasonic systems avilable on the market are often developed on empirical know-how (Nickel and Neis, 2007).

Ultrasonic disintegration presents the following particular advantages compared to other mechanical, thermal or chemical treatments (Nickel and Neis, 2007):

– compact dimensions of the treatment equipment
– ease of installation
– simple management.

However, the treatment is associated with significant energy consumption along with periodic replacement of the sonotrodes due to the erosion induced by ultrasonic cavitation.

10.3 EVALUATION OF ENERGY APPLIED IN ULTRASONIC TREATMENT

The main parameters used for describing operational conditions in applying ultrasonic disintegration are:

– *power intensity*: power per unit area of the sound-emitting surface (i.e. transducer area), expressed as W/cm^2;
– *power density*: power per unit volume of sludge treated, expressed as W/L;
– *specific (volumetric) energy* or *ultrasonic dose*: energy per unit volume of sludge treated, expressed as Ws/L or J/L;
– *specific energy*: energy per unit of TS in sludge treated, expressed as kJ/kgTS.

The specific energy (E_s) expressed per unit of TS (or TSS) represents the most widely used parameter.

In particular, E_s is calculated as follows:

$$E_s = \frac{P \cdot t}{V \cdot 1000} \quad (J/L)$$

$$E_s = \frac{P \cdot t}{V \cdot x \cdot 1000} \quad (kJ/kgTS)$$

where:
P = applied power (W);
t = treatment time (s);
V = treated volume (m^3);
x = sludge concentration (kgTS/m^3).

The parameter E_s is useful for the comparison of the results of sonication obtained under different operating conditions, simultaneously taking into account the power, treated volume and treatment time of sonication.

Although the expression above is widely used in the literature for calculating E_s and referring to the electrical power (P), the amount of energy effectively transferred to the working fluid would be a better parameter instead of gross (electrical) power. This value would ensure a better characterisation of the performance of ultrasound reactors, and facilitates accurate comparisons between different devices.

To measure the actual power effectively transferred, a method based on calorimetry can be applied (Mason *et al.*, 1992). This method is based on the assumption that ultrasounds that enter the sample are transformed into heat causing an increase in temperature (Gibson *et al.*, 2009). Thus, the power measured by calorimetry is independent of reactor geometry, materials, and other configuration aspects. The measurement refers to adiabatic conditions, i.e. without heat losses from the reactor, but, in practice, reactor insulation is not necessary if the measurement is performed within the first 30 s. The procedure is synthesised as follows (Gibson *et al.*, 2009):

(1) real-time measurement of temperature (T) in the initial period of sonication (t)
(2) calculation of the rate of temperature rise at the start of sonication $\left(\frac{\Delta T}{\Delta t}\right)_{t=0}$
(3) conversion to power using the expression:

$$P_{cal} = m \times C_p \times \left(\frac{\Delta T}{\Delta t}\right)_{t=0}$$

where P_{cal} is the calorimetrically determined power (kJ/L), m is the mass of water in the reactor, C_p is the specific heat capacity of liquid (kJ kg^{-1} °C^{-1}).

Finally, the value of E_s can be recalculated considering P_{cal} instead of P in the above expression.

During sonication at high E_s, an increase in sludge temperature is usually observed, which can reach as much as 60°C when energy over 100,000 kJ/kgTS is applied. In lab-scale tests, the sludge temperature increase is often controlled (i.e. by water bath) in order to evaluate only the effect of ultrasounds rather than the combination of ultrasounds + thermal effect. In full-scale applications the temperature does not always rise, because less E_s than the range tested at lab-scale are usually applied.

Considering the expression for calculating E_s, the same level of E_s can either be obtained by applying high power for a short duration, or lower power for a longer duration. An example of this concept, considering a suspension of *E. coli* is shown in Figure 10.4. By multiplying the power density (kW/L) by the treatment time (s), E_s can be calculated (kJ/L). The dynamic of cells correlates very well with E_s, and we suggest that it can be effectively used as a robust operating reference parameter for the evaluation of sonication performance. Therefore, to compare the different experiments reported in the literature, we chose to use E_s as a comprehensive parameter which was calculated by us when not explicitly indicated.

However, some authors suggest that at the same E_s, it is more economic to operate with a higher power for shorter time, because it is more efficient for COD solubilisation. This assumption is also supported by theoretical considerations as described by Show *et al.* (2007). In some experiences a "threshold effect" was observed, which is a power level beneath which ultrasonic disintegration is not effective even when applied for a long period.

10.4 THE INFLUENCE OF ULTRASOUND FREQUENCY

Ultrasound frequency is a very important parameter in this process, because it defines the sound field and affects sonication efficiency. The higher ultrasound frequencies favour hydroxyl free radical production and promote chemical reactions, while lower ultrasound frequencies produce stronger shockwaves favouring disintegration effects (Zhang *et al.*, 2008).

The influence of ultrasound frequency on sludge disintegration measured as DD_{COD} was evaluated between 41 and 3217 kHz on sludge with approximately 20 gTS/L (Tiehm *et al.*, 2001). The results are indicated in Figure 10.5.

Sludge disintegration is most effective at a frequency of 41 kHz, which allows a greater reduction of particle size to be obtained. On the basis of these results, better performance is expected from a further frequency decrease, to 20–30 kHz, which today represents the most widespread choice in full-scale applications.

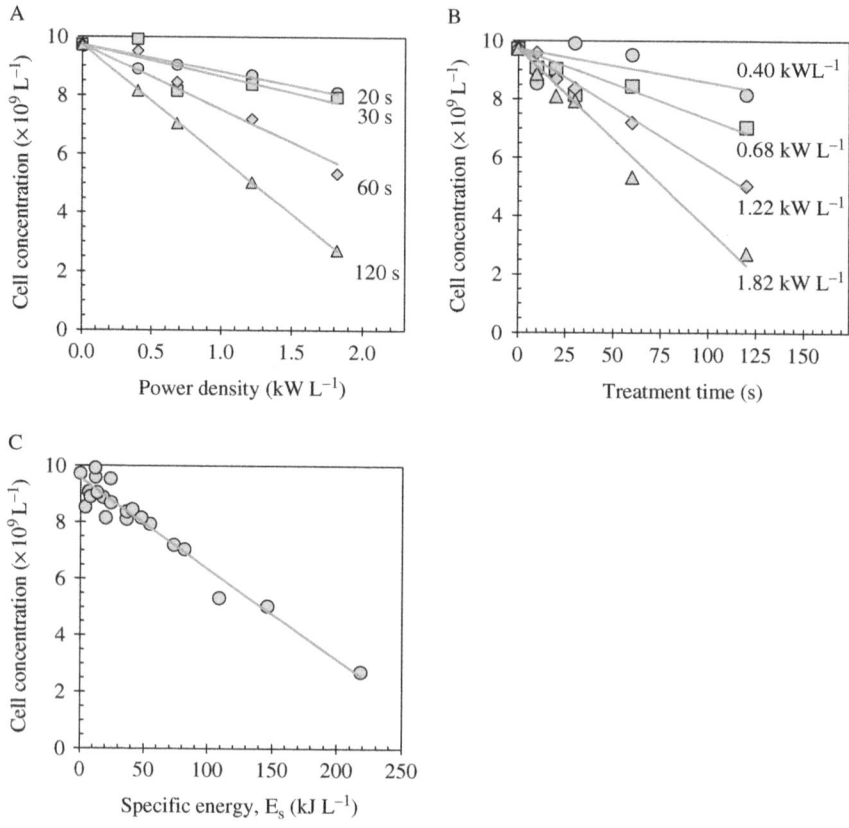

Figure 10.4. Dynamic of *E. coli* cells as a function of: (A) power density for selected treatment time; (B) treatment time for selected power density; (C) specific volumetric energy, E_s.

This dependence on frequency is also supported by theoretical considerations, as described by Tiehm *et al.* (2001), who indicated the following expression for describing the ultrasonic cavitation bubble radius in the case of air bubbles in water at atmospheric pressure:

$$R_r = 3.28 \cdot f_r^{-1}$$

where R_r is expressed in mm and f_r is the resonance frequency expressed in kHz.

The collapse of cavitation bubbles occurs when the expanding bubbles reach their resonant radius, which is a function of ultrasound frequency. The bubble radius is inversely proportional to the ultrasound frequency and therefore the application of low frequencies causes larger cavitation bubbles, which collapse and generate intense mechanical actions. It may be assumed that the energy released by this mechanical action is a function of the bubble size at the moment of collapse.

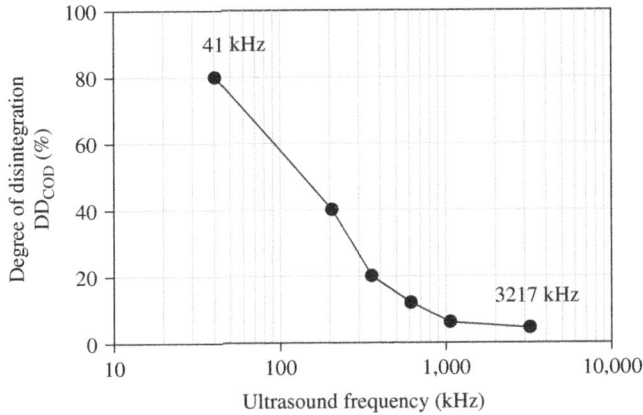

Figure 10.5. Degree of disintegration (DD_{COD}) as a function of ultrasound frequency (*from* Tiehm *et al.*, 2001).

In the case of sludge, the number and size of cavitation bubbles may differ from that of pure water, due to the high content of solids, different density of the medium and the presence of dissolved gases (Tiehm *et al.*, 2001).

10.5 COD SOLUBILISATION

During the ultrasonic disintegration of sludge, total COD, TS and VS concentrations remain quite constant, indicating that mineralisation of organic matter does not occur.

The COD solubilisation and the degree of disintegration (DD_{COD}) increase with increasing levels of E_s. In Figure 10.6 and in Figure 10.7 the values of COD solubilisation and DD_{COD} are indicated, comparing some data of the literature. A linear relationship between COD solubilisation and the logarithm of E_s in Figure 10.6, indicates that the solubilisation rate decreases as E_s increases.

The difference between the various experiences may be related to the fact that in most cases the applied power (or energy) is indicated and not the actual power effectively transferred (or energy), as described in § 10.3. Furthermore the various experiences are carried out using different ultrasound frequency from 20 to 41 kHz. For example, in the study of Tiehm *et al.* (2001) who used an ultrasound frequency of 41 kHz, the E_s required to achieve a certain degree of disintegration was higher as compared to the performance obtained with 31 kHz (Neis et al., 2000) or 20 kHz (Benabdallah El-Hadj et al., 2007), using similar TS concentrations (approximately 25.9–34.4 gTS/L).

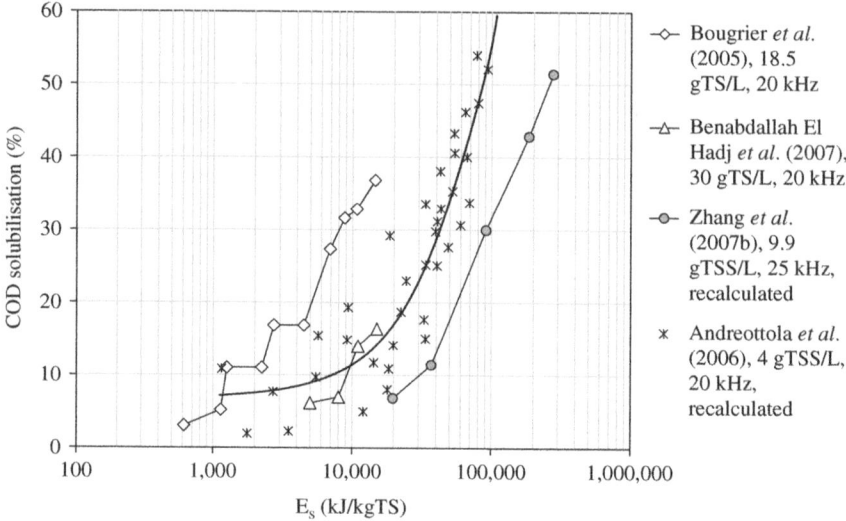

Figure 10.6. COD solubilisation as a function of E_s in the ultrasonic disintegration under various operational conditions.

Comparing DD_{COD} obtained at different TS concentration, it can be observed that the solid content favours the disintegration, as explained in § 10.5.1.

Zhang et al. (2007b) tested E_s up to 280,000 kJ/kgTSS (recalculated). At around 93,000 kJ/kgTSS, COD solubilisation was 30.1%, while solids reduction was 23.9%. At this E_s, the OUR (used as a measurement of biomass activity through respirometry) decreased significantly indicating an almost complete biomass inactivation of 95.5%.

During the ultrasonic disintegration of sludge flocs and solubilisation of EPS, proteins are easier to dissolve and resulted the main component released in the bulk liquid. Then, when the sludge has almost disintegrated, the dissolution of proteins slows down (Wang et al., 2006b). On the basis of their findings about COD, proteins and DNA solubilisation, these authors confirmed that that 50,000 kJ/kgTS gave the maximum level of solubilisation maintaining a fast rate, while over 50,000 kJ/kgTSS the release rate of proteins and DNA became slower.

Ultrasonic disintegration 199

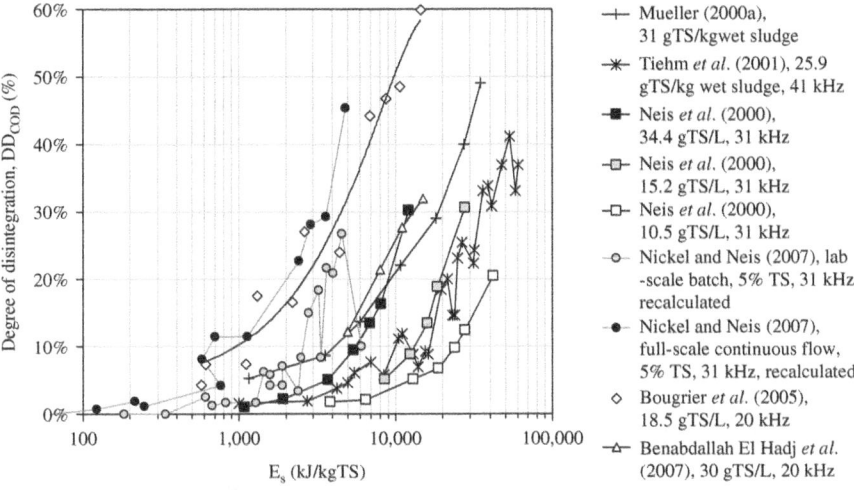

Figure 10.7. Degree of disintegration (DD_{COD}) as a function of E_s in the ultrasonic disintegration under various operational conditions.

10.5.1 The influence of sludge concentration

When ultrasound waves are applied to thickened sludge, the energy is mostly absorbed by water in sludge, which is almost exclusively composed of water (>95%).

The disintegration caused by ultrasonic treatment is more effective with higher TS concentrations in treated sludge. At the same applied E_s, the DD_{COD} is much lower for a concentration of 10.5 gTS/L compared to a concentration of 34.4 gTS/L (Neis *et al.*, 2000, see Figure 10.7).

In the presence of a higher TS concentration in the bulk liquid, a higher number of cavitation sites occurs and the probability that the solids come into contact with exploding cavitation bubbles increases.

Recently Mao *et al.* (2004) and Show *et al.* (2007) observed that a TS concentration limit exists, beyond which efficiency decreases. This value is posited by these authors at around 30 gTS/L, but it may change depending on the properties of the treated sludge. An explanation of the existence of a threshold could be as follows.

A liquid-solid system such as sludge must contain a sufficient quantity of liquid to allow the formation of micro-bubbles during cavitation. If the TS concentration is too high the bubbles formed cannot propagate and explode in the vicinity of the sonotrode causing high local temperature and pressure. It would thus not be feasible to use TS concentrations which are too high, because there is a risk of overheating and sonotrode erosion and plant breakdown.

Conversely, using diluted sludge (200 particles per mL), Gibson et al. (2009) observed that the fraction of particles disrupted by ultrasound at 20.3 kHz was not affected by the particle concentration and increased with increasing particle size.

Every ultrasonic reactor has a characteristic maximum value of solid content in the sludge to be treated, determined by its ability to ensure the propagation of ultrasound waves and thus sludge disintegration. The factors limiting the maximum solid content are: reactor size, transducer type, sludge viscosity, sludge temperature and polymer concentration if polymers are added to the sludge during flocculation (Grönroos et al., 2005).

10.6 INFLUENCE ON MICROORGANISMS

Recently, it was observed that ultrasound at low energy can stimulate the growth of bacteria, while only high energy is able to cause their disintegration. In fact, ultrasonic waves at low frequency and low energy were found to increase the concentration of microorganisms and to enhance bioreactor performance (Zhang et al., 2008). Furthermore, studies observed that low frequency was more effective in stimulating bacteria activity than high frequency, demonstrating that the mechanism is mechanical action rather than radical reactions. Ultrasound waves cause vibration of water molecules, microorganisms and other solutes in water at a frequency equal to the ultrasound frequency. This vibration enhances mixing and favours contact between substances, microorganisms and enzymes. Furthermore, ultrasound also improves cell membrane permeability, facilitating the transport of substances into the cells. The combination of these factors favours microbial growth and biodegradation (Zhang et al., 2008).

However, for these purposes the applied energy must be very low. For example, the optimal sonication conditions found by Zhang et al. (2008) correspond to around 670 kJ/kgTS, which caused an increase of sludge OUR (*Oxygen Uptake Rate*) of 28%. Furthermore, sonication applied in an SBR system treating synthetic wastewater, increased the COD removal efficiency by 5–12%, compared to the control (Zhang et al., 2008).

At much higher energy levels, sludge disintegration and cell lysis occur, as described below.

At $E_s > 15,000$ kJ/kgTSS, an immediately appreciable effect of ultrasonic disintegration is floc size reduction, easily seen under the microscope (Figure 10.8).

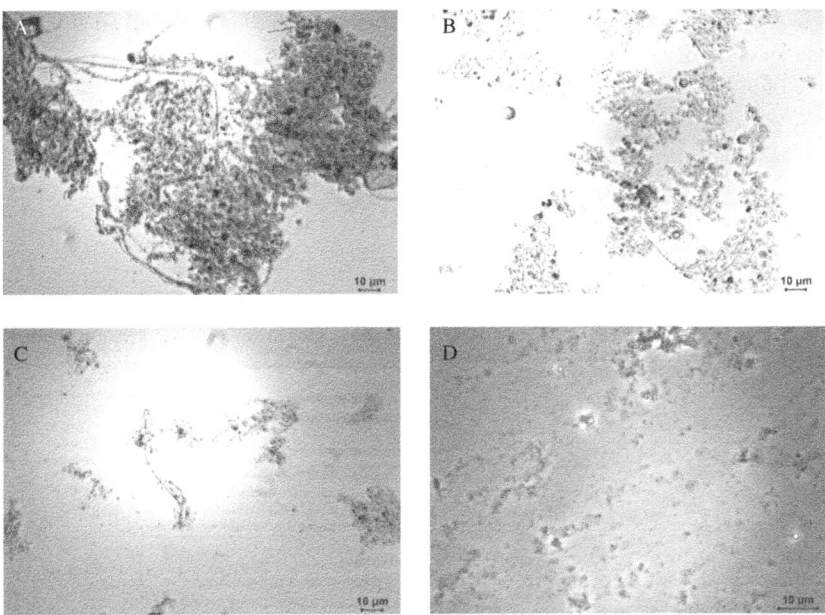

Figure 10.8. Microscope images of the following sludge samples: (A) untreated sludge; (B) sludge disintegrated at E_s of 15,000 kJ/kgTS; (C) sludge disintegrated at E_s of 40,000 kJ/kgTS; (D) sludge disintegrated at E_s of 108,000 kJ/kgTS.

As well as sludge solubilisation, bacterial cell damage is expected as a result of sludge disintegration. In order to investigate the effect of ultrasonic disintegration on the integrity or death of bacteria, advanced microbiological techniques were applied (Foladori *et al.*, 2007; Andreottola *et al.* 2006). These are based on the direct detection of cells after the fluorescent staining of their nucleic acids which allows the limits of the conventional cultivation based methods to be overcome. In particular the flow cytometry technique was used according to the approach described in § 7.7. The following aspects were evaluated:

(1) the number of disaggregated bacteria released in the bulk liquid during ultrasonic treatment;
(2) the number of intact, dead or completely disrupted bacteria[1].

---

[1] Intact bacteria present an intact cellular membrane; dead bacteria have a damaged and permeabilised membrane, but they still maintain their original shape; disrupted bacteria are no longer detectable because their fragments are too small to be detected by flow cytometry.

In untreated sludge (4 kgTS/m^3) most bacteria are aggregated in flocs; as a consequence of ultrasonic disintegration the number of bacteria released in the bulk liquid increased as shown Figure 10.9.

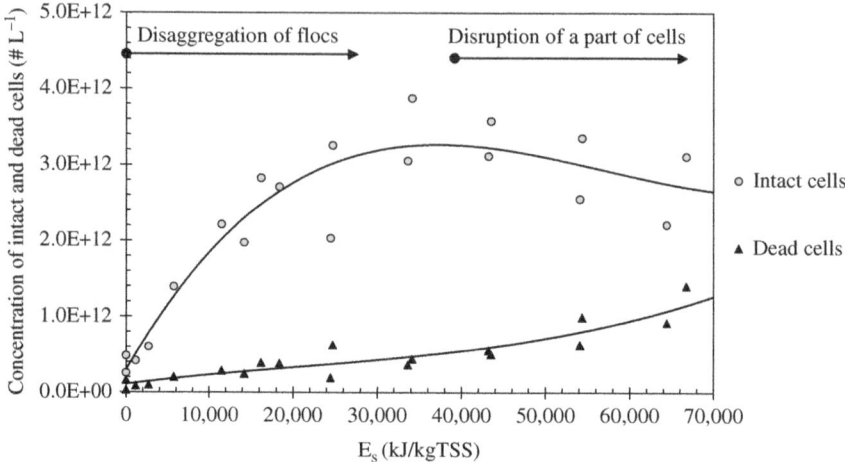

Figure 10.9. Concentration of intact and dead bacteria released in the bulk liquid as a function of E_s.

At $E_s < 32,000$ kJ/kgTS the concentration of intact and dead cells released in the bulk liquid increased significantly, up to 10 fold the concentration measured before sludge treatment. At this energy level the main phenomenon is floc disaggregation, while cell rupture is negligible. At $E_s > 32,000$ kJ/kgTS the number of intact bacteria starts to decrease, due to the progressive rupture of cells and at E_s of 68,000 kJ/kgTS a net reduction of intact cell numbers occurs.

From these results it can be observed that low E_s levels mainly cause floc size reduction without significantly damaging bacteria cells. Higher E_s levels are required to cause cell death/lysis.

However, even at low E_s levels benefits can be observed, such as an increase in the specific surface of sludge flocs and the dispersion of dense flocs, favouring contact among bacteria, substrates and enzymes. This enhances overall sludge biodegradability: for example, in the case of anaerobic digestion coupled to the ultrasonic disintegration pre-treatment, even at low energy levels a certain increase in biogas production is generally observed as demonstrated by several pratical experiences.

10.7 INFLUENCE ON SLUDGE SETTLEABILITY AND DEWATERABILITY

Because ultrasonic disintegration alters sludge floc properties (such as size, specific surface, shape, density, water content, porosity, charge, etc...), it also modifies its settling and filtering properties.

In particular, ultrasonic disintegration integrated in the wastewater handling units increases the settling rate and reduces SVI (Gonze *et al.*, 2003). At high E_s however a worsening of effluent water quality is observed, due to an increase in fine unsettleable particles (Gonze *et al.*, 2003).

At very high E_s (hower too high to be suitable in full-scale applications), sonication lyses microbial cells into small fractions, which do not settle well and enter the effluent, increasing effluent turbidity. The compactness of the EPS which constitute the floc structure decreases, causing an increase of the SVI (Zhang *et al.*, 2007a).

Resistance to filtration increases significantly when high E_s is applied; the filterability reduction is associated with worsened dewaterability and a reduction of the dry content of dewatered sludge (Gonze *et al.*, 2003).

10.8 INTEGRATION OF ULTRASONIC DISINTEGRATION IN THE WASTEWATER HANDLING UNITS

The decision to operate with a high sludge concentration allows energy consumption to be reduced and less energy to be lost in the bulk liquid. In the case of integration in wastewater handling (results indicated in Table 10.1), Müller and Strünckmann (2007) indicate the addition of a pre-thickening unit (with a centrifuge) before the ultrasonic disintegration application, according to the flow diagram indicated in Figure 10.1A. Due to this thickening a higher TSS concentration in the sludge is reached, which allows energy costs to be reduced.

Applying this process configuration, Müller and Strünckmann (2006) obtained the following reduction of sludge production, applying 105,000 kJ/m^3 of treated sludge taken from the secondary settler:

- 41% of TSS (41% of VSS) in the case of activated sludge fed with raw wastewater;
- 68% of TSS (69% of VSS) in the case of activated sludge fed with pre-settled wastewater.

Table 10.1. Performances obtained in ultrasonic pre-treatment integrated in the wastewater handling units.

Scale	Type of ww**	Frequency	Equipment	E_s*	Results	Reference
lab	rw	–	Stress frequency = 0,2 contact time = 25 min	105,000 kJ/m^3	Sludge reduction in activated sludge integrated with sonication = 41–68%	Müller and Strünckmann (2007)
lab	sw	25 kHz	Probe (surface area 2.12 cm^2)	108,000* kJ/kgTS	Sludge reduction in SBR integrated with sonication = 90%	Zhang et al. (2007a)
full	rw	31 kHz	Probe	(–) 8 W/cm^2 90 s	Sludge reduction in activated sludge integrated with sonication = 25%	Neis et al. (2008)

*some values recalculated.
**sw = synthetic wastewater; rw = real wastewater.

As explained in some experiences at pilot-scale (Ginestet, 2007a) the reduction of sludge production is higher in the case of activated sludge fed with pre-settled wastewater (68%) compared to the case of activated sludge fed with raw wastewater (41%).

In a lab-scale SBR system, fed with synthetic wastewater, and integrated with an ultrasonic treatment, a reduction of sludge production of up to 90% can be achieved, applying E_s of 108,000 kJ/kgTS (Zhang et al., 2007a). In this case the applied E_s is very high, causing an equally very high energy consumption of 3.21 kWh/m^3 (in term of wastewater). This operational cost is not economic viable because very high compared to expected sludge disposal costs. Moreover the authors observe that the 90% reduction of sludge production is only possible in systems fed with synthetic wastewater, while with real wastewater it is not possible to obtain such results – even applying high energy levels – because of the unavoidable progressive accumulation of inert solids in sludge (Zhang et al., 2007a).

A full-scale application of ultrasonic treatment was reported by Neis et al. (2008) at the WWTP Bünde (54,000 AE). In an activated sludge system operating with intermittent aeration, 30% of the daily thickened sludge stream was sonicated and the lysate recirculated in the activated sludge tank where it was used as an internal carbon source to improve denitrification. Excess sludge was reduced by 25% and the dewaterability of the sludge was improved by 2%.

10.9 INTEGRATION OF ULTRASONIC DISINTEGRATION IN THE SLUDGE HANDLING UNITS

Ultrasonic disintegration can be applied to the thickened sludge stream before entering the digester, as per Figure 10.1B. The introduction of this pre-treatment allowed foaming problems in the digesters to be solved, biogas production to be increased and an improvement of VS reduction (results indicated in Table 10.2). In some cases, the enhanced biogas production contributed to self-sufficiency with regard to energy supply.

Among the earlier studies, the enhancement of anaerobic digestion efficiency (36°C, seed:treated = 3:1) and methane production through ultrasonic pre-treatment was investigated by Wang et al. (1999) at lab-scale. At E_s of approximately 185,000 kJ/kgTS (recalculated) the total quantity of methane generated increased by 64% compared to the control. The solubilisation ratio and corresponding methane production depended linearly on E_s up to 185,000 kJ/kgTS, while a pre-treatment at higher E_s did not lead to further increases in methane production. Thus the authors indicated this value as the optimal pre-treatment for upgrading the anaerobic digestion process. Simultaneously, organic solids reduction passed from 27% (control) to 38% (increase of +38%). However, these E_s levels are too high to be economically viable considering the costs for sludge disposal.

Bougrier et al. (2005) investigated ultrasonic disintegration as a pre-treatment before mesophilic anaerobic digestion utilising thickened biological sludge with a concentration of 18.5 gTS/L (VS/TS = 81%). The ultrasonic disintegration was applied at 20 kHz and at E_s up to 15,000 kJ/kgTS. Biogas production in the anaerobic digester, as expected, increases for increasing levels of E_s in the ultrasonic disintegration pre-treatment, as demonstrated in Figure 10.10. Applying E_s of 7,000–15,000 kJ/kgTS biogas production increases by more than 40% (Bougrier et al., 2005).

Benabdallah El-Hadj et al. (2007) evaluated ultrasonic disintegration applied to mixed sludge (primary sludge + biological excess sludge) before mesophilic and thermophilic anaerobic digestion, using thickened sludge with a concentration of around 30 gTS/L. The enhancement of biogas production, after ultrasonic disintegration at 11,000 kJ/kgTS, was higher under mesophilic conditions (+30.88%±3.27) than under thermophilic conditions (+16.44%±1.76). Mesophilic digestion, being more limited by hydrolysis, is more enhanced by ultrasonic disintegration than thermophilic digestion.

Table 10.2. Performances obtained in ultrasonic disintegration pre-treatment prior to anaerobic digestion.

Scale	Frequency	Equipment/plant specification	E_s (kJ/kgTS)*	Results	Reference
Lab	31 kHz	Probe. Anaerobic digestion with SRT = 22 d	(−) 64 s	VS reduction = 50.3% (45.8% in the control)	Tiehm et al. (1997)
Lab	31 kHz	Probe. Anaerobic digestion with SRT = 8 d	(−) 64 s	Increase of biogas production (+120%)	Tiehm et al. (1997)
Lab	9 kHz	Probe (surface area of contact 100 cm^2)	185,000*	Increase of methane production (+64%) VS reduction = 38% (27% in the control)	Wang et al. (1999)
Lab	41 kHz	Disk transducers (area 25 cm^2)	25,000–30,000	VS reduction = 33.7% (21.5% in the control)	Tiehm et al. (2001)
Lab	20 kHz	Probe	7,000–15,000	Increase of biogas production (> +50%)	Bougrier et al. (2005)
Lab	20 kHz	Probe	11,000	Increase of biogas production (+31%)	Benabdallah El-Hadj et al. (2007)
Pilot	31 kHz	Probe. Anaerobic digestion with SRT = 16 d	(−) 10 W/cm^2 90 s	DD_{COD} = 20% VS reduction = 42.4% (32.3% in the control)	Neis et al. (2000); Nickel and Neis (2007)
Pilot	31 kHz	Probe. Anaerobic digestion with SRT = 8 d	(−) 10 W/cm^2 90 s	DD_{COD} = 20% VS reduction = 38.1% (27% in the control)	Neis et al. (2000); Nickel and Neis (2007)
Full	31 kHz	Probe. WWTP Meldorf (65,000 AE)	(−) 8 W/cm^2 90 s	Increase of biogas production (+30%)	Neis et al. (2008)
Full	31 kHz	Probe. WWTP Bamberg (330,000 AE)	(−) 8 W/cm^2 90 s	Increase of biogas production (+30%) VS reduction = 54% (42% in the control)	Neis et al. (2008)

* some values recalculated.

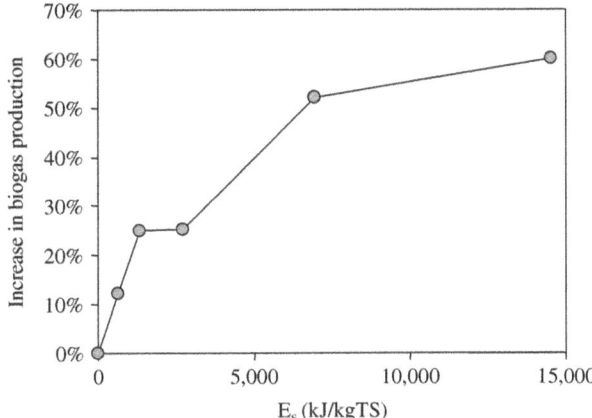

Figure 10.10. Increase in biogas production in anaerobic digestion as a function of E_s applied in the ultrasonic disintegration pre-treatment (*from* Bougrier *et al.*, 2005).

Tiehm *et al.* (2001) evaluated the reduction of sludge production in terms of VS comparing an anaerobic digestion unit fed with untreated sludge (control) and a unit fed with sonicated sludge (sludge concentration of 26 gTS/kg wet sludge). VS reduction was 21.5% in the control and 33.7% in the digester fed with sludge sonicated at E_s of 25,000–30,000 kJ/kgTS. Reducing E_s to more sustainable values, the benefit related to the VS mass reduction is reduced, but the increase in biogas production is always observed.

Full-scale applications of sonication + anaerobic mesophilic digestion have been applied since 2004 in Germany and some cases are reported by Neis *et al.* (2008). In all cases, the ultrasonic treatment was applied using operational parameters obtained in previous pilot-plant experiments (31 kHz, 8 W/cm^2, 90 s). Unfortunately the exact applied E_s is not available.

11
Thermal treatment

11.1 INTRODUCTION

The thermal treatment of sludge consists of heating to moderate ($<100°C$) or high temperatures up to $220°C$ or more, with contact times of minutes or hours, at the required pressure. The application of a thermal treatment produces various effects in the sludge such as:

- breakdown of the sludge structure, disaggregation of biological flocs,
- high level of sludge solubilisation,
- lysis of bacterial cells, release of intracellular constituents and bound water.

Therefore the water phase of sludge after a thermal treatment is characterised by a high content of dissolved organic compounds. Furthermore, as intracellular bound water is released in the bulk liquid by hydrolysis, the sludge viscosity changes considerably. A thermally treated sludge with a dry content of 12%

appears as liquid and can be handled in a similar manner to raw sludge with 5–6% (Kepp et al., 2000). This aspect can be exploited advantageously to reduce digester volume.

The effects induced by the thermal treatment can be used to increase biogas production in anaerobic digestion, improving dewaterability, pathogen inactivation and reduction of sludge produced.

Significant growth of interest in the thermal treatment of sludge dates back to the '60s and '70s, when the aim was to combine some of the benefits of dewaterability with improved digestion and full-scale installations began in the late '60s (inter alia Haug, 1977; Haug et al., 1978, 1983). The examples are the Zimpro process and the Porteous process (developed from 1939). Most of the early installations were operated at 180–250°C, to obtain optimum dewatering before incineration. Historically, the escalation of energy costs coupled with technical, operational and odour problems led to the early closure of these plants in the '70s and some as late as the '80s (Kepp et al., 2000; Neyens and Baeyens, 2003).

During these last years, other new thermal hydrolysis processes have however been developed, are currently commercially available and many are in full scale operation for a number of years. They are designed to thermally treat sludge for obtaining its reduction and to obtain a resulting sludge more amenable to dewatering and thus easier and cheaper to manage and dispose of.

The thermal treatment of sludge was also applied to generate an internal carbon source for denitrification (inter alia Smith and Göransson, 1992; Barlindhaug and Ødegaard, 1996). Thermal treatment was also used to reduce foaming in activated sludge reactors and digesters. Barjenbruch et al. (2000) observed that foaming in an anaerobic digester could effectively be stopped by thermal pre-treatment of sludge (121°C for 60 min).

The input of thermal energy is achieved by heat exchangers or by the application of steam to the sludge.

The main parameter for thermal treatment is temperature, whilst the duration of treatment generally has less influence.

Some experiences in the literature indicated that temperatures below 100°C can already contribute to partial sludge reduction, by increasing sludge biodegradability.

Temperatures above 150°C (and pressures of 600–2500 kPa combined with these temperatures) must be reached to liquefy sludge, which contributes significantly to sludge reduction, but the process is costly and requires demanding maintenance standards. In this temperature range contact time has little effect compared to the temperature range. For example, Dohányos et al. (2004) even proposed extremely short contact times in thermal treatment, lasting just 1 min at 170°C, to improve anaerobic digestion. Conversely, treatments at moderate temperatures, below 100°C, require a longer contact time (from some hours to one day).

Temperatures above 180°C do not cause a further appreciable increase of sludge biodegradability; indeed a gradual decrease is observed due to the formation of refractory compounds linked to Maillard reactions (Bougrier et al., 2007b).

11.2 COD SOLUBILISATION

Thermal treatment partially solubilizes sludge, but does not mineralize the organic matter, as demonstrated by the very slight decrease in total COD of sludge even after treatment at 175°C (Graja et al., 2005). When quantified, mineralisation is very low, resulting in a percentage of 0–5% of initial VS after treatment at 130°C for 60 min (Carballa et al., 2006).

Results obtained for COD solubilisation as reported in the literature are compared in Figure 11.1. It can be observed that COD solubilisation increases in an almost linear way in relation to temperature.

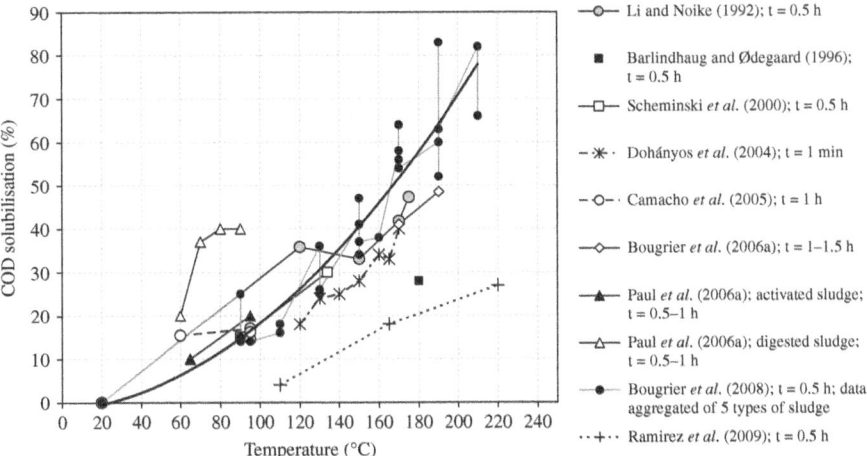

Figure 11.1. COD solubilisation as a function of temperature applied in the thermal treatment. In legend: t = contact time.

Many authors agree that COD solubilisation increases as the temperature rises, whereas contact time has little impact especially at higher temperatures.

In Figure 11.1, different types of sludge are included, but the origin of sludge (municipal) is significant in COD solubilisation, but not as important as the range of temperatures applied. This observation has been confirmed considering the data reported by Bougrier et al. (2008), which compared the efficiency of thermal treatment applied to 5 types of waste activated sludge: the dependence of this data on temperature is similar and therefore the data was aggregated in a single series in Figure 11.1. However, in an other investigation (Paul et al., 2006a), a significant difference was observed in the COD solubilisation between sludge of different origins: activated sludge and anaerobically digested sludge (see data in Figure 11.1). For digested sludge, the COD solubilisation was higher (40% at 80°C) and more rapid than that observed for activated sludge (20% at 95°C) (Paul et al., 2006a).

Li and Noike (1992) investigated COD solubilisation between 120 and 175°C at contact time of 30 min and the results obtained are reported in Figure 11.1. These authors noted that the solubilisation increase levelled off after the first 30 minutes. Heat treatment was effective in transforming sludge and solubilising carbohydrates, proteins and lipids, and converting them into low molecular weight VFA. However, not all organic compounds react in the same way (Li and Noike, 1992): according to Wilson and Novak (2009), thermal hydrolysis at 130°C has a larger impact on soluble polysaccharide release than on soluble protein release. The relative susceptibility of polysaccharides to solubilisation was also observed by Bougrier et al. (2008). For temperatures below 130–170°C (depending on sample) carbohydrates appeared to be more easily solubilised than proteins, while at higher temperatures, protein solubilisation also started to increase. The authors suggested that carbohydrates were mainly located in the exopolymers of sludge structure whereas proteins were mainly located inside the cells. At relatively low temperatures, exopolymers and carbohydrates were solubilised, while higher temperatures are needed to lysate cell walls and solubilise proteins. Furthermore, the low ammonia level found in the thermally treated sludge demonstrated that proteins were solubilised and only slightly hydrolysed to ammonia (Bougrier et al., 2008).

Bougrier et al. (2006b) observed that organic solids were more affected by thermal treatment. In fact, slightly higher solubilisation levels were observed for volatile solids than for total solids and thus VSS/TSS ratio decreases at increasing temperatures and particles become more inorganic (Bougrier et al., 2008).

11.2.1 COD solubilisation at moderate temperatures (<100°C)

The effect of thermal treatment at T<100°C was investigated by Camacho et al. (2005) and Paul et al. (2006a) on activated sludge and digested sludge.

They found that the solubilised COD after thermal treatment increased as a function of contact time at temperatures between 60°C and 95°C, as indicated in Figure 11.2.

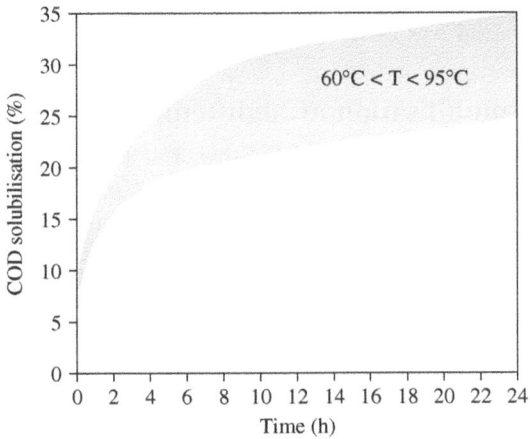

Figure 11.2. Solubilisation of COD as a function of contact time, at 60 and 95°C (*modified from* Camacho et al., 2005).

The lowest temperature (40°C) is not enough to produce a significant release of COD. Beyond 60°C, rapid COD solubilisation occurs immediately after the start of the treatment (~flash temperature) and continues with a lower rate.

At 60°C the maximum release of COD was around 25% after 24 h contact time, but was already at 15% after 2 h.

COD solubilisation of activated sludge is also rapid at 95°C, reaching 6–12% of total COD in the first minutes of contact time. A much slower COD solubilisation continues for longer contact times, reaching solubilisation of 15% after 2 h, and a maximum release around 30–35% at 95°C after 24 h of contact time (Camacho et al., 2002b; 2005). Therefore, COD solubilisation apppears to be dependent on both the temperature and the contact time. However, because the COD solubilisation rate decreases significantly with time, it is not worth prolonging the heating time beyond 30–60 minutes.

Paul et al. (2006a) observed that heating sludge from 25 to 95°C transfers an energy of 447 kJ/mol of VSS (weight of biomass 113 gVSS/mol, assuming the composition $C_5H_7NO_2$), which is higher than the energy of a non-covalent link that can be assumed to be 20–30 kJ/mol. Therefore, at this energy level the non-covalent links of biological flocs are expected theoretically to be broken up and the sludge structure destroyed. Conversely, the complex structure of biological flocs is characterised by the effects of electrostatic, ionic and hydrogen bonds, and the global non-covalent energy may reach values near that of covalent links (Paul et al., 2006a). This is probably the reason why temperatures below 100°C do not destroy flocs completely and only a partial sludge destructuration was observed (Paul et al., 2006a).

11.2.2 COD solubilisation at high temperatures (>150°C)

Thermal treatment at temperatures higher than 150°C covers the applications of hydrothermal oxidation, such as wet air oxidation (§ 6.17) or supercritical water oxidation (§ 6.18). A useful curve of the results reported in the literature was prepared by Camacho et al. (2002b) considering a wide range of temperatures up to 420°C. Figure 11.3 shows the amount of COD solubilized (including mineralisation) by the simple thermal treatment and the hydrothermal oxidation as a function of reaction temperature, indicated as a continuous curve in the figure and compared with experimental data already shown in Figure 11.1.

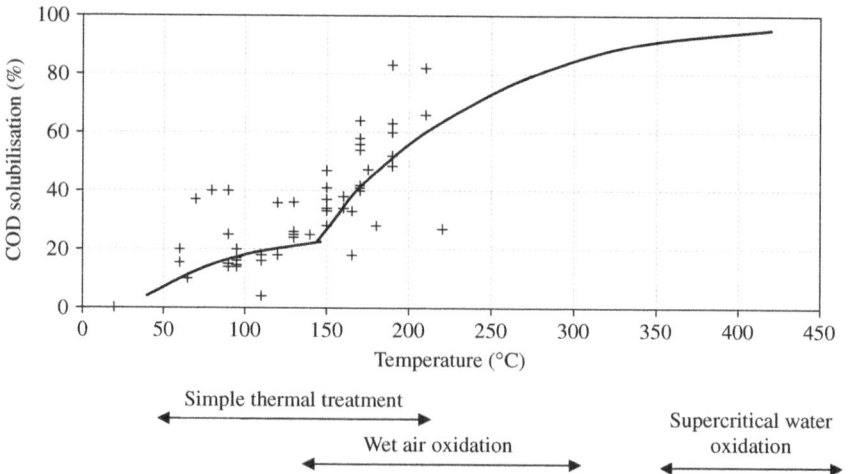

Figure 11.3. Approximate curve of COD solubilisation (*modified from* Camacho et al., 2002b); points refer to data in Figure 11.1.

Hydrothermal oxidation is more efficient than simple thermal treatment for organic matter solubilisation and so very high excess sludge reduction is expected.

A threshold value of solubilised COD was proposed at around 150°C, which is probably characteristic of different physico-chemical modifications (Camacho et al., 2002b).

In the case of simple thermal treatment at temperatures around 150°C, sludge heating causes liquefaction, desorption and cell lysis which leads to COD solubilisation. As a consequence, the accessibility of the material for subsequent biodegradation may be increased but there is little change in the sludge molecular chemistry. Thus, the effect of the thermal treatment may simply improve the availability of the already biodegradabile matter (Paul et al., 2006a). Conversely, hydrothermal oxidation greatly increases the sludge solubilisation yield and reactive oxygen is probably incorporated in some molecules thus increasing their biodegradability.

The mineralisation of the organic matrix is dependent on the temperature applied and on the oxidation level during hydrothermal oxidation, while no mineralisation is observed at moderate temperatures with simple thermal treatment, where only COD solubilisation occurs.

11.3 INCREASE OF BIODEGRADABILITY

Paul et al. (2006a) investigated the biodegradability of solubilised COD after thermal treatment at T < 100°C. The biodegradability of soluble organic matter in sludge was measured by respirometric experiments (batch sequenced respirometer, according to Spérandio and Paul (2000), see § 7.4, and SAPROMAT BOD-measuring unit). The authors observed that the biodegradability of the solubilised COD was low, around 30–35% and decreased as temperatures rose from 40°C to 95°C. This unexpected phenomenon is probably due to the release of part of the sludge material by desorption or floc destructuration while no increase in its intrinsic biodegradability occurs (Paul et al., 2006a).

On the basis of these observations, Paul et al. (2006a) concluded that thermal treatment at T < 100°C does not increase the intrinsic biodegradability of sludge, while it does improve the availability of the already biodegradable matter through molecule desorption and floc destructuration, which significantly improves the availability of organic matter to enzymes and microorganisms.

Sludge biodegradability increases with pre-treatment temperatures in the range 90–210°C, as observed by Carrère et al. (2008), and sludge biodegradability enhancement is linearly correlated to COD solubilisation for temperatures up to 190°C. At 210°C, biodegradability was lower than at 190°C, due to the formation of recalcitrant compounds. In fact, at high temperatures the phenomenon of sugar "caramelisation" (pyrolysis) and Maillard reactions (reaction of carbohydrates with aminoacids) occur and the characteristic "tea-coloured" liquor is produced, due to the presence of new compounds, such as Amadori compounds and melanoidins which are refractory and potentially inhibitory to subsequent anaerobic digestion (Ramirez *et al.*, 2009).

It has often been thought that thermal treatment would not be advantageous for primary sludge and also, historically, this treatment has been applied to excess biological sludge rather than primary sludge. Recently, Wilson and Novak (2009), revisited the effect of thermal treatment at 130–220°C, and demonstrated that primary and secondary sludge responded similarly in the breakdown of proteins, lipids and polysaccharides during thermal treatment. Furthermore thermal treatment of primary sludge can increase its dewaterability (Kepp *et al.*, 2000).

Due to its enhanced biodegradability, the thermal hydrolysate – which is the supernatant of thermally treated sludge – produced after a treatment at 180°C for 30 min, was also used as a carbon source to support post-denitrification in a moving bed biofilm reactor (Barlindhaug and Ødegaard, 1996).

11.4 NITROGEN AND PHOSPHORUS SOLUBILISATION

Nitrogen – Nitrogen compounds are released during the thermal treatment of sludge and the following characteristics were observed (Xue and Huang, 2007):

- significant solubilisation of organic nitrogen, which is the major component of the total N solubilised;
- very low ammonia, nitrite and nitrate concentrations;
- negligible variations of total nitrogen concentration in sludge.

In a thermal treatment at 175°C on thickened sludge, the solubilisation of total nitrogen was 32%, but only a fifth of it was converted to NH_4 (Graja *et al.*, 2005).

Solubilisation of N for temperatures in the range 40–70°C for a contact time of 1–3 h is shown in Figure 11.4 (Xue and Huang, 2007).

Phosphorus – With regard to P solubilisation, the results reported in the literature do not completely agree.

During thermal treatment P is solubilised and increasing concentrations of PO_4 and total P are measured in the supernatant for temperatures above 50°C. Whereas the released P is moderate at 40°C, at 50–70°C the release rates were much higher, especially in the first hour and then gradually slowed for longer contact time. A temperature of at least 50°C is thus needed for an appreciable P release from sludge (Xue and Huang, 2007). For example, a maximum PO_4

release of approximately 90 mg/L (about 95% of total P released) was observed in the first hour at 50°C after the thermal treatment of sludge having around 8 gTSS/L and 3–4% of total P content (Xue and Huang, 2007).

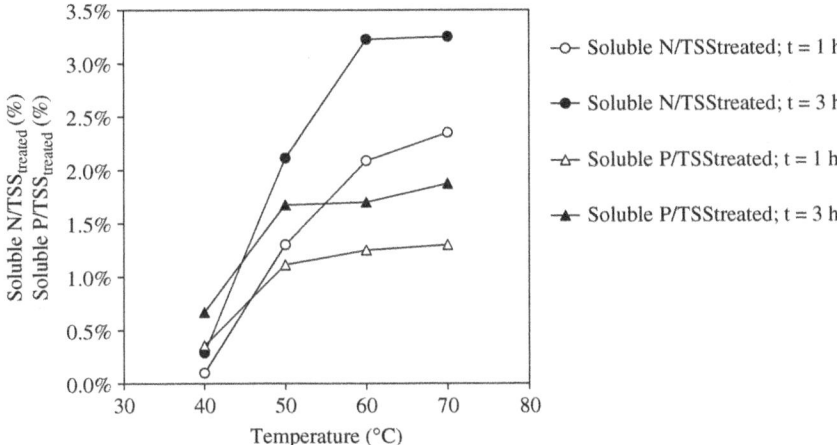

Figure 11.4. Specific P and N solubilisation (per unit of TSS treated) during thermal treatment (*modified from* Xue and Huang, 2007). In legend: t = contact time.

However, Kuroda *et al.* (2002) observed only a small amount of P released in 1 h at 50°C, while the optimal conditions for P release were 1 h at 70°C (polyphosphate around 85–90 mg/L from a sludge with 3.1 gTSS/L and 4.3% of total P content).

The release of metal cations, including Mg^{2+}, K^+ and Ca^{2+}, during thermal treatment were investigated by Xue and Huang (2007) and it was found that concentrations of Mg^{2+} and K^+ increased with contact time and temperature, but Ca^{2+} decreased, probably due to a spontaneous precipitation (as calcium phosphate) occurring during thermal treatment (a similar hypothesys to explain the disappearance of P was drawn by Graja *et al.*, 2005). The release of Mg^{2+} and K^+ agreed well with total P release. The metal cations are generally considered to play important roles in biological phosphorus removal in activated sludge systems, because, besides phosphorus, they are components of polyphosphate granules.

Some investigations evaluated the release of P from thermally treated sludge with the aim of recovering P from sludge (*inter alia* Xue and Huang, 2007; Kuroda *et al.*, 2002). In fact nearly all of the polyphosphate accumulated in large quantities in the activated sludge (in enhanced biological phosphorus removal processes) can be released from the sludge when it is heated to 70°C for about 1 h. The addition of Ca^{2+} ions allows approximately 75% of the total P to be recovered in a reusable form, without pH adjustment (Kuroda *et al.*, 2002).

11.5 INFLUENCE ON MICROORGANISMS

The influence of heating on microorganism viability in lab cultures is commonly evaluated through cultivation methods. The heating of biomass for 10 min at 90°C resulted in a rapid viability loss, nearly 100%, explained as cell death followed by cell lysis (Canales *et al.*, 1994).

With regards to enteric bacteria, at the pasteurisation temperature of 70°C, the maximum survival time is around few minutes or seconds and then organisms are killed (Lang and Smith, 2008). At 55°C, 90% inactivation of enteric is reached between 2 and 8 min. These results demonstrated that enteric bacteria, including thermotolerant strains, were rapidly destroyed by the time × temperature exposure conditions required for the pasteurisation or thermophilic treatment of sludge.

Yan *et al.* (2008) demonstrated that after 1 h of thermal treatment at 60°C nearly 98% of the mesophilic bacteria died, leaving 2% of thermophilic bacteria, which were able to secret protease and grow. The protein content in sludge is typically 20–60% w/w and the proteolytic cleavage of peptide bonds by protease generally plays an important role in the digestion or lysis of excess sludge during heat-treatment. Usually, protease is contained in the cells of sludge and these intracellular proteases would not contribute to sludge reduction, but once they are released from the cells (as a consequence of thermal treatment), they contribute to the hydrolysis of proteins in the sludge, providing the substrates for microbial cryptic growth.

However, it should be considered that cultivation methods are very restrictive for evaluating bacteria viability. In fact, it is well known that culture-dependent methods cannot detect all the bacteria in activated sludge and wastewater, due to the presence of the viable-but-not-culturable state of most cells. Therefore conventional measurement should be combined with other advanced microbiological investigations (see § 7.7).

After thermal treatment at 95°C for 5–45 min, biological inactivation of activated sludge (evaluated by respirometry) was higher than 93% (Camacho *et al.*, 2005). In any case, a complete recovery of initial activity was systematically observed after a certain time (reactivation time). However, in this process, additional maintenance energy requirements were observed with a significant reduction in the Y_{obs}, which passed from 0.65–0.77 to 0.42–0.55.

In summary, the thermal treatment of sludge at temperatures < 100°C induces:

- cell lysis by increasing membrane fluidity, just after the beginning of the thermal treatment;
- release of intracellular enzymes (protease) from the cells, which contributes to sludge lysis;
- thermal hydrolysis of intra- or extra-cellular components;
- groups of thermophilic bacteria may become dominant in the microbial population in thermal treatment around 60–70°C;
- modifications in chemical structure of exopolymers.

In practice, bacterial inactivation levels observed at full-scale would be expected to be lower than those determined at lab-scale, because of operational factors and heat transfer kinetics in sludge (Lang and Smith, 2008).

11.6 INFLUENCE ON SLUDGE SETTLEABILITY AND DEWATERABILITY

It is already well known from earlier studies on thermal treatment of sludge at high temperatures, that this treatment improves the dewatering properties of primary and secondary sludge. Sludge structure is significantly modified by thermal treatment.

Thermal treatment enhances sludge settleability: for example, the SVI decreased with the rise in temperature, passing from 140 mL/gTSS for untreated sludge to 47 mL/gTSS for T = 150°C, and then stable around 36 mL/gTSS for higher temperatures (Bougrier et al., 2008). Neyens et al. (2004) explained this by the fact that EPS are hydrated compounds able to absorb huge quantities of water and EPS solubilisation induced by thermal treatment causes the release of a part of this water.

The filterability of sludge, evaluated by CST, varies with temperature of treatment with a threshold value of 150°C: CST values increase for T < 150°C (deterioration of filterability may be due to sludge solubilisation and the release of small particles), while they strongly decrease for higher treatment temperatures (improvement of filterability, due to the release of linked water). Anderson et al. (2002) observed that at 132°C sludge became more difficult to dewater.

Experiments by Neyens et al. (2004) demonstrated that thermal hydrolysis carried out at 120°C and contact time of 1 h was effective to reduce the amount of dewatered sludge to be disposed off (by solubilisation and passage of soluble solids in the water phase) and to enhance solid content in the dewatered sludge cake. The dry content of sludge to be dewatered was reduced to 58% of the initial untreated amount and the solid content in the dewatered cake passed from 30% (untreated sludge) to approximately 43% (Neyens *et al.*, 2004).

11.7 INTEGRATION OF THERMAL TREATMENT IN THE BIOLOGICAL PROCESSES

The reduction of sludge by thermal treatment is based on the mechanisms of cell lysis – cryptic growth. The lysated sludge treated thermally is applied or recirculated in the biological reactors where a part is consumed for the catabolism and finally emitted as CO_2.

Thermal treatment can be theoretically integrated both in the wastewater handling units and in the sludge handling units, but this latter option is generally preferred, due to the higher solids concentration in thickened sludge. In fact, several investigations have proved that solid content of sludge is an important factor in thermal treatment, because an increase of solids concentration (and then a water content decrease) means an energy saving by avoiding the use of most of the heat energy for water heating and thus smaller volumes for the hydrolysis reactor (Xue and Huang, 2007).

Although some studies have been aimed at testing the integration of thermal hydrolysis in the wastewater handling units, full-scale applications are mostly based on the combination of thermal pre-treatment + anaerobic digestion, such as the Cambi process (Kepp *et al.*, 2000) and the BioThelys® process (Chauzy *et al.*, 2008), currently available commercially. The basic idea is to enhance hydrolysis, the kinetically limiting step for sludge digestion, by causing it thermally rather than biologically.

11.7.1 Integration of thermal treatment in the wastewater handling units

Very few studies regarding the integration of thermal treatment in the wastewater handling units have been published. For example, thermally treated sludge has been used to give biodegradable COD to support pre-denitrification (Barlindhaug and Ødegaard, 1996). The main limitation to the application of this configuration is the high costs for heating sludge at low solid concentrations. One option to reduce costs associated with heating, is to apply thermal treatment at temperatures below 100°C, which reduces the heating and pressure requirement.

It was demonstrated that a thermal treatment, even at $T < 100°C$ integrated in the activated sludge stages causes a significant reduction of excess sludge production. This reduction is directly linked to an immediate decrease in biological activity, an increase of maintenance requirements and a partial increase of the biodegradability of organic compounds (Camacho et al., 2005).

In earlier studies (Canales et al., 1994), the thermal treatment was integrated in a lab-scale MBR fed with synthetic wastewater (the thermally lysated biomass is recirculated in the MBR) and a 60% reduction in the excess sludge production was observed.

Sludge production in thermal treatment + activated sludge process (SRT = 12–15 d) was investigated by Camacho et al. (2005) at pilot-scale, treating sludge at 95°C, with contact times of 5–45 minutes, and with various stress frequencies (as defined in § 7.11). For each operational condition the sludge production was systematically lower than in the untreated biological process, and the highest reduction was achieved at 45 minutes, demonstrating that the contact time becomes important for T below 95°C. The stress frequency of the thermal treatment was also important and the optimal value was found to be $0.3\ d^{-1}$. A maximum sludge reduction of 55% was reached at 95°C, 45 min and stress frequency of $0.3\ d^{-1}$. No deterioration of effluent quality was observed, showing that the biomass had maintained its efficiency in carbon removal and nitrification (Camacho et al., 2005; Paul et al., 2006b).

11.7.2 Integration of thermal treatment in the sludge handling units

Thermal hydrolysis + mesophilic anaerobic digestion

The combination of thermal pre-treatment + mesophilic anaerobic digestion has been widely investigated in the literature, aimed at improving sludge biodegradability, sludge reduction and higher biogas production.

In the earlier studies, Li and Noike (1992) found that the optimal conditions for thermal pre-treatment were contact time of 30–60 min and temperatures of around 150°C. Gas production (evaluated during 30 d batch tests) increased as the contact time (in the thermal treatment) became longer, but there was only a marginal additional increase at contact times over 30 min. Methane gas production reached the maximum for thermal treatment at around 150°C (30 min), with an increased efficiency of 35% compared to the control. The VSS degradation during anaerobic tests was higher for sludge thermally treated at 170°C (VSS degradation of 60%) compared to the control (30%). These results demonstrate the potential of reducing the retention time necessary for anaerobic digestion to 5 days.

Carbohydrates are solubilised in higher proportion compared to proteins and lipids, and the soluble carbohydrates produced by thermal pre-treatment were almost completely degraded during anaerobic digestion (to a greater extent than in the digestion of untreated sludge), demonstrating that the degradability of carbohydrates was markedly increased by thermal hydrolysis (Li and Noike, 1992).

Optimal conditions for thermal pre-treatment + anaerobic mesophilic digestion are thus generally 160–180°C for 30–60 min, as confirmed in several studies. Thermal treatment at around 175°C, combined with anaerobic digestion, can significantly reduce sludge production and this reduction can reach 50–70% according to the process applied (as referred to by Bougrier et al., 2007b).

With regard to the enhancement of biogas production, the use of $T < 100°C$ may lead to an average 20–30% increase in biogas production, while the increase is much higher at temperatures of 160–180°C (40–100% biogas production increase).

Treatment time is often mentioned as having a very low influence, and Dohányos et al. (2004) proposed a very fast thermal pre-treatment, lasting only 1 minute, at a high pressure and temperature (120–170°C, 0.8 MPa, pilot-scale). Due to the short retention time in the reactor, the cells are disrupted by the shock pressure, but an inactivation of enzymes does not take place to any great extent.

In thermal pre-treatment from 90°C to 210°C the biogas volume enhancement was linked to sludge COD solubilisation and the lower the initial biodegradability, the higher the impact of thermal treatment (Bougrier et al., 2008). At temperatures above 175°C, in spite of the enhancement of sludge solubilisation, biogas production may not be enhanced or may even decrease, because of the formation of products of Maillard reactions (Pinnekamp et al., 1989), which are difficult or impossible to degrade and it is associated with the brown color of the supernatant (Bougrier et al., 2007b, 2008).

With a COD solubilisation of 34% and 46% at 135°C and 190°C respectively, the TS reduction in anaerobic digestion was 35% at 135°C and 49% at 190°C, compared to the untreated sludge (TS reduction of 31% without thermal treatment) (Bougrier et al., 2007b). However, the authors further stated that long-term sludge storage may have caused an underestimate of the effect of thermal pre-treatment in this study.

TS reduction passed from 25% (control) to 43% (150°C) and 52% (170°C) in the study by Bougrier et al. (2006b).

Thermal hydrolysis + thermophilic anaerobic digestion

Few studies have analysed thermal pre-treatment + thermophilic anaerobic digestion (Gavala et al., 2003; Skiadas et al., 2005; Ramirez et al., 2009).

With regard to thermal pre-treatment carried out at moderate temperatures (<100°C), the effect of 70°C (contact time of 2 d) on thermophilic anaerobic digestion (55°C, HRT = 15 d) of primary and secondary sludge in lab-scale digesters was investigated by Gavala *et al.* (2003) and Skiadas *et al.* (2005). The process with the thermal pre-treatment step resulted in higher sludge reduction (43% of VSS) than the control without thermal pre-treatment (6% of VSS), as a result of the enhancement of the solubilisation of secondary sludge in the first case. In particular, in the thermally pre-treated line, 33% of VSS was removed in the thermal pre-treatment, while 10% took place in the anaerobic digester. The thermal pre-treatment also completely destroyed the faecal streptococci (chosen as a pathogen indicator) compared to the control which had a lower pathogen reducing effect.

Passing to a thermal pre-treatment at higher temperatures coupled with anaerobic thermophilic digestion, a maximum value of sludge biodegradability and methane production was observed at 165°C (Ramirez *et al.*, 2009). This result is also confirmed by other studies which showed that a pre-treatment temperature of 170°C seems to be the limit for the improvement of methane production. At 220°C the COD solubilisation reached 27% and carbohydrates in the soluble phase reacted with other components to form products (Maillard reactions) which are slow or difficult to biodegrade.

A synthesis of the results obtained with thermal treatment and mesophilic or thermophilic anaerobic digestion is indicated in Table 11.1.

Thermal pre-treatment reduces sludge mass and considerably increases methane production in mesophilic anaerobic digestion, while the increase is to a lesser extent for thermophilic anaerobic digestion. Therefore, reduced benefits are expected when a thermal pre-treatment is applied before the thermophilic digestion (Appels *et al.*, 2008).

11.7.3 Full-scale applications

Thermal pre-treatment requires the input of a considerable amount of energy for heating. However, thermal pre-treatment + anaerobic digestion may result in a net energy production from the system due to:

(1) increased biogas production, as a result of increased biodegradability;
(2) reduced digester heating requirements.

In practice, sludge heating costs can also be lowered by working with higher solid concentrations in feed sludge.

Table 11.1. Performances obtained in thermal pre-treatment prior to anaerobic digestion.

Scale	Optimal conditions		Results	Reference
	T (°C)	Contact time (h)		
lab	150°C	0.5 h	Increase of methane production (+35%)	Li and Noike (1992)
lab	170°C	0.5 h	VSS reduction (+30%)	Li and Noike (1992)
lab	150°C	0.5 h	TS reduction 43% (25% for the untreated sludge)	Bougrier et al. (2006b)
lab	170°C	0.5 h	TS reduction 52% (25% for the untreated sludge)	Bougrier et al. (2006b)
full	160–180°C	0.5 h	VSS reduction 62% (42% for the untreated sludge)	Pickworth et al. (2006)
lab	135°C	0.5 h	TS reduction 35% (31% for the untreated sludge)	Bougrier et al. (2007b)
lab	190°C	0.25 h	TS reduction 49% (31% for the untreated sludge)	Bougrier et al. (2007b)
full	150–180°C	0.5 h	VSS reduction (+45%)	Chauzy et al. (2008)

Odorous compounds generally caused by the thermal treatment can be reduced during digestion of thermally pre-treated sludge (Neyens and Baeyens, 2003).

An example of a thermal treatment at around 190°C, integrated in the conventional mesophilic anaerobic digestion is shown in Figure 11.5 (Bougrier et al., 2007b). The thickening of sludge to reach a TS concentration of around 50 g/L is important to reduce heating energy consumption. Sludge passes through two heat exchangers and undergoes heating to the desidered temperature (for example 190°C). The aim of the first heat-exchanger is also to cool thermally treated sludge from 190°C towards 35°C, suitable for anaerobic digestion, recovering heat to help increase the inlet sludge temperature to 175°C. The hot fluid is water vapour, which condenses in the heat-exchanger and is heated again in the boiler by burning methane. Bougrier *et al.* (2007b) demonstrated that that energy required for heating in the process can be positively balanced by biogas production.

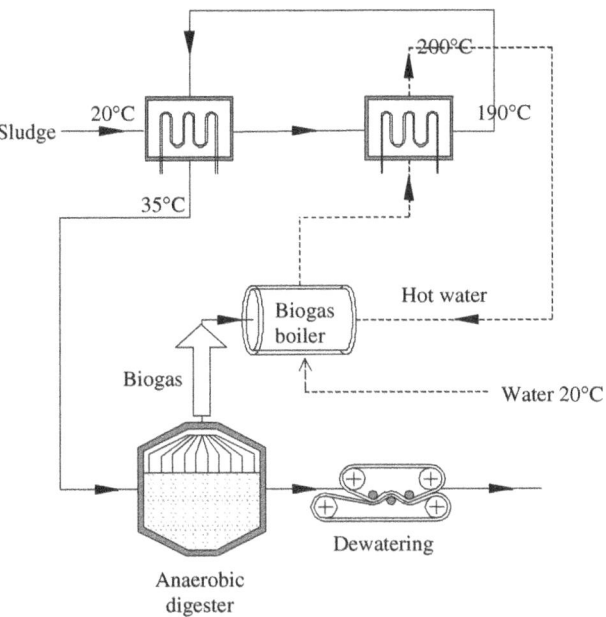

Figure 11.5. Scheme of a thermal treatment + anaerobic digestion (Bougrier *et al.*, 2007b).

The combination of thermal pre-treatment + mesophilic anaerobic digestion (fixed-film reactor), aimed at methane production, was studied in 2003–2004 by Chauzy *et al.* (2005) in a small full-scale plant (Witry-les-Reims; 2,500 PE; activated sludge SRT ~15 d). In this process thermal hydrolysis replaces the hydrolytic step in the anaerobic mesophilic process, which is performed at HRT <3 d. The thermal treatment is applied to a portion of the return flow, coming from the settler to the biological tank (Figure 11.6). Sludge has been previously thickened up to 60–80 gTS/L to reduce lysis reactor volume. The treatment takes place in (Graja *et al.*, 2005):

- thermal unit composed of a heat exchanger and a thermal reactor (V = 0.23–0.4 m³), running continuously and mixed; gas flow injection (nitrogen or oxygen) is provided to maintain the pressure inside the reactor marginally higher than the saturated vapour pressure in order to

keep sludge in liquid phase; temperature is set around 175°C and contact time is 40 min;
- solid/liquid separation of the hydrolysed sludge in a centrifuge; only the liquid fraction containing most of the biodegradable and soluble matter enters the fixed-film anaerobic digester to avoid clogging; the centrifuged solids are recirculated several times in the thermal unit to optimise the overall solubilisation, before being drawn out of the system;
- rapid anaerobic digestion unit: the liquid fraction (centrate) from the dewatering unit, containing soluble compounds and low TSS, is fed into a fixed-bed anaerobic mesophilic digester (37°C, V = 5.5 m^3, HRT = 3 d) where conversion to biogas take place; the treated effluent returns to the activated sludge process.

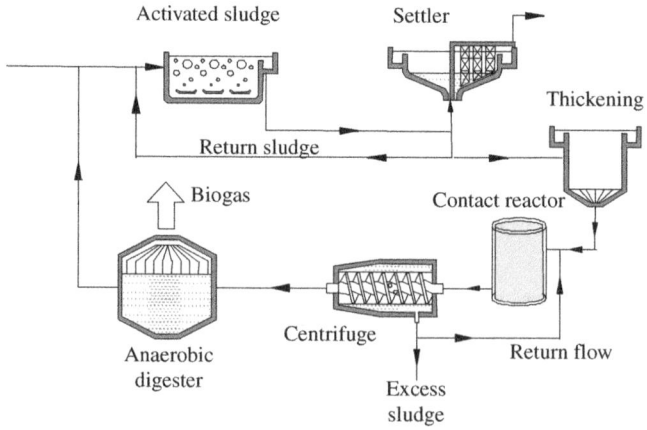

Figure 11.6. Scheme of the thermal treatment coupled with anaerobic digestion of the centrate.

The thermal treatment can be performed at temperatures in the range 150–185°C, pressures of 12–15 bar and contact time of 30–60 min. In this study 87% of sludge production without thermal treatment was treated thermally.

The centrate entering the anaerobic digester was characterised by 14,000 mg/L of soluble COD and its removal was 55%. Oxygen demand in the activated sludge stage increased by 13%; concentration of solubile COD and solubile organic N in the effluent increased very slightly due to the refractory COD produced in thermal hydrolysis and anaerobic digestion (Chauzy et al., 2005).

With respect to an Y_{obs} of 0.45 kgTSS/kgCOD in the control line, in the thermal line the yield was 0.31 kgTSS/kgCOD (a reduction of 38% calculated by the authors).

The overall solids removal obtained in the thermally treated line achieved a 52% reduction of TSS according to Graja *et al.* (2005), to be compared with the usual 40% TSS reduction in 15–20 d obtained by conventional anaerobic digesters.

A further advantage, is the improvement in sludge dewatering, obtaining a dry content of more than 35% (Chauzy *et al.*, 2005), due to the lower viscosity of thermally hydrolysed sludge (photographically demonstrated by Graja *et al.*, 2005), which makes it very easy to manage.

The process described above is considered as a prototype of the following commercial system.

The BioThelys® process (developed by Veolia Water) is a thermal hydrolysis process, proposed for enhancing the mesophilic anaerobic digestion of municipal or industrial sludge (Chauzy *et al.*, 2008). The first BioThelys was installed at Saumur in France (60,000 AE; 2 reactors for thermal hydrolysis of 4.6 m³), and started operations in 2006. The thermal hydrolysis of sludge is realised by batch, by steam injection at a temperature of 160°C (150–180°C) for about 30 min (Figure 11.7). The subsequent mesophilic anaerobic digestion (500 m³) of hydrolysed sludge is done with an HRT of 15 d and enhances VS reduction by 45% compared to the system without thermal hydrolysis. The increase of biogas production can even make anaerobic digestion suitable for the treatment of secondary sludge alone instead of (as conventionally done) mixed (primary + secondary) sludge.

Despite the advantages (greater biodegradability of thermally treated sludge, pathogen inactivation, enhanced dewaterability), one drawback is related to the high strength return flows recirculated to the wastewater handling units after dewatering.

Another similar process, available at full-scale (thermal hydrolysis + anaerobic digestion) is the Cambi process (developed by the Norwegian company Cambi AS), which aims to give a safe, stockable and stable final product. The Cambi process was first installed at the start of 1993 as a pilot plant. The first full-scale application of the Cambi as a thermal hydrolysis pretreatment before anaerobic digestion (HRT of 15 d) was built in Hamar (Norway) and has been in operation since 1995. Today the digested and disinfected sludge produced in this plant is used 100% in agriculture and land reclamation (Kepp *et al.*, 2000).

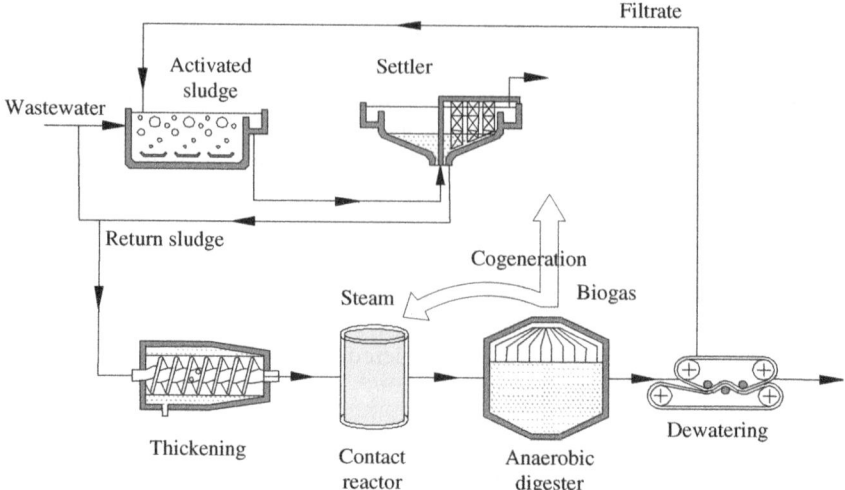

Figure 11.7. Scheme of the thermal hydrolysis process (such as BioThelys®) integrated in the sludge handling units.

The process operates at 160–180°C by means of steam, at 10 bar and 30 min contact time. Hydrolysis takes place in three steps (Figure 11.8; Kepp et al., 2000):

(1) pulper
(2) contact reactor
(3) flash-tank

and then, the hydrolysed sludge is pumped into the anaerobic digester at 10–12% dry content and this solid concentration allows 50% of digester volume to be saved.

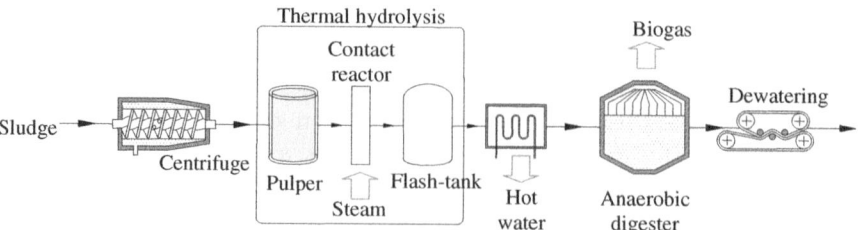

Figure 11.8. Scheme of the thermal hydrolysis process (such as Cambi) integrated in the sludge handling units (Kepp et al., 2000).

Considering thermal pre-treatment + anaerobic digester, it results in solid solubilisation of around 30%, 1.5-fold increase in biogas production and a 50% reduction of volume of dewatered sludge, ensuring good disinfection (Weemaes and Verstraete, 1998; Ødegaard, 2002; Kepp et al., 2000). The degree of stabilisation (as COD) reached 59% compared to a typical 40% in anaerobic digesters without thermal-treatment (Kepp et al., 2000).

Pickworth et al. (2006) found a VSS reduction in the Cambi system of 62% compared to 42% in anaerobic digestion alone.

11.8 MICROWAVE TREATMENT

Interest in the use of microwaves for heating instead of conventional thermal methods, has grown in recent years due to the significant reduction in reaction times and energy requirements (as a result of much lower thermal losses in transferring energy).

In fact, in conventional heating, in which heat is transferred from the heating device to the medium, performance depends on thermal conductivity, temperature gradients and convection. Conversely, in microwave irradiation, heating occurs throughout the medium in a very rapid way, but localized superheated regions (hot spots) can occur in the medium heated (Eskicioglu et al., 2007c).

Microwaves are characterised by wavelengths of 1 cm–1 m and frequencies of 30 GHz–300 MHz. To avoid interference with telecommunications and cellular telephone frequencies (872–960 MHz), the heating applications generally use a frequency of 2450 MHz, which can be freely used in industrial applications without requiring special permits (Eskicioglu et al., 2007a, 2007c).

Microwave irradiation in the field of WWTPs has recently been proposed as an innovative sludge pre-treatment before anaerobic digestion, aimed at the enhancement of digestion performance, the improvement of dewaterability and pathogen inactivation. Analogously to conventional heating, microwave application to sludge has the following effects:

- disintegration of biological flocs,
- high level of sludge solubilisation, EPS solubilisation
- lysis of bacterial cells, release of intracellular constituents and bound water.

In the literature, the effect of microwaves on microorganisms is not yet fully understood. The damage and death of organisms can theoretically be due to two effects (Eskicioglu et al., 2007a):

(1) thermal effect due to the heat generated by the rotation of dipole molecules under the oscillating electrical field;
(2) athermal effect (also known as nonthermal or microwave effect), which is caused by polarized parts of macromolecules aligning with the poles of the electromagnetic field (orientation) potentially resulting in the breakage of hydrogen bonds.

Both the thermal (heating) and the athermal effect (orientation) break the polymeric network of sludge flocs and may cause cell death. Therefore, due to the potential athermal effect, cells undergoing exposure to microwave treatment may be theoretically subjected to greater damage compared to cells conventionally heated to the same temperature. However, these two effects are not always distinguishable from each other, and it is not yet clear which of them is prevalent (Eskicioglu *et al.*, 2007a).

In the literature there are varying and divergent opinions on the extent of sludge solubilisation, biogas production rate and VS reduction obtained through microwave treatment compared to conventional heating at the same temperature. Pino-Jelcic *et al.* (2006) found better performance for microwave applications, while Eskicioglu *et al.* (2006; 2007b) reported higher solubilisation and greater biodegradation with conventional heating due to the longer exposure times required to reach the desired temperature.

In fact, Eskicioglu *et al.* (2006) observed that the level of increase in soluble COD, soluble sugar, soluble protein and soluble VFA was different for sludge treated by microwave and conventional heating (carried out up to the same maximum temperature of 96°C but with different contact time), and, in particular, higher COD solubilisation and higher cumulative biogas production was obtained from conventional heating.

In a subsequent work, Eskicioglu *et al.* (2007a) suggested that a correct comparison between microwave treatment and conventional heating has to be performed imposing identical temporal heat temperature profiles. This comparison was carried out by the authors with the aim of correctly evaluating the potential athermal effects of microwaves. Applying 50–96°C, similar COD solubilisation was observed for the two types of treatment, resulting in no discernable microwave athermal effect. However, the improvement of biogas production for microwave treated sludge seems to be related to an athermal effect on the mesophilic anaerobic biodegradability of sludge.

Sludge solubilisation due to microwave irradiation has been reported in several investigations and a comparison of some results is shown in Figure 11.9, compared to data of conventional thermal treatment in the same range of temperature (data of Figure 11.1).

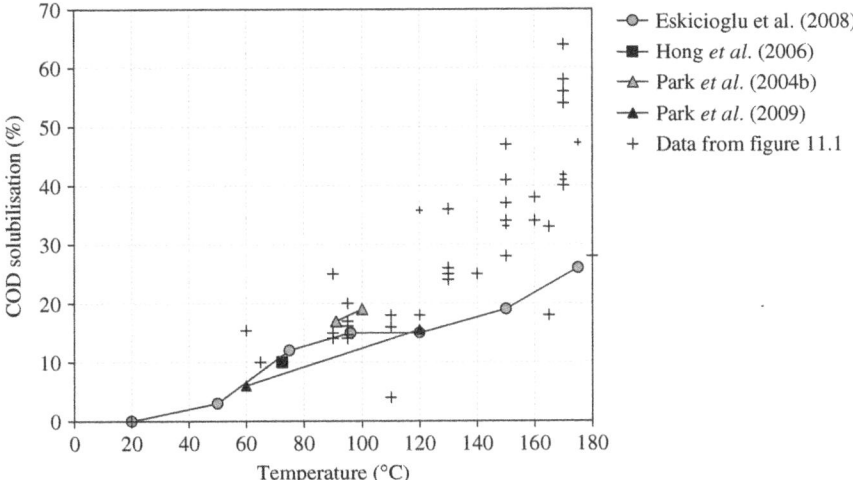

Figure 11.9. COD solubilisation by microwave heating, compared to data from conventional thermal treatment in the same temperature range shown in Figure 11.1.

There is a quite linear relation between microwave temperature and the solubilisation level. Microwave induced a COD solubilisation up to 25% at temperatures of up to 180°C (the initial soluble COD was substracted in the calculation of COD solubilisation).

COD solubilisation of excess sludge increased from 8% in the control to 18% after microwave pre-treatment carried out at 72.5°C (+10%) (Hong et al., 2006).

COD solubilisation of +17% and +19% was reported for excess sludge microwave irradiated to 91°C and boiling temperatures, respectively (Park et al., 2004b). Park et al. (2009) indicates that COD solubilisation increase from 6.9% (control) to 8.0–17.5% after microwave pre-treatment at 60–120°C.

However, only in the temperature range of 50–120°C, a similar level of solubilisation between microwave heating and conventional thermal treatment can be observed, while for higher temperatures this latter is more efficient.

Although data is available in the literature on sludge solubilisation, as far as we know, microwave technology has not yet been successfully applied at full-scale scale.

12
Chemical and thermo-chemical treatment

12.1 INTRODUCTION

As well as simple thermal treatment, chemical and/or thermo-chemical techniques, based on alkaline, acid reagents or combinations, have also been proposed. Coupling an increase in temperature with a strong change in pH, away from the optimal values for microorganisms, cell breakage occurs and therefore these types of treatment enhance the process of cell lysis-cryptic growth.

When compared with simple thermal treatment, thermo-chemical treatment has a higher efficiency in sludge solubilisation when applied at the same temperature, with additional costs for reagents, however. Acid or alkaline agents can be considered as "catalysts" in the thermal hydrolysis of organic macromolecules, and temperatures above 170–200°C, which cause the formation of soluble

© 2010 IWA Publishing. *Sludge Reduction Technologies in Wastewater Treatment Plants.* By Paola Foladori, Gianni Andreottola and Giuliano Ziglio. ISBN: 9781789065305. Published by IWA Publishing, London, UK.

refractory compounds, can be reduced. (Thermo)-chemical treatment can also be employed at lower temperatures of 50–90°C (Vlyssides and Karlis, 2004), or around ambient temperature, both under alkaline or acidic conditions.

The application of a thermo-chemical treatment to sludge is aimed at:

- promoting hydrolysis and solubilisation of complex polymeric substances into smaller molecules;
- improving biodegradability
- improving settling/dewatering properties.

For sludge reduction, the process can be theoretically integrated either in the wastewater handling units or in the sludge handling units. However, in practice, the integration in the wastewater handling units is rare due to uneconomic reasons associated with a low solid content, while more common is the use to enhance anaerobic digestion or sludge dewaterability, contributing to sludge volume reduction.

The main drawback is related to corrosion probems of the equipment used for thermo-chemical hydrolysis at full-scale, induced both by increased temperatures (coupled with high pressure for T > 100°C) and by the pH alterations.

In the mid '80s a growing interest in technologies based on acid or alkaline, or either combined with thermal hydrolysis led to the development of the Synox, Protox and Krepro processes, although some of them suffered from high costs and poor-quality products (Neyens and Baeyens, 2003).

12.2 TYPES OF ACIDIC OR ALKALINE REAGENTS

The thermo-chemical treatment of sludge can be performed by using:

(a) alkaline reagents, such as NaOH, KOH, CaO, $Mg(OH)_2$ or $Ca(OH)_2$;
(b) acidic reagents, such as HCl or H_2SO_4.

In the literature, many investigations report that sludge solubilisation significantly increases with alkaline treatment, and to a lesser extent by acid treatment (Cassini *et al.*, 2006; Chen *et al.*, 2007). Most authors indicate that alkaline reagents may be more effective, since an alkaline treatment avoids acid-corrosion effects thus allowing the use of common steel in construction. Moreover the sludge treated with alkaline reagents tends to have a less negative effect on subsequent biological treatments. Other authors propose acid treatments, such as acid thermal hydrolysis based on the use of H_2SO_4, applied in the sludge handling units to thickened sludge before dewatering, to reduce the dewatered cake volume (Neyens et al., 2003b).

Among acidic reagents, the capacity of HCl to induce cell lysis is reported to be lower than H_2SO_4 (Rocher et al., 1999). Furthermore, to obtain the same level of organic carbon solubilisation, the quantity of acid required may be higher than for alkaline reagents, resulting in higher chemical costs for the treatment (Rocher et al., 1999).

Alkali added to a cell suspension react with the cell walls in several ways, including the saponification of lipids in the cell walls, which leads to solubilisation of the membrane. Disruption of microbial cells then leads to leakage of intracellular material out of the cell. In addition, as the pH increases, the bacterial surfaces become negatively charged, which causes electrostatic repulsion and desorption of some parts of EPS (Neyens et al., 2003a).

Among alkaline reagents, NaOH was found to be the most efficient reagent for inducing cell lysis and cause sludge solubilisation (Rocher et al., 1999). For producing the same organic carbon solubilisation (pH 10.5), the quantity of NaOH required is lower than KOH (Rocher et al., 1999).

Compared to NaOH, the use of $Ca(OH)_2$ in the thermo-chemical treatment of sludge was less effective in sludge solubilisation (Li et al., 2008). The solubilisation level decreased gradually when $Ca(OH)_2$ dosages were over 0.02 mol/L, due to a reflocculation phenomenon. In fact, the bivalent cations (including Ca^{2+} and Mg^{2+}) have a key role in connecting cells with EPS and thus their presence can obstruct the solubilisation of the organic polymers enhancing the re-flocculation of fragments produced by alkaline treatment.

According to some studies, Na^+ may induce an inhibitory effect towards microorganisms and, in particular, to methanogenic bacteria (inter alia Feijoo et al., 1995; Rocher et al., 1999). The influence of Na^+ on the anaerobic biodegradability of a thermo-chemically pre-treated sludge was investigated by Delgenès et al. (2000) on industrial sludge. In particular, anaerobic acetate biodegradability tests run with increasing Na^+ concentrations (up to 15 gNa/L added as NaCl) showed that no inhibition occurred below 8 gNa/L, while a slight influence was observed when 15 gNa/L were added.

A chemical pre-treatment was also perfomed by using lime (CaO) to operate at pH over 12 (Carballa et al., 2006).

12.3 COD SOLUBILISATION

The efficacy of the thermo-chemical treatment depends mainly, as expected, on the temperature and pH applied. Many investigations aimed at evaluating sludge solubilisation during thermo-chemical treatment have been carried out at lab-scale, testing wide ranges of temperature (both above and below 100°C) and pH from 4 to 13. Analogously to the findings described in Chapter 11 (for simple thermal treatment) the application of temperatures above 100°C generally requires contact times of 30–60 min, since longer times do not effectively improve solubilisation.

In the case of thermo-chemical treatment carried out at temperatures lower than 100°C, the equipment is simpler since it can be carried out at atmospheric pressure. (Thermo)-chemical treatment at ambient temperature can significantly improve solid solubilisation, but the efficiency is higher only when coupled with a significant pH change, either in the acid or alkaline range. One benefit of the application of high temperatures is the shorter contact time required for the thermo-chemical treatment.

In the following sections some results obtained from thermo-chemical treatment of sludge are indicated, referred to contributes of the literature published in the last decade. Although very many contributes were published in the '70s and '80s in the field of thermo-chemical treatments – aimed to the study of sludge solubilisation, biogas production and increase of solid content in dewatered cake – they are not included in this review, because in general no indications about sludge mass reduction were given. Up to now, although the investigations on sludge solubilisation are very numerous, rare are the investigations on the entity of sludge reduction and therefore we only can image an expected entity of sludge reduction on the basis of the level of solubilisation.

12.3.1 Effect of temperature

In general, as can be seen in Figure 12.1, the higher the temperature the higher the extent of COD solubilisation and VSS solubilisation.

Experimental data of VSS solubilisation due to sludge solubilisation are reported by Vlyssides and Karlis (2004) during hydrolysis process at moderate temperatures (50–90°C), pH 11 and contact time of 1–10 h. A rapid increase of soluble COD was observed after 1 h of treatment, obtaining VSS solubilisation in the band 10–20%. For more intensive hydrolysis conditions (pH 11, 90°C, 10 h) VSS solubilisation increased up to 40%.

A thermo-chemical treatment based on NaOH was used by Delgenès *et al.* (2000) on industrial sludge, characterised by initial 56 gCOD/L, of which 20% was soluble. pH value was raised to 12 and treatment lasted 30 min at temperatures ranging from 90 to 200°C. Particulate COD solubilisation increased up to 140°C, while no further increase was observed at higher temperatures (Figure 12.1).

Chemical and thermo-chemical treatment 237

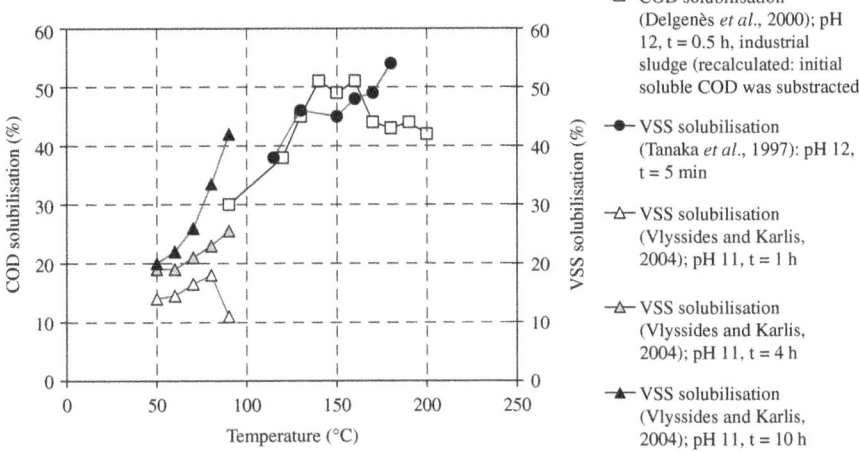

Figure 12.1. Influence of temperature on COD solubilisation and VSS solubilisation (pH and contact time are indicated).

VSS solubilisation (15%) can also be obtained at ambient temperature, but in this case the required dosage of alkaline reagents is very high (Tanaka et al., 1997).

12.3.2 Effect of pH

The effect of pH on COD solubilisation and VSS solubilisation is shown in Figure 12.2. In the experience of Chen et al. (2007) – who investigated the whole range of pH from 4 to 11 – COD solubilisation was significantly higher at alkaline pH (range 9–11) than at neutral pH (range 6–8) or at acidic pH (range 4–6). Thus the sludge hydrolysis rate appears accelerated under alkaline conditions.

The effects of pH in the range 8–13 on particulate COD solubilisation at 140°C in the case of an industrial sludge was investigated by Delgenès et al. (2000). COD solubilisation increased from 36.9% at pH 8 to 76.1% at pH 13. Taking into account the soluble COD initially present in the sludge, the net increase of COD solubilisation was 16% at pH 8 and 56% at pH 13 (Figure 12.2). The authors suggested optimal conditions (pH 12, 140°C, 30 min) for obtaining the maximum solubilisation rate leading to a net increase of COD solubilisation of 51%.

In alkaline treatment using dosages up to 0.25 gNaOH/gVSS (corresponding to pH around 12) at 130°C for 5 min, the VSS solubilisation increased significantly, up to 70%, while no improvement was seen above this dosage (Tanaka et al., 1997).

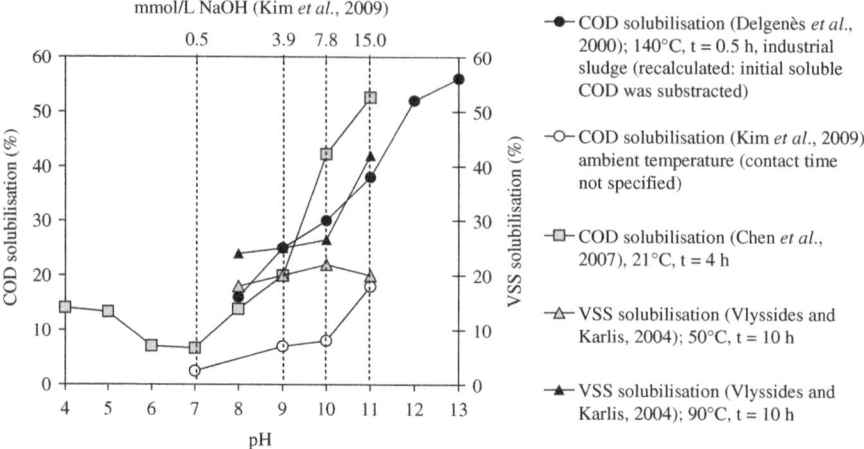

Figure 12.2. Influence of pH on COD solubilisation and VSS solubilisation (temperatures and contact time are indicated). Horizontal axis at the top of the graph is relative only to the data of Kim *et al.* (2009).

The efficiency of (thermo)-chemical hydrolysis was evaluated as a function of H_2SO_4 and NaOH concentration at a contact time of 8 h and at ambient temperature, by Cassini *et al.* (2006). The best COD solubilisation was achieved in the alkaline hydrolysis using 20–60 meq/L of NaOH. At a dosage of 20 meq/L, COD solubilisation was 11% for acid treatment and 60% for NaOH.

12.3.3 Effect of contact time

The effects of contact time on thermo-chemical treatment in terms of COD and VSS solubilisation are shown in Figure 12.3. In general, there are two distinct phases in this hydrolysis process:

- an initial short phase (minutes or one hour) with a rapid hydrolysis rate,
- a subsequent longer phase, until 10 h or more, with a slower hydrolysis rate.

Among authors there is agreement on this two-step behaviour and on the short duration of the first phase (Rocher *et al.*, 1999; Li *et al.*, 2008; Hao *et al.*, 2003).

This two-step behaviour is also confirmed by Tanaka *et al.* (1997), who observed that a contact time over 5 minutes (130°C, pH 12) did not produce a further significant VSS solubilisation.

Chemical and thermo-chemical treatment 239

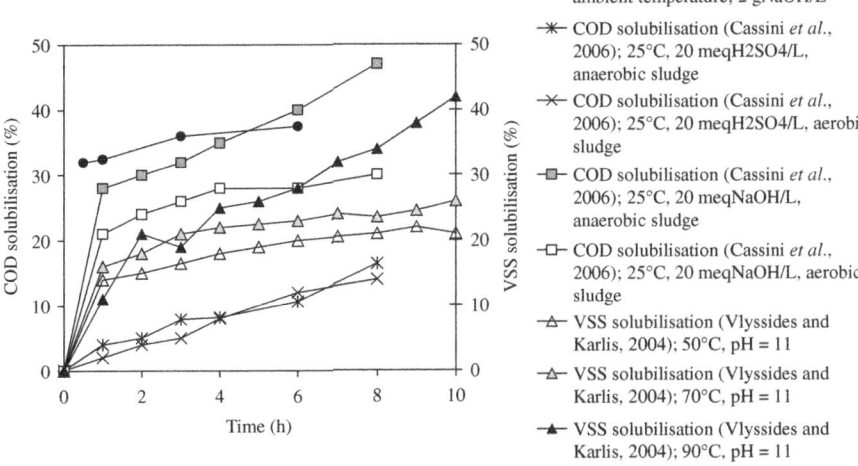

Figure 12.3. Influence of contact time on COD solubilisation and VSS solubilisation (pH and temperatures are indicated).

The influence of contact time is more important at lower temperatures (50–90°C), as observed in the investigation of Vlyssides and Karlis (2004) and compared in Figure 12.3. VSS solubilisation increased rapidly during the first 2 h. At pH 11 and 70°C, VSS solubilisation in the initial rapid hydrolysis phase was about 17%, which corresponded to 60–70% of the final value obtained after 10 h of hydrolysis retention time (VSS solubilisation of 25%).

This behaviour is valid also for chemical treatment carried out at ambient temperature. For example Li et al. (2008) observed, in the first 30 min and using 2 gNaOH/L, a COD solubilisation level of 60–71% of the final value after 24 h. Therefore, the authors suggested that the optimal treatment duration was 30 min.

In the investigation of Heo et al. (2003) the COD solubilisation also increased rapidly during the first 30 min and the particulate hydrolysis was nearly completed within 4 h.

Comparing acid and NaOH treatment for two types of sludge (aerobic and anaerobic), Cassini et al. (2006) observed that alkaline hydrolysis causes a rapid increase of soluble COD which levelled off after just 1 h, while soluble COD increased linearly during acid hydrolysis (Figure 12.3). The authors did not give any explanation of these results which need further investigation. However, the results of Cassini et al. (2006) confirm again that alkaline hydrolysis is more efficient in COD solubilisation that the acid treatment.

12.3.4 Comparison of solubilisation levels under different conditions

Table 12.1 summarises the optimal conditions for (thermo)-chemical treatment as proposed in the literature, and divided in two categories:

- chemical treatment at ambient temperature (without heating);
- thermal + chemical treatment (with heating).

For example, Tanaka *et al.* (1997) presented a comparison between the following treatments: chemical, thermal + chemical and thermal, obtaining the following results:

- *chemical* – at ambient temperature, pH 12, VSS solubilisation was 15% (see first part of Table 12.1);
- *thermal + chemical* – at 130°C, pH 12, VSS solubilisation was 45% (see second part of Table 12.1);
- *thermal* – at 180°C, neutral pH (without alkali dosage) VSS solubilisation was 30%.

This comparative study demonstrates the weak effect of alkaline alone without the addition of a thermal effect. The highest COD and VSS solubilisation is obtained with the combination thermal + chemical, even when using a temperature (130°C) lower than the simple thermal treatment (180°C).

ORP measurement during alkaline hydrolysis may have a potential as an indicator for monitoring the process. In this case, the ORP curve decreased correspondingly as soluble COD increased sharply; when soluble COD tended to level off, the ORP increased gently (Chiu *et al.*, 1997; Chang *et al.*, 2002).

12.4 NITROGEN AND PHOSPHORUS SOLUBILISATION

During thermo-chemical treatment the release of N and P depends both on temperature and pH value. The influence of temperature is described in § 11.4 regarding thermal treatment.

With regard to the effect of pH, Chen *et al.* (2007) indicate that the concentration of ammonia (and also orthophosphate) was highest in the acidic pH (range 4–5) rather than under alkaline conditions (pH 11). The low ammonia release at pH 11.0 is due to the toxicity of strong alkaline conditions which decreases the activity of hydrolytic sludge enzymes, such as protease, peptidase, etc.

Table 12.1. Comparison of the conditions applied in thermo-chemical treatments.

Scale	Reagent	Optimal conditions pH or dosage	T (°C)*	Contact time (h)	Results	Reference
		Chemical treatment at ambient temperature (without heating)				
lab	NaOH	12	amb.	1 h	VSS solubilisation (15%)	Tanaka et al. (1997)
lab	NaOH	40 meq NaOH/L	amb.	24 h	COD solubilisation (36.3%)	Chiu et al. (1997)
lab	NaOH	20–80 meq/L NaOH	25°C	2 h	COD solubilisation (30–40%)	Chang et al. (2002)
lab	NaOH	40 meq/L NaOH	25°C	10 h	COD solubilisation (45%)	Chang et al. (2002)
lab	NaOH	pH 12 (45 meq/L)	25°C	4 h	COD solubilisation (27.7%)	Heo et al. (2003)
lab	H_2SO_4	20 meq/L	25°C	8 h	COD solubilisation (11%)	Cassini et al. (2006)
lab	NaOH	20 meq/L	25°C	8 h	COD solubilisation (60%)	Cassini et al. (2006)
lab	CaO	12	amb.	24 h	VSS solubilisation (0–13%)	Carballa et al. (2006)
lab	NaOH	11	21°C	4 h	COD solubilisation (53%)	Chen et al. (2007)
lab	NaOH	pH 11 (22.3 meq/L)	amb.	3 h	COD solubilisation (14%)	Oh et al. (2007)
lab	NaOH	0.05 mol/L	amb.	0.5 h	COD solubilisation (32%)	Li et al. (2008)
lab	NaOH	pH 10 (7.8 mM NaOH)	25°C	–	COD solubilisation (9.8%)	Kim et al. (2009)

(continued)

Table 12.1. Continued

Scale	Reagent	Optimal conditions			Results	Reference
		pH or dosage	T (°C)	Contact time (h)		
Thermal + chemical treatment (with heating)						
lab	NaOH	12	130°C	0.083 h	COD solubilisation (45%)	Tanaka et al., 1997
lab	NaOH	0.01N	60°C	24 h	Increase of TOC solubilisation: • +8% compared to 60°C without NaOH • +30% compared to 37°C without NaOH	Saiki et al. (1999)
lab	NaOH	10	60°C	0.33 h	Increase of 150 mgDOC/gTSS	Rocher et al. (1999, 2001)
lab	NaOH	12	140°C	0.5 h	COD solubilisation (51%)	Delgenès et al. (2000)
lab	NaOH	12	130°C	0.5 h	COD solubilisation (53%)	Tanaka and Kamiyama (2002)
lab	NaOH	pH 11 (45 meq/L)	55°C	4 h	COD solubilisation (38.3%)	Heo et al. (2003)
lab	–	11	50°C	10 h	Increase of VSS solubilisation: • +17% compared to untreated • negligible compared to 50°C without NaOH	Vlyssides and Karlis (2004)
lab	–	11	90°C	10 h	Increase of VSS solubilisation: • +45% compared to untreated • +25% compared to 90°C without NaOH	Vlyssides and Karlis (2004)
lab	NaOH	11	75°C	3 h	COD solubilisation (25%)	Banu et al. (2009)

*amb. = ambient temperature

Alkali reagents, NaOH or $Ca(OH)_2$, cause a significant release of soluble TKN, but NaOH causes a higher release of soluble organic P and orthophosphate compared to $Ca(OH)_2$. The reason is probably due to the fact that calcium interferes with phosphate to form calcium phosphate.

The solubilisation of nutrients after thermo-chemical treatment applied using HCl and H_2SO_4 was investigated by Smith and Göransson (1992). The solubilisation of nutrients was quite high for N but very low for P, obtaining N/P ratio in the hydrolysate of 72. The authors suggest that iron hydroxide is released during the acidic treatment of sludge, which react with phosphate ions.

12.5 INFLUENCE ON SLUDGE DEWATERABILITY

In general, the acid and alkaline hydrolysis combined with a thermal treatment resulted in an improvement of the rate of filtration (measured from CST values), the rate of mechanical dewatering and an increased dry content in the filter cake (Neyens *et al.*, 2004).

Treatment with acid reagents (sulphuric acid) improves dewaterability because EPS leave the activated sludge surface, which help to pack the sludge aggregates and to reduce the water content of dewatered sludge (Neyens and Baeyens, 2003).

Li *et al.* (2008) observed that sludge dewaterability deteriorated when NaOH dosage was lower than 0.1 mol/L, due to disruption of sludge flocs. Increasing dosages to 0.5 mol/L disrupted floc fragments were re-flocculated and the dewaterability improved gradually.

Although high dosages of $Ca(OH)_2$ can enhance dewaterability and sludge compactibility, it causes an increase in the weight of sludge cake, because of the addition of inorganic matter. For example, when the $Ca(OH)_2$ dosage was 0.05 mol/L, the weight of sludge cake increased by 24.3%, while it decreased by 26% for NaOH treatment at the same dosage (Li *et al.*, 2008).

12.6 INTEGRATION OF THERMO-CHEMICAL TREATMENT IN THE WASTEWATER HANDLING UNITS

The thermo-chemical treatment of excess sludge increases its biodegradability and induces the mechanism of cell lysis-cryptic growth when lysate is recirculated in the wastewater handling units, favouring reduction in excess sludge production.

The entity of sludge reduction induced by an alkaline thermal treatment, carried out at 60°C, pH 10 (NaOH) for 20 min, was evaluated in a lab-scale plant (chemostat) fed with synthetic wastewater (Rocher et al., 2001). Working at relatively low alkaline conditions eliminates or reduces the need for expensive chemical neutralisation. With a treatment frequency of 4/SRT (the biomass is exposed 4 times to the alkaline treatment during SRT), a 37% reduction of sludge production was obtained, compared to the control. The efficiency of organic matter removal in the bioreactor remained stable, around 93% (Rocher et al., 2001).

However, the integration of a thermo-chemical treatment in the wastewater handling units at full-scale is rare and as far as we are aware, it is difficult to find successful results in the literature. The reason is the unfavourable economic balance of this application due to the wasted energy for heating a low-concentrated sludge, such as in the return flow of activated sludge processes, and the increased reagent dosage for changing pH of a great sludge volume.

The raising or lowering of pH of sludge requires the addition of chemicals to increase the ionic strength of sludge, but a subsequent pH neutralisation is required when high pH values (more efficient in sludge reduction) are applied.

12.7 INTEGRATION OF THERMO-CHEMICAL TREATMENT IN THE SLUDGE HANDLING UNITS

Thermo-chemical hydrolysis + anaerobic digestion

The most widely proposed application of thermo-chemical treatment in the sludge handling units is a pre-treatment prior to anaerobic digestion. The aim is to obtain an increase in biodegradable COD leading to a reduction of the digester volume, an increase in biogas production and contribution to sludge reduction. Part of the heat used for the thermo-chemical pre-treatment can be recovered to heat the mesophilic or thermophilic anaerobic process, thus leading to saving in operational costs.

It has to be underlined that methanogenic bacteria activity and thus the production of methane is strongly affected by pH which should be maintained in the range of 6.6–7.6 with an optimum near pH 7.0. When the pH decreases from 6.0 to 4.0, or increases from 7.0 to 10.0, methane production falls (Lay et al., 1997; Chen et al., 2007). No methane at all was generated at pH 10 and 11 (Chen et al., 2007). Several authors highlighted that the activity of methanogens decreased or was lost in very alkaline or acidic environments.

This factor has to be taken into account in the application of a thermo-chemical pre-treatment before conventional anaerobic digestion so as not to worsen methanogenesys and biogas production, limiting the benefits of the sludge solubilisation and biodegradability induced. Conversely, further neutralisation of pH may be performed, for example, passing from 12 to 7 by adding HCl before feeding digesters. Regarding this latter aspect, it has to be taken into account that the pH tends to neutralise spontaneously after the hydrolysis of sludge therefore it becomes suitable for biological processes. For example, the addition of 7.78 mM of NaOH (pH 10) in the thermo-chemical treatment resulted in a final pH of 7.3 and sludge became again suitable for biological treatment (Kim *et al.*, 2009).

An example of the pH dynamic, indicating alkali consumption during the treatment with 45 meq/L of NaOH at ambient temperature (25°C) is indicated in Figure 12.4. It can be observed that the pH value falls during hydrolysis, eliminating the need for chemical neutralisation before the sludge enters the anaerobic digestion (Heo *et al.*, 2003).

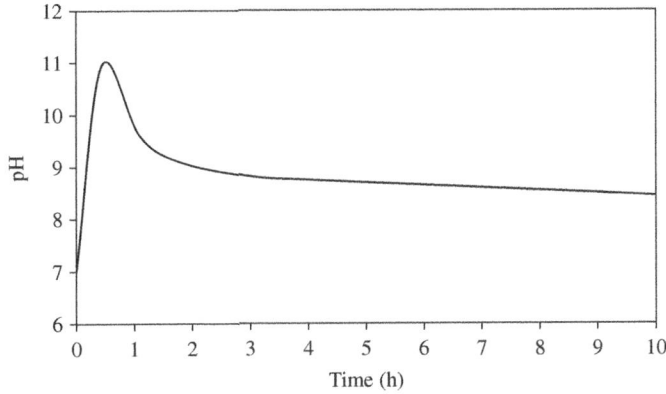

Figure 12.4. Dynamic of pH during a thermo-chemical treatment (45 meqNaOH/L, 55°C, 3% TS) (*modified from* Heo *et al.*, 2003).

Table 12.2 summarises the results of biogas enhancement in anaerobic digestion (most cases at lab-scale), after a thermo-chemical treatment.

For example, Tanaka *et al.* (1997) pre-treated municipal and industrial sludge, applying 130°C, NaOH dosage of 0.3 g/gVSS (corresponding to pH 12), contact time of 5 min obtaining a VSS solubilisation of 45%. After anaerobic digestion a 120% increase in methane production was observed. Under similar pre-treatment conditions (pH 12), but at ambient temperature, the increase in methane production was only 50%, demonstrating the lower effect of alkaline alone without the combination of a thermal effect. Compared to the simple thermal treatment even at 180°C, the increase in methane production was only 80%, which again highlights the advantageous role of alkali addition (Tanaka *et al.*, 1997).

Table 12.2. Performances obtained in thermo-chemical pre-treatment prior to anaerobic digestion. Other data from these references are also indicated in Table 12.1.

scale	reagent	Optimal conditions			Results	Reference
		pH	T (°C)*	Contact time (h)		
lab	NaOH	12	amb.	1 h	Increase of methane production (+50%)	Tanaka et al., 1997
lab	NaOH	12	130°C	0.083 h	Increase of methane production (+120%)	Tanaka et al., 1997
lab	NaOH	12	230°C	0.5 h	HRT or digester volume reduced to 1/3. Increase of methane production (+200%)	Tanaka and Kamiyama (2002)
lab	NaOH	pH 12 (45 meq/L)	25°C	4 h	Increase of methane production (+66%)	Heo et al. (2003)
lab	NaOH	pH 11 (45 meq/L)	55°C	4 h	Increase of methane production (+88%)	Heo et al. (2003)

*amb. = ambient temperature

In the lab-scale investigation by Tanaka and Kamiyama (2002), removal rates of COD in anaerobic digesters increased from 31% to 48% at HRT of 6 d, indicating that HRT or digester volume could be reduced to one third (Tanaka and Kamiyama, 2002). Methane production was greatly enhanced by the pre-treatment, especially at shorter HRT. Specific methane production rates in the pre-treated process were three times as high as the normal process.

Thermo-chemical hydrolysis + dewatering of thickened sludge

Acid and alkaline hydrolysis performed at high temperatures (120°C) was also proposed for treating thickened sludge (5–6% TS) before dewatering, in order to improve dewaterability and contribute to reduce solids to be disposed. The scheme of this configuration applying acidic conditions, is shown in Figure 12.5: water phase separated from dewatering and typically rich in solubilised organic compounds, N and P, is recirculated to the wastewater handling units.

Neyens et al. (2003b) applied acid hydrolysis to thickened sludge by using H_2SO_4, at pH 1, 3, 5 and temperatures above 120°C and contact time of 1 h. The results are summarised in Table 12.3.

Working at pH 2–3 with H_2SO_4, a significant quantity of sulphate is added during the process and subsequently transferred to the water phase. Furthermore, heavy metals, which are normally integrated in the organic complex molecules, destroyed during acid hydrolysis, are released as soluble salts. The water phase need neutralisation, metal hydroxides will be precipitated and eliminated as hydroxide-sludge (Neyens et al., 2003b).

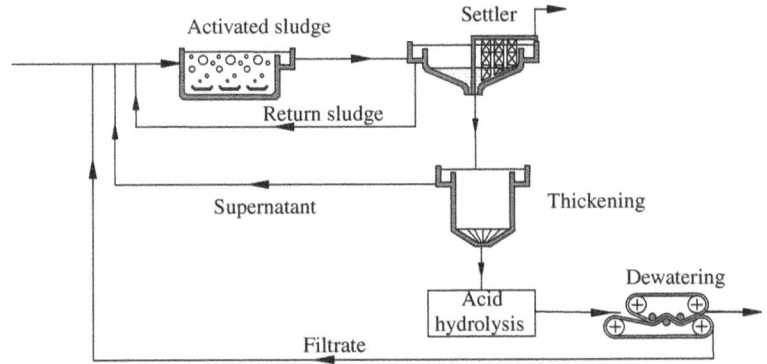

Figure 12.5. Scheme of the acid hydrolysis at high temperatures applied to improve dewaterability (*modified from* Neyens et al., 2003b).

Table 12.3. Thermo-chemical treatment of thickened sludge (5–6% TS), prior to dewatering.

scale	reagent	Optimal conditions			Results		Reference
		pH	T (°C)	Contact time (h)	TS solubilised*	Solid content in dewatered cake	
lab	H_2SO_4	<5	120°C	1 h	70%	>52% (compared to 22.5% in untreated sludge)	Neyens et al. (2003b)
lab	$Ca(OH)_2$	10	100°C	1 h	40%	46% (compared to 28% in untreated sludge)	Neyens et al. (2003a; 2003b)

*TS solubilised = percentage of dry solids solubilised as a fraction of the initial untreated solids (this fraction is no longer to be dewatered).

Alkaline treatment performed under similar conditions but pH 10 (Table 12.3) performs less well than acid conditions, but it avoids acid-corrosion effects, thus allowing the use of common steel in reactor construction (Neyens *et al.*, 2003a).

Pathogen inactivation and a high load (COD, BOD, N, P) water phase separated during dewatering are other effects.

Krepro process

This process was initially developed in the mid '80s by the Swedish company Kemira Kemwater, aimed to produce organic matter to be used in the wastewater handling units as a carbon source for denitrification and to recover phosphorus to be recycled in agriculture (Ødegaard *et al.*, 2002).

In this process the thickened sludge (4–6% TS) is acidified by addition of H_2SO_4 to reach pH 1–2. Then the sludge undergoes a thermal treatment at 140°C and 3.5 bar for 30–40 min, obtaining a sludge solubilisation of 40%. The sludge is cooled in a heat exchanger in order to recover heat to help increase the inlet sludge temperature. Then the hydrolysed sludge pass in a centrifuge for the separation of solids from the phosphorus-rich centrate, which react further with added ferric salts and alkali (to reach pH 3) for enhancing the precipitation of a very pure ferric-phosphate. The liquid phase after separation of the ferric-phosphate fraction (with 35% TS, 15% P) contains organic carbon (11,000–13,000 mgCOD/L) which is recirculated in the wastewater handling units to support denitrification. This return flow contains also ammonia which causes an additional load in the activated sludge process. Furthermore the return flow contains iron (coagulant) which can be exploited in the primary settler to enhance solid separation from raw wastewater which reduces the organic load applied to nitrification (Ødegaard *et al.*, 2002).

13
Ozonation

> "...partial oxidation (by ozone) as low as possible
> and biological oxidation as high as possible"
> (Ried et al., 2007)

13.1 INTRODUCTION

Ozone (O_3) is a gaseous compound made up of three atoms of oxygen, produced at industrial level by means of silent electrical discharges. Due to the high instability of the O_3 molecule, its production is realised on-site, by means of ozone generators fed with air (filtered and dehumidified) or pure oxygen. An electric energy supply of approximately 9–15 kWh/kgO_3 is necessary for the production of ozone, depending on whether pure oxygen or air is used and on the small or large scale of application.

Ozone is well-known as a strong chemical oxidant, used in advanced oxidation processes for the reduction of anthropogenic, inhibitory or recalcitrant organic compounds (e.g. endocrines, hormones, pharmaceuticals, ...) from wastewater, due to the enhancement of the biodegradability of these substances, for tertiary treatment of biologically effluents (e.g. decolourisation) or for deodorisation. Ozone oxidizes organic compounds either by direct oxidation or through the

© 2010 IWA Publishing. *Sludge Reduction Technologies in Wastewater Treatment Plants.* By Paola Foladori, Gianni Andreottola and Giuliano Ziglio. ISBN: 9781789065305. Published by IWA Publishing, London, UK.

generation of hydroxyl radicals, or both, transforming organic compounds into more soluble and more biodegradable oxygenated intermediates. In the ozonation process important factors are gas absorption and the chemical reaction, affected by the reaction kinetics and the gas-liquid mass transfer rate, which is low due to the low solubility of ozone (Zhou and Smith, 2000).

In recent decades the ozonation of sludge aimed at its reduction has received increasing attention, as a process able to affect the growth of microorganisms and sludge characteristics (Ried et al., 2007).

During ozonation several processes occur, causing disintegration of the sludge complex matrix, solubilisation and mineralisation, involving the transformation of soluble compounds and particulates (Figure 13.1). These phenomena demonstrate the competition for ozone between soluble and particulate matter during sludge ozonation (Cesbron et al., 2003).

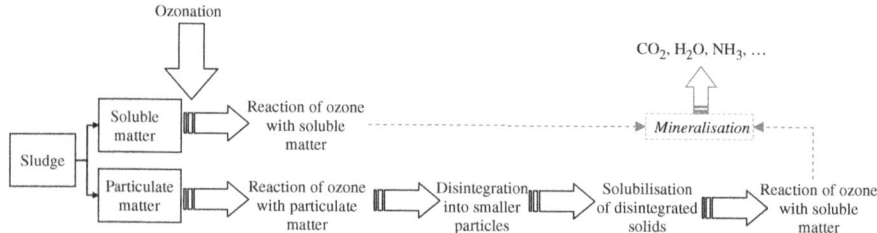

Figure 13.1. Decomposition of sludge caused by ozonation.

Furthermore, during ozonation the following effects occur:

- damage and inactivation of microorganisms present in sludge; permeabilisation and cell death and the release of intracellular compounds;
- conversion of organic matter to biodegradable molecules which can be more susceptible to decomposition by microorganisms in subsequent biological treatment;
- to a lesser extent, mineralisation of the soluble organic matter into CO_2, H_2O, NH_3,..., due to chemical oxidation.

During ozonation the effect of solubilisation appears to be the most important for sludge reduction at relatively low ozone dosages. On the contrary sludge reduction by mineralisation requires unrealistically high ozone dosages. Mineralisation can be usually calculated as the difference between total COD before and after ozonation and causes an immediate sludge mass reduction.

To reduce excess sludge, the ozonation process has been integrated either into the wastewater handling units (see § 13.9) or in the sludge handling units (both in aerobic or anaerobic digestion, see § 13.10).

The organic-rich solubilized or the unsettled fraction of ozonated sludge can be used in the wastewater handling units, to produce an internal carbon source for denitrification thus avoiding additional costs for other carbon sources (Ahn *et al.*, 2002a). During biological treatment, part of the ozone-treated sludge is oxidized to CO_2, while another part is turned into new microbial cells, due to the lysis-cryptic growth process (see § 4.2).

While ozonation has also been applied to control sludge bulking in activated sludge stages or digestors (Collignon *et al.*, 1994), a technical drawback in sludge ozonation is the occurrence of foaming in ozonation reactors (caused by sludge disintegration) which is hard to avoid.

The high cost of ozone production is the main limitation in the application of ozonation for sludge reduction. It is well known that higher ozone dosages produce less excess sludge, but excessive dosages lead to very high operational costs and potentially (but not confirmed) to excessive inhibition/decrease of microbial biomass which plays a fundamental and complementary role in carbon and nitrogen utilisation.

To obtain a more cost-effective process, moderate ozone consumption is imperative, exploiting biological processes such as activated sludge or anaerobic digesters to remove the biodegradable compounds produced, on the philosophy, "partial oxidation as low as possible and biological oxidation as high as possible" (Ried *et al.*, 2007). Coupling ozonation and biological treatment, the objective of ozonation should be only to convert solids into more easily biodegradable matter in order to facilitate their removal in the cheaper biological reactors. Therefore it is desirable to increase the biologically degraded portion of the sludge and to decrease the ozone used, applying multi-step ozonation (Ried *et al.*, 2007). The synergistic effect of combining ozonation/biological treatment has yet to be optimised with the aim of establishing the lowest possible quantity of O_3 used for sludge reduction.

13.2 PARAMETERS INVOLVED IN OZONATION

A number of processes occur simultaneously during sludge ozonation and influence the overall performance of the process (*inter alia* El-Din and Smith, 2001):

- convection and backmixing processes of the liquid and gas phases flowing through the ozonation reactor,
- ozone gas mass transfer process,

252 Sludge Reduction Technologies in Wastewater Treatment Plants

- ozone auto-decomposition process,
- competitive reactive processes between dissolved ozone and organic/inorganic compounds,
- particle disintegration, solubilisation,
- substrate biodegradation, mineralisation.

The efficiency of sludge ozonation may depend on the reaction of ozone with organic constituents of sludge and on the following factors:

- wastewater or sludge quality,
- configuration of ozonation reactor,
- flow rate and concentration of ozone in the gas-phase,
- flow rate and solid concentration in the treated sludge,
- efficiency of ozone transfer in the ozonation reactor,
- contact time between ozone and sludge in the ozonation reactor,
- ozone dosage: this is the most important parameter and is expressed as mass of ozone applied per mass of TSS treated or per mass of TSS removed, or other definitions (see § 13.4),

Since ozone is an expensive reagent, optimisation of the ozonation process is a crucial factor and therefore, up to now, efforts have been made to improve ozonation efficiency, both in the reactor configuration and in the gas transfer systems, with the aim of minimising ozone dosages and maximising sludge reduction (*inter alia* Yan *et al.*, 2009).

To evaluate the effect of ozonation, the following parameters are taken into account:

- sludge mineralisation (§ 13.5);
- COD solubilisation (§ 13.6);
- TSS disintegration: this is due to mineralisation and solubilisation, as defined in § 7.2 (§ 13.6);
- sludge reduction when ozonation is integrated in the wastewater handling units and in the sludge handling units (§ 13.8, § 13.9, § 13.10).

13.3 CONFIGURATION OF OZONATION REACTORS

An example of a fully equipped sludge ozonation system is indicated in Figure 13.2. It consists of an ozone generator, an ozone contactor and a residual ozone destruction unit. The mass of ozone input can be adjusted by varying the gas flow and the converter power. The gas absorption system can have different features, using either diffusers or injectors. In the example in Figure 13.2 – in which a full-scale installation is shown (Sievers *et al.*, 2004) – the gas absorption system is installed in a loop equipped with a booster pump and an injector; the flow containing ozone is mixed with the sludge inlet flow pumped at 0.5–10 m^3/h and then fed into a reaction tank of 6 m^3 to extend the reaction time. In this configuration the gas absorption system is independent of the flow rate of the feed pump. Ozonation reactors are generally equipped with a residual ozone destructor to eliminate ozone in the off-gas.

With regard to the possible escape of some volatile organic compounds with the off-gases, studies refer that a minimal level of volatile organics were detected in a trapping solution and by means of conventional COD determination.

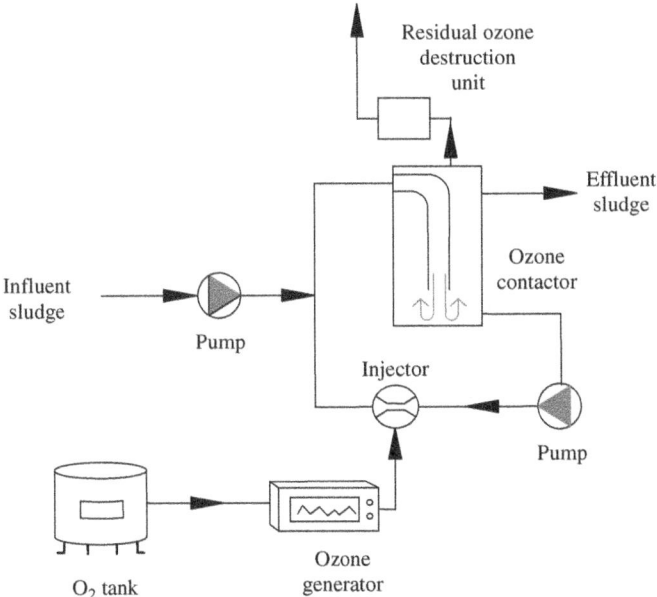

Figure 13.2. Scheme of an ozonation system (*modified from* Sievers *et al.*, 2004).

Although ozone can be produced from air, in most cases it is produced by an ozone generator converting pure O$_2$ in O$_3$ by silent electrical discharge.

Often, serious problems of foaming are observed in the ozonation reactor. Spray nozzles and mechanical mixers can be employed in the ozone reactor in order to hydraulically reduce foam generated during the ozonation of sludge. An example is indicated by Park *et al.* (2003; 2008) in Figure 13.3, where the ozone was supplied by diffusers and the ozonation of sludge was operated in a semi-batch mode.

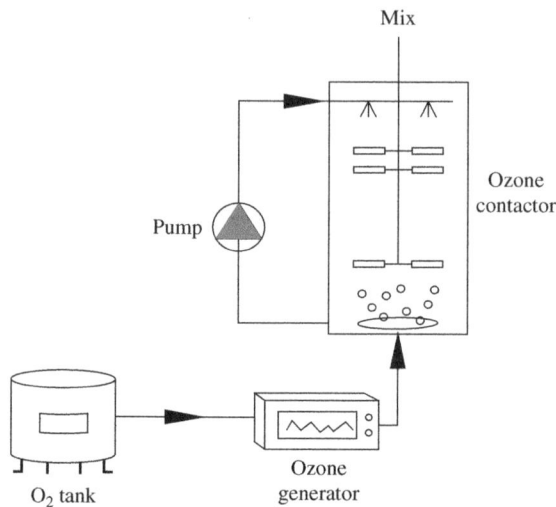

Figure 13.3. Scheme of an ozonation reactor equipped with devices for reducing foam (*modified from* Park et al., 2003).

Gas/liquid contact systems, such as bubble columns with porous diffusers or mixed reactors equipped with injectors (Venturi-like systems), can be chosen for ozone transfer. A system has to be chosen considering that activated sludge may be difficult to handle because of its viscosity and foaming tendency.

The rate of ozone mass transfer can be increased through the generation of smaller bubbles or microbubbles (Mitani *et al.*, 2005; Chu *et al.*, 2007). In fact, the larger surface area to volume ratio of microbubbles favours the ozone mass transfer. A study of the mass transfer from very small oxygen bubbles in tap water showed that the best oxygen transfer rate came from bubbles with a diameter in the range 300–1000 μm (Qiu *et al.*, 2001). Microbubbles are defined as having a diameter of up to a few hundred μm, compared to conventional bubbles with diameters of several mm. In addition to the enhancement of mass transfer, microbubbles, which have low rising velocity in liquid phase and higher inner pressure, can accelerate the formation of hydroxyl radicals and may thus contribute to sludge solubilisation (Chu *et al.*, 2007, Chu *et al.*, 2008). With the use of a microbubble generator – made up of a recycling pump, a gyratory accelerator and an injector – the sludge disintegration step was faster (Chu *et al.*, 2007), but this system did not give a net benefit in sludge disintegration when more than 0.16 gO₃/gTSS of ozone was applied (Chu et al., 2008; Yan et al., 2009).

In a semi-industrial ozonation plant, applying ozone dosages between 0.01 and 0.035 gO₃/gTSS, Manterola et al. (2008) observed that the gas flow rate was a more significant factor affecting ozone mass transfer, rather than the ozone gas concentration. Improved sludge solubilisation was observed in the case of higher gas flow rates (and lower ozone gas concentrations) with the same ozone dosages, due to increased ozone mass transfer from gas to liquid.

The question of whether the depth of the ozonation column affects the sludge solubilisation was investigated by Egemen et al. (2001): they concluded that the depth of their ozonation columns (from 2 to 6 ft) did not significantly affect the solubilisation efficiency.

13.3.1 Ozone transfer in sludge

In general, low off-gas concentrations of ozone were observed in several ozonation reactors. This could be due to the fact that initial chemical reactions of ozone transferred to the liquid with the various dissolved organics are fast enough to be completely depleted at the gas-liquid interface (El-Din and Smith, 2001). As a consequence, a total absence of dissolved ozone within the bulk liquid is often observed and a "fast kinetic regime" or an "instantaneous-reaction kinetic regime" occurs, where the apparent rate of ozone mass transfer can exceed the maximum rate of physical gas-liquid mass transfer.

This mass transfer rate enhancement is characterized by the enhancement factor, E, defined as the ratio between the actual flux of ozone (at numerator) and the maximum flux corresponding to physical absorption (at denominator) (Deleris et al., 2000; Paul and Debellefontaine, 2007):

$$E = \frac{\text{actual flux of ozone}}{\text{maximum flux due to physical absorption}} = \frac{r_O}{k_L a \times C_O^*}$$

The actual flux of ozone (r_{O3}) has to be defined as the ozone effectively transferred to the liquid and it is quantified considering the gas flow rate (Q_{O3}) and the ozone concentrations at the inlet and outlet. The maximum flux is calculated multiplying the equilibrium concentration of ozone in the liquid phase, C_O^*, and the overall mass transfer coefficient, $k_L a$.

The profile of dissolved O_3 in the liquid film is indicated in Figure 13.4 (Paul and Debellefontaine, 2007). The reaction between ozone and sludge occurs only near the gas-liquid interface, a liquid film having thickness to the order of μm (El-Din and Smith, 2001).

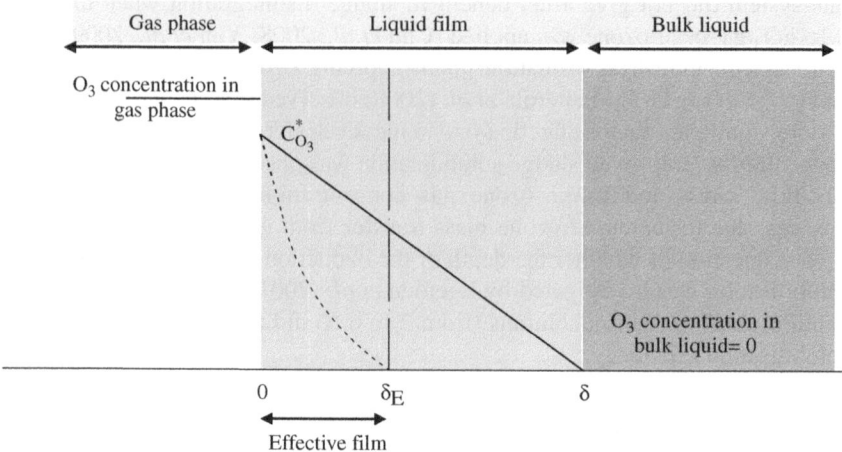

Figure 13.4. Scheme of the profile of dissolved O_3 in the liquid film (*modified from* Paul and Debellefontaine, 2007).

Without any enhancement by chemical reactions, the thickness of the liquid film is δ, and the ozone concentration decreases from C_O^* to 0. In the presence of organic compounds leading to an enhancement of the transfer, the profile of ozone corresponds to the dotted line and reaches 0 at thickness δ_E, which is smaller than δ:

$$\delta_E = \frac{\delta}{E}$$

The fastest reacting compound, with high reactivity and/or in high concentration, is preferentially attacked by ozone and it determines the E value. Consequently an effective film, having a reduced thickness, δ_E, can be identified, beyond which no more dissolved ozone is present. The parameter E can reach up to 10 depending on the type of reactor and on the conditions, leading to a film thickness of a few µm, which is very thin compared to the size of activated sludge flocs or colonies of active bacteria (Paul and Debellefontaine, 2007). The large size of flocs (several hundred µm) causes the protection of the active biomass – made up of bacteria aggregated in clusters or forming microcolonies inside the flocs – against oxidation and thus ozone is preferentially directed to oxidizing the other organic materials. Part of the sludge floc reacts with ozone in the effective film, producing soluble species and small sized colloids, which are partially oxidized by ozone near the gas-liquid interface, mainly migrate towards the bulk liquid for further biological degradation (Paul and Debellefontaine, 2007).

Déléris et al. (2000) observed a high difference obtained during experiments carried out with both very diluted sludge (0.2 gTSS/L) and normally concentrated sludge (2 gTSS/L), at the same applied ozone dosages. This phenomenon can be explained as follows: because of the high concentration of oxidizable compounds, the E factor observed during the ozonation of concentrated sludge was higher than that observed with diluted sludge, which results in an increased ozone transfer rate and greater efficiency of the applied ozone. However TSS concentration used in the practice are not so diluted, but in the range 0.4–6% TS.

13.4 DEFINITION OF OZONE DOSAGE

In various studies, different methods have been used to evaluate the amount of ozone given to sludge in an ozonation process.

The main parameters to be considered are the following:

- volume of sludge in the ozonation reactor, to be considered in the case of batch reactors (V),
- flow rate of sludge fed into the ozonation reactor, to be considered in the case of continuously fed reactors (Q_s),
- initial TSS concentration of sludge fed into the ozonation reactor (x),
- flow rate of gas containing ozone (Q_{gas}),
- initial ozone concentration in the gas flow (O_3),
- time of ozonation (t).

Some parameters useful in ozonation processes are defined as follows:

(1) *mass of ozone* (M_{O3}): defined as the total amount of ozone applied in the reactor:

$$M_{O3} = Q_{gas} \cdot O_3 \cdot t$$
$$\left[m^3/h\right] \cdot \left[gO_3/m^3\right] \cdot [h] = [gO_3]$$

(2) *ozone dosage* (D_{O3}): defined as the mass of ozone applied in the reactor per mass of TSS in the sludge:

$$D_{O3} = \frac{\text{mass of ozone}}{\text{mass of sludge}}$$

$$D_{O3} = \frac{M_{O3}}{V \cdot x} = \frac{Q_{gas} \cdot O_3 \cdot t}{V \cdot x} \quad \text{in batch reactors}$$

$$\frac{[gO_3]}{[m^3] \cdot [gTSS/m^3]} = [gO_3/gTSS]$$

$$D_{O3} = \frac{M_{O3}}{Q_s \cdot x \cdot t} = \frac{Q_{gas} \cdot O_3 \cdot t}{Q_s \cdot x \cdot t} \quad \text{in continuously fed reactors}$$

$$\frac{[gO_3]}{[m^3/h] \cdot [gTSS/m^3] \cdot [h]} = [gO_3/gTSS]$$

(3) *specific ozone dosage*: defined as the mass of ozone applied in the reactor per mass of COD removed from the influent wastewater in the wastewater handling units (Paul et al., 2006b; Paul and Debellefontaine, 2007);

(4) *ozonation rate*: defined as the ozone dosage per hour of ozonation: it is obtained by dividing the ozone dosage by the time of ozonation;

(5) *ozone factor*: defined as the ozone mass required for the elimination of a unit of TSS (or COD); in the calculation of TSS removed, the biological degradation (enhanced by ozonation) may also be taken into account.

Commonly the ozone dosage is expressed per mass of TSS ($gO_3/gTSS$), but sometimes it is expressed per mass of VSS ($gO_3/gVSS$) or mass of COD ($gO_3/gCOD$).

The mass of ozone can be expressed in terms of "applied" ozone or "transferred" (net) ozone. In particular the net mass of ozone used in a reactor ($M_{O,net}$) is calculated considering the difference between the concentration of ozone at the inlet of the reactor ($O_{3,in}$) and leaving the reactor ($O_{3,out}$) measured directly in the effluent gas flow. Thus the net mass of ozone is expressed as:

$$M_{O3,net} = Q_{gas} \cdot (O_{3,in} - O_{3,out}) \cdot t$$

$$[m^3/h] \cdot [gO_3/m^3] \cdot [h] = [gO_3]$$

The approach for measuring the mass of ozone in the gas flow is generally based on the use of potassium iodide traps.

In the literature, authors refer either to the net ozone dosage or to the applied ozone dosage, when the ozone leaving the system is not determined. In several experiences on sludge ozonation, the ozone concentration at the contactor outlet was very low, showing that sludge completely absorbed the ozone flux (*inter alia* Zhang et al., 2009).

Ozonation

In the case of ozonation integrated in the wastewater handling units or sludge handling units, the ozonated line is often compared to a control line (not ozonated): in this case the approaches for calculating the ozone dosage may differ among the experiences cited in the literature; a proposal to clarify some calculations is indicated in the scheme of Figure 13.5.

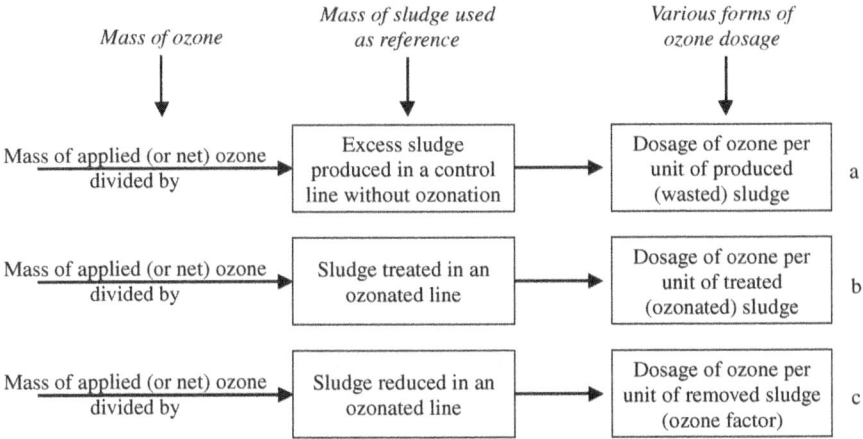

Figure 13.5. Various calculations of ozone dosages found in the literature.

The mass of ozone can be expressed in terms of applied or net ozone, as indicated above. With regard to the mass of sludge used as reference for the calculation of ozone dosage, the following terms can be considered:

(1) the mass of excess sludge produced in a control line operating without ozonation (cases "a" of Figure 13.5);
(2) the mass of sludge treated in an ozonated line (case "b" of Figure 13.5);
(3) the mass of sludge reduced in an ozonated line, calculated as the difference between the sludge production in the ozonated line and in the control line without ozonation (case "c" of Figure 13.5).

An example of the various calculations of ozone dosages comparing an ozonated activated sludge line and a control line is the following:

- ozonation of sludge in a contact reactor using an ozone dosage of 0.02 $gO_3/gTSS_{treated}$ (value "b" of Figure 13.5);
- production of excess sludge of 500 kgTSS in the control line in the entire monitoring period;

- production of excess sludge of 250 kgTSS in the ozonated line in the entire monitoring period;
- mass of applied ozone during the entire monitoring period, 23.6 kgO$_3$;
- calculation of value "a" of Figure 13.5: ozone dosage = 23.6 kgO$_3$/500 kgTSS = 0.047 gO$_3$/gTSS$_{produced}$;
- calculation of value "c" of Figure 13.5: ozone dosage = 23.6 kgO$_3$/(500–250) gTSS = 0.094 gO$_3$/gTSS$_{removed}$.

In general the ozone dosage expressed as gO$_3$/gTSS$_{treated}$ (value "b" of Figure 13.5) is relative to each passage through the ozonation reactor depending on the treatment frequency (see § 7.11).

Another example of the relationship between the different forms of ozone dosage and the treatment frequency is indicated in § 13.9.1, pag. 271.

The values of ozone dosages used in the various studies cover a very wide range, from less than 0.01 gO$_3$/gTSS to much high dosages around 1 gO$_3$/gTSS.

The optimal ozone dosage is normally a compromise between ozonation costs, excess sludge disposal costs and the maintenance of the effluent quality required for wastewater discharge.

In order to calculate an approximate cost for sludge ozonation, operational costs for the production of 1 kg of ozone[1] can be assumed to be roughly 20 kWh (Böhler and Siegrist, 2007; Goel et al., 2003a) or in the range of 1.5 to 2.0 Euros (Ried et al., 2007). An energy consumption cost of about US$ 1.8 is indicated by He et al. (2006) to produce 1 kg of ozone gas from air.

13.5 EFFECT OF SOLIDS MINERALISATION

Total solids or total COD before and after ozonation can be measured to assess the solids mineralisation during the ozonation stage. Data available in the literature regarding the level of mineralisation during ozonation do not always agree.

At ozone dosages up to 0.1 gO$_3$/gTSS the effect of mineralisation is very small resulting in approximately 5% of total solids mass, compared to 24% of solubilisation. Goel et al. (2003a; 2003c) referred that about 5% of TS is mineralized during ozonation at an ozone dosage of 0.05 gO$_3$/gTS.

At higher dosages, the entity of mineralisation rises, resulting in the decrease of solubilized organics (Yeom et al., 2002). The mineralisation of sludge

---

[1] Energy to produce ozone of approximately 9–15 kWh/kgO$_3$, energy to produce liquid oxygen around 0.5 kWh/Nm^{3}O$_2$, energy to transfer ozone to the sludge (2.5 kWh/kgO$_3$).

reached 14% and 26% when the ozone dosage was increased to 0.18 and 0.27 gO$_3$/gTSS respectively (Yan et al., 2009).

Considering the variation of total COD, no significant mineralisation was observed by Bougrier et al. (2007a) for ozone dosages up to 0.18 gO$_3$/gTS, as demonstrated by the fact that total COD remained almost constant. A low mineralisation effect of 5% on carbon was observed with an ozone dosage of 0.2 g O$_3$/gTS (Ahn et al., 2002b).

A significant decrease in suspended solids concentrations was observed by many authors, but this variation can not be considered when evaluating mineralisation, because ozonation causes the conversion of suspended solids into a soluble form.

13.6 COD SOLUBILISATION AND TSS DISINTEGRATION

Ozonation is expected to generate soluble organic matter by the disintegration of solids and the oxidation of organic polymers. Production of soluble COD during ozone treatment is in fact generally correlated with TSS disintegration.

Sludge solubilisation depends strongly on the ozone dosage, as shown in Figure 13.6 and Figure 13.7 for COD solubilisation and TSS disintegration respectively.

In the case of COD solubilisation (Figure 13.6), the data refers to the release of soluble COD after ozonation, excluding the contribution of COD mineralisation and biodegradation in biological stages.

TSS disintegration (Figure 13.7) is instead due to solubilisation + mineralisation (potential mineralisation of suspended solids, according to the definition in § 7.2), while biodegradation in the biological stage is not taken into account.

In Figure 13.6 and Figure 13.7 cases of excess sludge originated from both real and synthetic wastewaters are shown. The sludge solubilisation in the case of synthetic sludge (continuous curves in the figures) is generally higher than that obtainable with real sludge, because of the higher presence of non organic solids in real wastewater compared to synthetic and different resistance to ozone (Nagare et al., 2008).

Sludge produced in a plant fed with synthetic wastewater was used by Nagare et al. (2008), Saktaywin et al. (2005) and Yasui et al. (1994). With regard to TSS disintegration, 30% was reached at an ozone dosage of 0.03–0.05 gO$_3$/gTSS. This value is higher than the solubilisation level reported for real sludge (Figure 13.7).

From Figure 13.6 and Figure 13.7 it can be noticed that the data of COD solubilisation and TSS disintegration are very scattered over 0.04–0.05 gO$_3$/gTSS. The same ozone dosage gives different efficiency levels in the various investigations because other parameters affect the ozonation process, such as:

(a) size of sludge flocs which influences the floc surface and the diffusion of ozone;
(b) presence and concentration of soluble organic compounds which react with ozone;
(c) efficiency of ozone transfer and the hydrodynamic of the ozonation reactor.

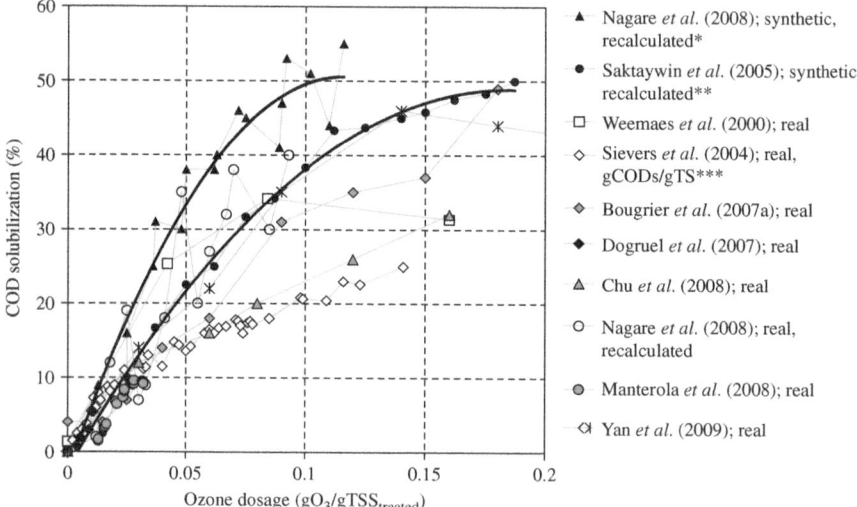

* in the data of Nagare et al. (2008) the mineralisation effect was excluded (recalculated as 5% on average);
** in the data of Saktaywin et al. (2005) the mineralisation effect was excluded (recalculated using the authors' data);
*** in the data of Sievers et al. (2004) solubilisation is expressed as g solubleCOD/gTStreated;
Real = sludge produced in a plant fed with real wastewater;
Synthetic = sludge produced in a plant fed with synthetic wastewater;
Continuous curves are referred to synthetic wastewater (Saktaywin et al., 2005; Nagare et al., 2008).

Figure 13.6. COD solubilisation as a function of ozone dosage.

TSS disintegration increases in an almost linear way at the initial stage, up to a dosage of around 0.04–0.05 $gO_3/gTSS$, where a 20–35% disintegration is reached.

At an ozone dosage of up to 0.04–0.05 $gO_3/gTSS_{treated}$, the degree of solubilisation was in the range from 2 to 6 $gTSS_{solubilised}/gO_3$ (Nagare et al., 2008; Sakai et al., 1997).

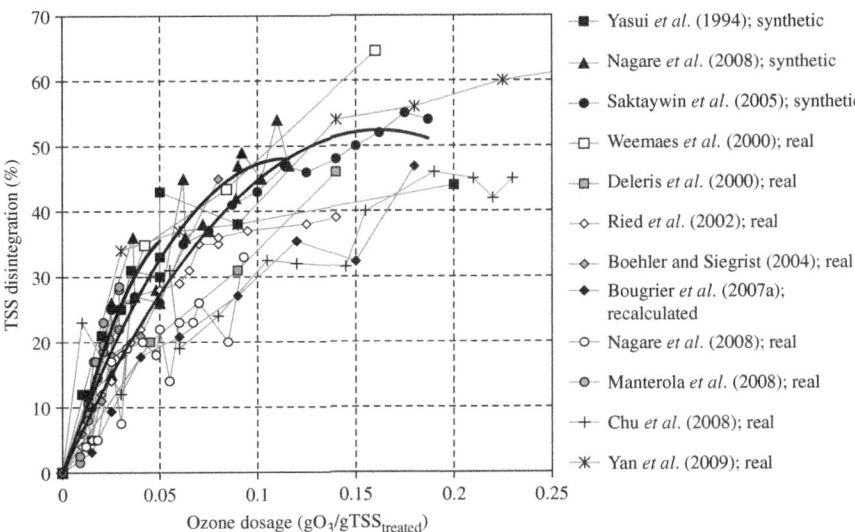

Figure 13.7. TSS disintegration as a function of ozone dosage.

For ozone dosages higher than 0.05 $gO_3/gTSS$ the COD solubilisation rate generally increases slowly: in this range the amount of COD solubilized by ozone is of the same order of magnitude as the amount of dissolved organic matter oxidized by ozone, thereby resulting in a moderate change in the soluble COD concentration.

Zhang et al. (2009) indicated that when the ozone dose was high (0.080 gO_3/gTS), a high portion of ozone reacted with dissolved organic matter.

When the ozone dosage is above 0.14 $gO_3/gTSS$, the ozone failed to oxidize the sludge matrix efficiently due to the release of radical scavengers such as lactic acid and SO_4^{2-} from the microbial cells in the sludge (Yan et al., 2009).

For the maximum ozone dosages tested, higher than 0.2 $gO_3/gTSS$, the efficiency of sludge disintegration and solubilisation does not change significantly (range 45–65%). These findings indicate that the efficiency of sludge disintegration by ozonation has its limits, even at high ozone dosages.

Competition between soluble and particulate – The competition for ozone in reaction with soluble and particulate organic matter was investigated by Cesbron et al. (2003). The applied ozone may react first with soluble compounds and then attack the particulate solids, thus the soluble fraction may have a screening effect on the particulate matter attacked by the ozone (Cesbron et al., 2003). Even at low solubilisation levels, hydroxyl radicals react quickly with solubilised compounds which act as scavengers of particulate solids. Therefore, using very high ozone dosages may result in little improvement of sludge disintegration and solubilisation (Chu et al., 2008).

Similar observations were found by Sievers et al. (2004) in full-scale WWTP, who observed that the solubilised COD per unit of ozone (gCOD/gO$_3$) decreased with increasing ozone dosages as shown in Figure 13.8. The lower values of this ratio at high ozone dosages demonstrated the progressive oxidation of the solubilised COD, which confirms the observations of other authors as described above. This aspect is important for an economic use of ozone. In Figure 13.8, also the solubilised COD per unit of TSS treated are indicated; for example, 110 gCOD were solubilised per 1 kg TSS treated at ozone dosage of 0.03 gO$_3$/gTSS$_{treated}$.

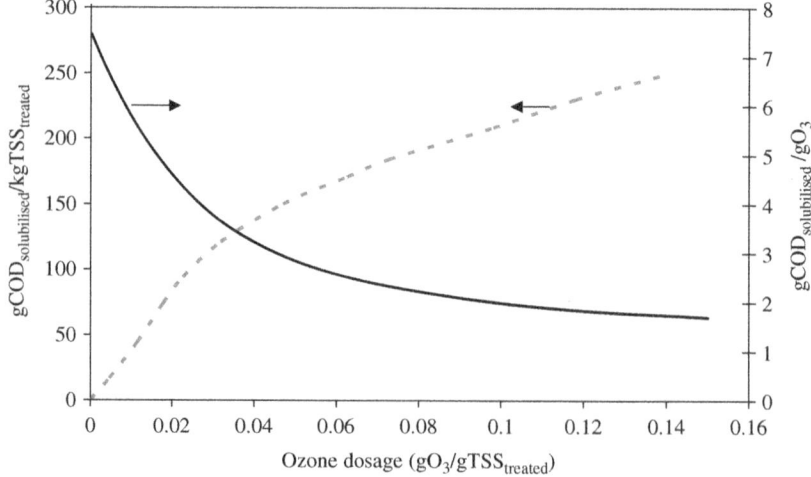

Figure 13.8. Solubilised COD per unit of TSS treated (gCOD/kgTSS) as a function of ozone dosage. Solubilised COD per unit of ozone (gCOD/gO$_3$) as a function of ozone dosage (*modified from* Sievers et al., 2004).

Contact time – Due to the fast reaction rates between sludge compounds and ozone, Manterola et al. (2008) suggested that a contact time in the ozonation reactor of 10 min can be enough to achieve the maximum value of COD solubilisation for the ozone dosages tested (up to 0.035 gO$_3$/gTSS).

Biodegradability – With regard to the biodegradability of the solubilised compounds, Saktaywin *et al.* (2005) found that the percentage of biodegradable compounds compared to solubilised compounds decreases with increasing ozone dosages. For COD solubilisation below approximately 20%, around 60% of soluble COD was biodegradable, while the remaining soluble organic matter was refractory. This percentage decreases with higher ozone dosages and in fact for COD solubilisation around 60%, soluble biodegradable COD dropped to 30%. Therefore, increasing ozone consumption did not lead to an increase of the biodegradability of solubilized sludge (Saktaywin *et al.*, 2005). This phenomenon suggests that part of the ozone was probably used to oxidize biodegradable compounds produced during ozonation (Nishijima *et al.*, 2003) or soluble biodegradable compounds can be immediately biologically consumed.

Solids concentration – With regard to how the TSS concentration may affect ozonation, it appears that:

- the absolute value of solubilisation (concentration of the solubilised COD) effectively depends on the initial TSS concentration of sludge,
- the solubilisation level does not change significantly; a small decrease may be observed in sludge solubilisation at higher solid concentrations.

The data shown in Figure 13.6 and Figure 13.7 are relative to various investigations carried out at different solid concentrations of sludge, but it appears that the solid concentration does not affect the solubilisation percentage. The VSS solubilisation was reduced minimally for sludge concentrations increasing from 1.8 to 2.6% at an ozone dosage of 0.05 gO_3/gTS (Goel *et al.*, 2003a).

In the investigation of Manterola *et al.* (2008) a lower COD solubilisation level was observed in the case of an initial TSS concentration of 9 gTSS/L, compared to 4 gTSS/L, both ozonated at ozone dosages of 0.01–0.02 $gO_3/gTSS$.

VSS/TSS ratio – No significant change in VSS/TSS ratio was observed at ozone dosages below 0.04 $gO_3/gTSS$, whilst VSS/TSS ratio decreased quickly above 0.04 $gO_3/gTSS$ (Zhao *et al.*, 2007).

Mineral fraction – With regards to the solubilisation of the mineral fraction contained in the destroyed suspended solids, Paul and Debellefontaine (2007) reported that organic and mineral fractions of excess sludge were not reduced to the same extent. Bougrier *et al.* (2007a) reported that less than 18% of mineral solids were solubilized at an ozone dosage of 0.18 gO_3/gTS, while the solubilisation of organic solids was greater reaching 48%.

Development of solubilisation and mineralisation – The profiles of the particulate and soluble organic matter and CO_2 production during ozonation up to very high dosages is shown in Figure 13.9. The dosages indicated are very high, reaching 2.5 $gO_3/gTSS_{treated}$, and therefore are not significant in practice, but the wide range of ozone dosages allows us to understand the evolution of sludge solubilisation and mineralisation (associated with CO_2 production). The range of ozone dosages which are of interest in pratical applications is below 0.1 $gO_3/gTSS$ and thus only the first part of the curves in Figure 13.9 is relevant (where ozone dosage and sludge solubilisation are linearly correlated). Once above 1.0 $gO_3/gTSS$ most of the particulate organic matter is transformed into soluble compounds. Above this threshold a net decrease in soluble COD is observed, while at 2.5 $gO_3/gTSS$ a wide part of the particulate and solubilised compound are mineralised.

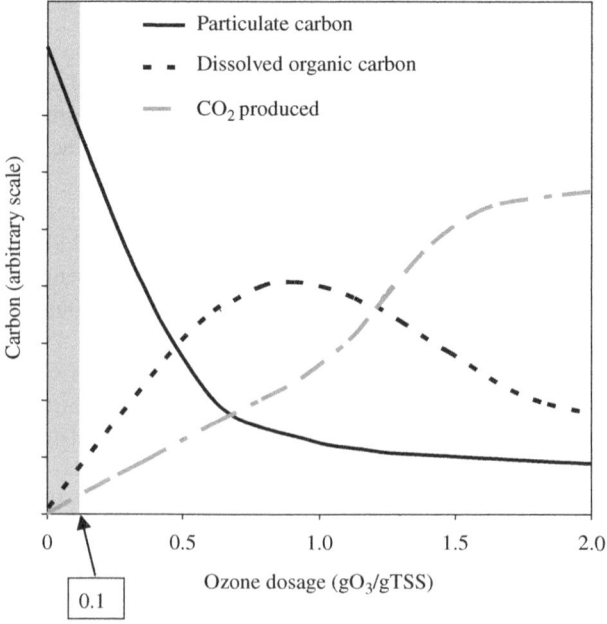

Figure 13.9. Profiles of particulate and soluble fraction of sludge and CO_2 production by mineralisation for increasing ozone dosages (*modified from* Camacho et al., 2002b).

13.7 NITROGEN AND PHOSPHORUS SOLUBILISATION

The production of soluble N forms can be used as an indicator of ozonation efficiency (Bougrier *et al.*, 2007a; He *et al.*, 2006; Chu *et al.*, 2008; Manterola *et al.*, 2008); in particular during ozonation the following dynamics were observed:

- a significant solubilisation of organic N
- a small increase in NH_4 concentration
- very limited or absent variation of NO_2-N
- an increase of NO_3-N only for long contact times (hours).

Protein hydrolysis – Proteins are an important compound in excess sludge. At an ozone dosage of 0.06–0.16 gO_3/gTSS, the fraction of protein-N in the supernatant of the ozonated sludge was 17–27% of total N (Chu *et al.*, 2008) (Figure 13.10).

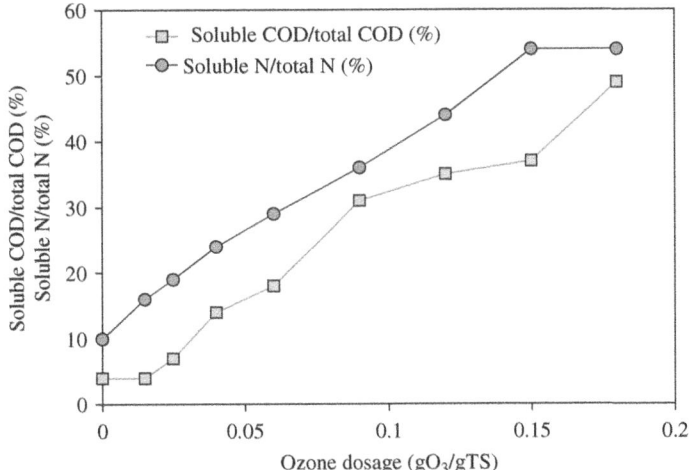

Figure 13.10. COD and N solubilisation as a function of ozone dosage (*modified from* Bougrier *et al.*, 2007a).

N solubilisation during sludge ozonation was found to be proportional to the COD solubilisation, as shown in Figure 13.10, explaining the similar action of ozone on both organic N and organic matter in sludge.

Ammonia concentration – In general ammonia generation indicates protein hydrolysis. Thus the low ammonia concentration generated by ozonation can be explained by the fact that although ozone reacts with proteins, but protein hydrolysis is not noticeable (Bougrier *et al.*, 2007a).

The slight increase in ammonia concentration was proportional to ozone dosage at values higher than 0.02 gO$_3$/gTSS, suggesting that a minimum dosage may be necessary to reach protein destruction and subsequent ammonia release.

Nitrate and nitrite concentration – The ozonation process can efficiently oxidise nitrite to nitrate, but its ability to convert ammonia in nitrite or nitrate is much less effective. In the investigation of Manterola *et al.* (2008) no significant increase in nitrite concentration was observed during sludge ozonation.

Phosporus concentration – With regards to P solubilisation during ozonation, He *et al.* (2006) found an increase of PO$_4$-P and total soluble P concentration. The increase of PO$_4$-P accounted for 45.4% of total soluble P at an ozone dosage of 0.036 gO$_3$/gTSS, which indicated that part of the organic phosphorus was oxidized to phosphate.

The solubilisation of total N and total P calculated per unit of TSS is indicated in Figure 13.11. The entity of P solubilisation among the different experiences are highly variable, probably because the different types of sludge, e.g. characterised by absence/presence of P accumulating microorganisms. For example, total P solubilised per unit of TSS treated was below 0.5% (up to ozone dosages of 0.2 gO$_3$/gTSS) according to Zhao *et al.* (2007) and Dogruel *et al.* (2007); conversely, in the experience of Chu *et al.* (2008), a very high P concentration was solubilised (up to 2.3%).

Due to the release of N and P in the supernatant after ozonation, special care should be taken in the use of ozonated sludge as carbon source, because also additional N and P loads are applied in the subsequent biological system (Zhao *et al.*, 2007).

13.8 INTEGRATION OF OZONATION IN THE BIOLOGICAL PROCESSES

For the reduction of sludge production, ozonation is integrated with biological processes by using various configurations. After a partial ozonation of sludge, the treated flow is recycled into the biological reactors, where the lysates produced by ozonation are treated biologically. Ozonation can be integrated both in the wastewater handling units or in the sludge handling units. Both alternatives have already been developed as industrial processes and have several full-scale applications, but improvements aimed at energy saving are the current objective.

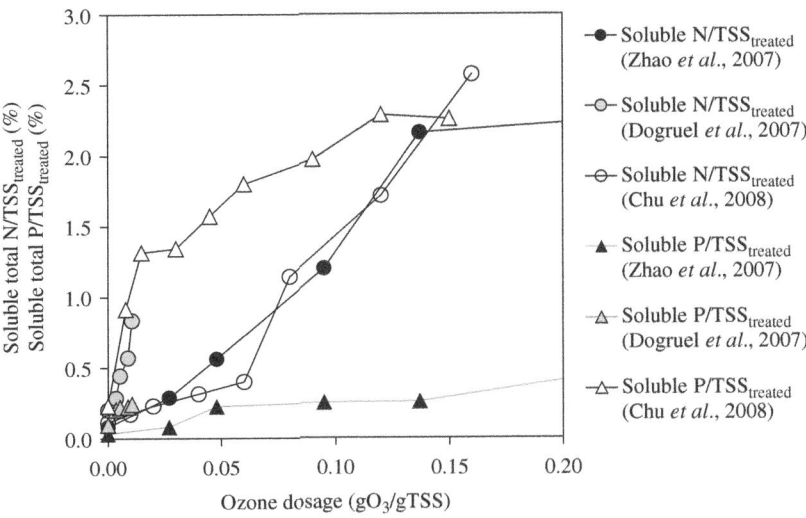

Figure 13.11. P and N solubilisation (per unit of TSS treated) as a function of ozone dosage.

In the case of ozonation integrated in the wastewater handling units (see § 13.9), the various configurations include traditional activated sludge reactors (Yasui and Shibata, 1994; Yasui et al., 1996), alternating nitrification/denitrification, SBRs (Sequencing Batch Reactors, Chiavola et al., 2007) or MBRs (Oh et al., 2007).

Regarding the integration in the sludge handling units (see § 13.10), the combination of ozonation + anaerobic digestion is based on the conversion of sludge into biodegradable matter, while mineralisation takes place in the subsequent anaerobic process (Battimelli et al., 2003). To optimise ozonation prior to anaerobic digestion, organic matter mineralisation should be minimised, while organic matter solubilisation should be maximised, in order to favour the release of biodegradable substrates to enhance methane production (Bougrier et al., 2007a). The combination of ozonation with aerobic digestion was also proposed (Caffaz et al., 2005).

13.9 INTEGRATION OF OZONATION IN THE WASTEWATER HANDLING UNITS

In the experiences proposed since the mid '90s, ozonation has been applied: (a) on activated sludge taken directly from the aeration tank as indicated in Figure 13.12A or (b) on the sludge return flow as shown in Figure 13.12, options

270   Sludge Reduction Technologies in Wastewater Treatment Plants

B1 and B2. In both cases sludge passes through the ozonation contactor, where solubilisation and mineralisation occur (Kamiya and Hirotsuji, 1998; Yasui and Shibata, 1994). When the ozonated sludge is recycled to the activated sludge system for further biological degradation, new biomass grows on the solubilised degradable organic fraction (mechanism of cell lysis-cryptic growth, see § 4.2) and an inert soluble organic fraction is also produced (Böhler and Siegrist, 2007).

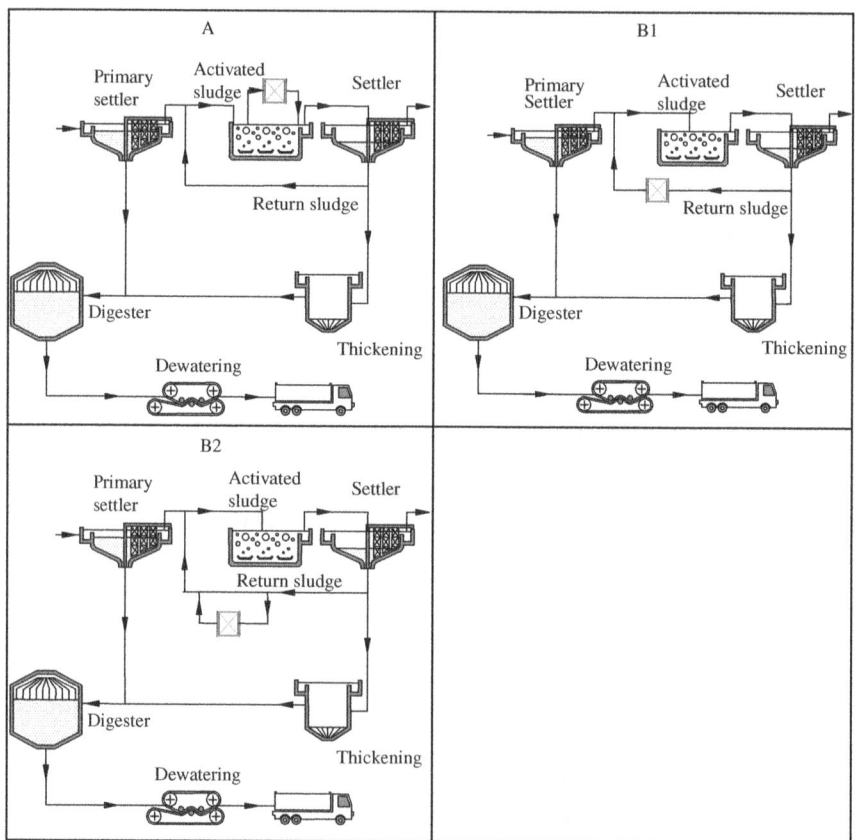

Figure 13.12. Alternatives for the integration of ozonation in the wastewater handling units. The symbol ⊠ indicates the ozonation equipment.

The ozonation process has already had full-scale applications for more than 15 years. The process called *Biolysis*® *O* (Ondeo-Degrémont, France) involves the ozonation of a portion of activated sludge extracted from the aeration basin. After sufficient time in the ozone contactor, the ozonated sludge is returned to the aeration basin (option A in Figure 13.12). Another process is called Bioleader™ (Kurita Water Industries Ltd, Japan) and other patents are developed by PRAXAIR, INC. (U.S.) and WEDECO Umwelttechnologie (Germany).

13.9.1 Initial studies and ozone dosage calculation

In these first experiences from the mid '90s, synthetic wastewater (based on peptone, yeast extract, meat extract, etc...) was fed into the activated sludge.

Operating on the recirculation rate and ozone dosage, Yasui and Shibata (1994) demonstrated no excess sludge production for 6 weeks and a net decrease of Y_{obs} from 0.40 gTSS/gBOD$_5$ (control value of a conventional wastewater treatment process) to nearly zero, when 30% of the activated sludge was recirculated daily from the biological stage to the ozonation stage (at ozone dosage of 0.05 gO$_3$/gTSS$_{treated}$ and treatment frequency at 0.3 d^{-1}). The lab-scale activated sludge plant was operated with SRT of 10 days, influent organic load of 1.0 kgBOD$_5$ m^{-3} d^{-1} and TSS concentration in aeration tank of 4 kgTSS/m^3.

As indicated in Figure 13.5 the ozone dosage differs when calculated per unit of treated (ozonated) sludge (case "b", Figure 13.5) or per unit of excess sludge initially produced without ozonation (case "a", Figure 13.5). In the case described by Yasui and Shibata (1994), the ozone dosage calculated per unit of initially produced sludge, taking into account the values indicated above, is the following:

$$\text{daily TSS production (without ozonation)} =$$
$$1.0 \text{ kgBOD}_5 \text{ m}^{-3}\text{d}^{-1} \times 0.40 \text{ kgTSS}_{produced}/\text{kgBOD}_5$$
$$= 0.4 \text{ kgTSS}_{produced}\text{m}^{-3}\text{d}^{-1}$$

Ozone dosage corresponds to 0.05 kgO$_3$/kgTSS$_{treated}$, obtaining a daily ozone dosage of:

$$0.05 \text{ kgO/kgTSS}_{treated} \cdot 4\text{kgTSS}_{treated} \text{ m}^{-3} \cdot 0.3 \text{ d}^{-1} = 0.06 \text{ kgOm}^{-3}\text{d}^{-1}$$

Finally, the ozone dosage calculated per unit of initially produced sludge (case "a", Figure 13.5) is:

$$\frac{0.06 \text{ kgO}_3 \text{ m}^{-3} \text{ d}^{-1}}{0.4 \text{ kgTSS}_{produced} \text{ m}^{-3} \text{ d}^{-1}} = 0.15 \text{ kgO}_3/\text{kgTSS}_{produced}$$

In this example the treatment frequency was 0.3 d^{-1}: this corresponds to the recirculation in the ozonation contactor of 33% of total biomass present in the aeration tank per day.

The first results from full-scale activated sludge processes where ozonation was applied in the sludge return flow were published by Japanese researchers in the mid '90s (Yasui et al., 1996; Sakai et al., 1997). Yasui et al. (1996) studied the effect of ozonation of sludge produced from several types of industrial and municipal wastewater at lab and full-scale, with the aim of achieving zero excess sludge production, following the previous experiences of Yasui and Shibata (1994). In particular, in a 3-year experience on a full-scale pharmaceutical WWTP, the optimal setting for obtaining the complete elimination of excess sludge was selected according to Yasui and Shibata (1994): the ozone dosage was set at 0.05 $kgO_3/kgTSS_{treated}$ (about 0.15 $kgO_3/kgTSS_{produced}$) and a percentage of 30% of total biomass present in the aeration tank per day was recirculated to ozonation (Yasui et al., 1996).

Another experiment aimed at zero sludge production was reported by Sakai et al. (1997) on a full-scale municipal WWTP, applying ozone to a part of the return sludge at a dosage of 0.034 $kgO_3/kgTSS_{treated}$ (case "b", Figure 13.5), corresponding to 0.136 $gO_3/gTSS_{produced}$ (case "a", Figure 13.5). Thus 25% of total biomass present in the aeration tank was recirculated to ozonation per day. In other words, the dosage of 0.034 $kgO_3/kgTSS$ was applied 4 times to the mass of sludge which would have been produced without ozonation.

While zero sludge production remains the goal, it has some significant drawbacks:

- increase of metals and phosphate in the treated effluents which can not be retained in the sludge;
- inert particulate COD and inorganic TSS present in the influent wastewater may accumulate in the sludge, resulting in a decrease of the VSS/TSS ratio and a consequent reduction of active biomass fraction (Huysmans, 2001);
- potential loss of plant performance, especially with regard to nitrification or COD removal;
- effluent TSS concentration may deteriorate;
- high ozone dosages do not always lead to corresponding economies in sludge disposal.

Therefore in the recent experiences on ozonation, moderate ozone dosages aimed at partial excess sludge reduction, instead of zero sludge production, have been preferred, as described in the investigations referred to in the next section.

13.9.2 Results on sludge reduction

Some case-studies at lab-, pilot- and full-scale aimed at sludge reduction are briefly described below and the main results are summarised in Table 13.1. In this section sludge reduction (TSS reduction) is the result of mineralisation + solubilisation + biological treatment.

Egemen *et al.* (1999) investigated two continuous flow lab-scale activated sludge systems fed with synthetic wastewater, operating at SRT of around 5 d and run in parallel: one as control system and one integrated with ozonation. The Y_{obs} of the ozonated system averaged 0.11 gTSS/g soluble COD removed, with respect to a value of 0.29 gTSS/g soluble COD removed, measured in the control system. Thus a yield reduction of 62% was observed, which equates to about a 40 to 60% reduction of excess sludge, although the exact dosage of ozone was not indicated.

Using intermittent ozonation on a part of the sludge return flow in a lab-scale SBR at ozone dosage of 0.048 $gO_3/gTSS_{produced}$ (correspondent to 0.019 $gO_3/gTSS_{treated}$), the Y_{obs} was 0.083 $gTSS/gCOD_{added}$, half the yield measured in a control SBR used as reference, which was 0.164 $gTSS/gCOD_{added}$ (Huysmans *et al.*, 2001). This reduction of the Y_{obs} indicated that a 50% decrease of the net sludge production could be achieved in the ozone-treated reactor.

In the experience at pilot-scale reported by Déléris *et al.* (2002) – operating with activated sludge at SRT of 10 d – ozonation was applied to a fraction of sludge pumped out from the aerated tank, as in Figure 13.12 option A. The Y_{obs} was 0.28 $gVSS/gCOD_{removed}$ (0.34 $gTSS/gCOD_{removed}$) in the untreated line and decreased to 0.09 g $VSS/gCOD_{removed}$ at applied ozone dosage of 0.062 $gO_3/gTSS_{treated}$ (0.05 $gO_3/gVSS_{treated}$). Thus ozonation resulted in a 70% reduction of sludge production. Previous results found by the same authors (cited by Déléris *et al.*, 2002) showed that an ozone dosage of 0.1 $gO_3/gVSS_{treated}$ was excessive, and led to a decrease of the biomass in the reactor.

Ried *et al.* (2002) – in a two-line full-scale activated sludge plant (WWTP Enger, 20,000 PE) without primary settling and with SRT of 15 d – measured in the ozonated line a sludge reduction of 30% at dosage of 0.052 $gO_3/gTSS_{treated}$ and a daily treatment of 10% of the activated sludge. This ozone dosage corresponds to about 0.08 $gO_3/gTSS_{produced}$ (calculated with reference to the initial excess sludge) (Ried *et al.*, 2002).

In an SBR system fed with synthetic wastewater, where ozone was supplied directly during the aerobic phase of the SBR cycle instead of the air supply, Egemen Richardson *et al.* (2009) used a daily net ozonation rate of 0.058 gO_3 $gTSS^{-1}$ d^{-1} (considering TSS of ozonated sludge) obtaining an average sludge reduction of 29%.

Table 13.1. Performance of ozonation integrated in the wastewater handling units for sludge reduction.

Scale	Type of wastewater*	Wastewater treatment process	O_3 dosage ($gO_3/gTSS_{produced}$)	O_3 dosage ($gO_3/gTSS_{treated}$)	Sludge reduction (TSS)	References
lab	sw	activated sludge (SRT = 10 d)	0.05	0.05	33%	Yasui and Shibata (1994)
lab	sw	activated sludge (SRT = 10 d)	0.10	0.05	67%	Yasui and Shibata (1994)
lab	sw	activated sludge (SRT = 10 d)	0.15	0.05	100%	Yasui and Shibata (1994)
full	rw	activated sludge	0.136	0.034	100%	Sakai et al. (1997)
lab	sw	activated sludge (SRT = 5 d)	–	–	40–60%	Egemen et al. (1999)
lab	rw	SBR	0.048	0.019	50%	Huysmans et al. (2001)
pilot	rw	activated sludge (SRT = 10 d)	–	0.062	70%	Déléris et al. (2002)
full	rw	activated sludge (SRT = 15 d)	0.08	0.052	30%	Ried et al. (2002)
lab	rw	MBR (SRT = 120 d)	–	0.1	100%	Song et al. (2003)
pilot	rw	activated sludge	$x = 0.02 \div 0.08$ $gO_3/gCOD_{removed}$**		$y\,(\%) = 1344\,x$	Paul et al. (2006b)
lab	sw	SBR	0.08	0.03	25%	Dytczak et al. (2007)
lab	sw	SBR (SRT = 20 d)	ozonation rate = 0.058 gO_3 $gTSS^{-1}\,d^{-1}$		29%	Egemen Richardson et al. (2009)

* sw = synthetic wastewater; rw = real wastewater.
** specific ozone dosage, defined at pag. 258.

In a lab-scale MBR system with a long SRT of 120 d and with Y_{obs} of approximately 0.15 gTSS/gCOD, ozonation was applied at dosage of 0.1 gO_3/gTSS according to the configuration of Figure 13.13. This application allowed excess sludge to be eliminated completely and a sludge production of almost zero to be obtained (Song *et al.*, 2003). In spite of the absence of sludge extraction, TSS concentration in the biological reactor remained around 8 gTSS/L and the effluent quality was kept at a satisfactory level both for COD and total nitrogen.

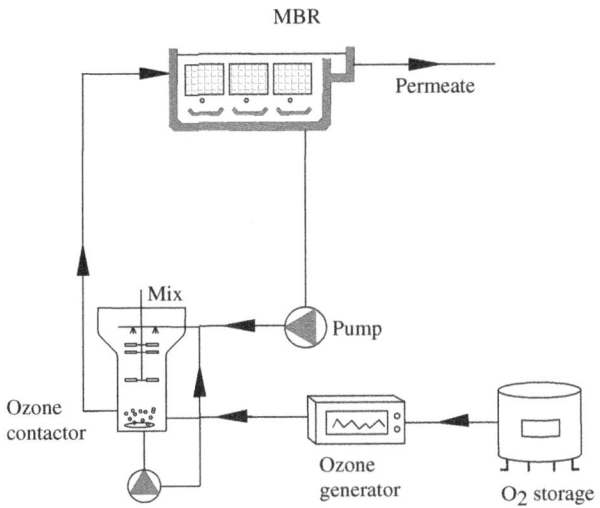

Figure 13.13. Ozonation integrated in the lab-scale MBR system (*modified from* Song *et al.*, 2009).

Dytczak *et al.* (2007) evaluated the effectiveness of partial ozonation of return activated sludge on a nitrifying SBR receiving partially ozonated sludge and applying alternating anoxic/aerobic conditions to achieve denitrification/ nitrification. The authors found that biomass in the alternating anoxic/aerobic reactor was easier to destroy than in the aerobic reactor, generating approximately twice as much soluble COD. Sludge reduction was up to 25% at ozone dosage of 0.08 gO_3/$gTSS_{produced}$ (corresponding to 0.03 gO_3/$gTSS_{treated}$). It is interesting to note that nitrification rates deteriorated much more in the aerobic than in the alternating anoxic/aerobic reactor. The extra supply of solubile COD, proportional to ozone dosage, greatly increased the denitrification rate in the alternating anoxic/aerobic reactor (Dytczak *et al.*, 2007). The results found in this research were affected by the denser, stronger and more compact structure of flocs in the aerobic reactor compared to the weak, thin and elongated flocs in the alternating reactors.

In a recent review on sludge ozonation, Paul and Debellefontaine (2007) referred to their own results obtained in a 6-year research programme. The excess sludge reduction (due to ozonation + biological effects) as a function of the specific ozone dosage expressed as $gO_3/gCOD_{removed}$ is indicated in Figure 13.14 (Paul et al., 2006b). The specific ozone dosage, as defined on pag. 258, takes into account the COD removed from the influent wastewater in the wastewater handling units. The original specific ozone dosage expressed as $gO_3/gCOD_{removed}$ was converted by us into $gO_3/gTSS_{produced}$ (referred to the untreated line) considering: $Y_{obs} = 0.33$ $gVSS/gCOD_{removed}$ in the untreated line and VSS/TSS = 0.86.

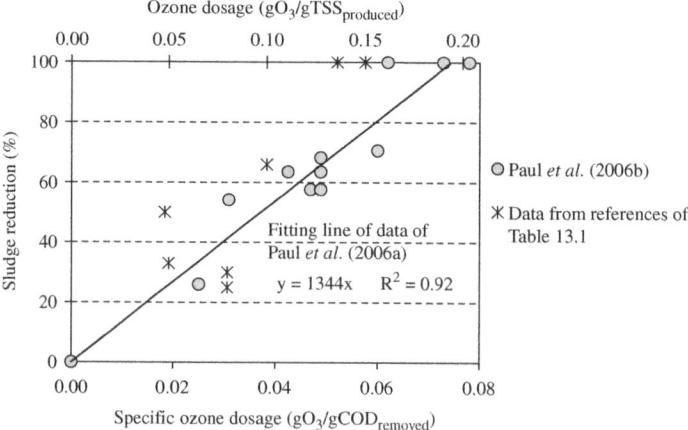

Figure 13.14. Relationship between specific ozone dosage and excess sludge reduction.

In Figure 13.14 the following data is also indicated:

- fitting line calculated by us from data in Paul et al. (2006b);
- data presented in Table 13.1 used for comparison and expressed as $gO_3/gTSS_{produced}$.

As expected, the reduction of sludge is proportional to the ozone dosage and it confirms that it is possible to obtain a 100% reduction with a dosage of around 0.07–0.08 $gO_3/gCOD_{removed}$ (calculated by us to be 0.16 $gO_3/gTSS_{produced}$; this ozone dosage agrees with other studies aimed at zero sludge production: 0.15 by Yasyui et al., 1996; 0.136 by Sakai et al., 1997; 0.1 by Song et al., 2003).

Furthermore, looking at the dependence of the Y_{obs} on ozone dosage (Figure 13.15), it can be noted that for increasing ozone dosages a progressive decrease of the Y_{obs} occurs: for ozone dosages of 0.05 $gO_3/gTSS_{produced}$ the reduction is significant, but the value reaches 0 for dosages higher than 0.15 $gO_3/gTSS_{produced}$.

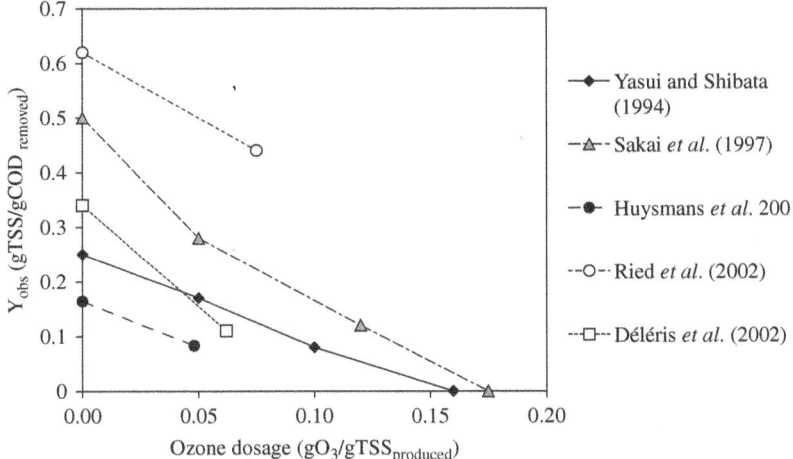

Figure 13.15. Reduction of observed sludge yield as a function of ozone dosage.

Table 13.1 summarises the main results from the investigations described above.

Ozonation integrated in the wastewater handling units may influence the biological treatment and its efficiency, including the following main aspects:

- effluent quality
- pH of sludge
- activity, respiration and integrity of microorganisms in sludge
- soluble and bound EPS content
- nitrification
- denitrification
- sludge settleability

as described in detail in the following sections.

13.9.3 Influence on WWTP effluent quality

Several experiences in the literature reported a slight decrease in effluent quality after sludge ozonation and recirculation of ozonated sludge into the activated sludge reactors.

During ozonation, inert soluble and colloidal COD is released to the bulk solution just as during the long-term stabilisation of activated sludge (Böhler and Siegrist, 2007). This effect leads to a slight increase in the soluble COD and TOC levels in the treated effluents, due to the fact that elimination of sludge generates some non-biodegradable soluble organics (Boero *et al.*, 1991; Chudoba, 1983; Déléris *et al.*, 2002; Egemen Richardson *et al.*, 2009; Orhon *et al.*, 1989; Yasui *et al.*, 1996).

Huysmans *et al.* (2001) indicate a slight increase in effluent concentration of COD and PO_4-P at ozone dosage of 0.048 $gO_3/gTSS_{produced}$ and a sludge reduction of 50%.

Also Egemen Richardson *et al.* (2009) reported a slight increase of soluble COD in the effluent of an SBR treated with ozonation rate of 0.058 gO_3 $gTSS^{-1}$ d^{-1}; soluble COD passed from 39 to 52 mg/L, which caused a reduction of removal efficiency of 2% compared to a control SBR reactor (not ozonated).

In a continuous-flow activated sludge system, effluent soluble COD passed from 50 to 69 mg/L after integration with an ozonation stage, but the soluble COD removal efficiency remained practically the same (Egemen *et al.*, 1999).

Böhler and Siegrist (2007) indicated a specific increase of inert COD in an ozonated process of about 0.15 $gCOD/gTSS_{reduced}$, which would lead to an approximate 25% increase of the effluent COD for a plant with 30% sludge reduction.

Soluble COD concentration in effluent from a pilot-scale activated sludge plant integrated with ozonation, increased by 15–40 mgCOD/L (Paul and Debellefontaine, 2007), but the total COD concentration was still far below the limit for discharge, compatible with the European regulations in force (125 mg COD/L). In general, the amount of COD discharged with the effluent could only account for 10 to 20% of the excess sludge reduction observed.

Table 13.2 summarises the increase of effluent COD concentration in activated sludge systems integrated with ozonation.

When high ozone dosages are applied, the concentration of P in effluent becomes higher as a result. In these cases the solubilized P has to be removed by further treatment to keep its concentration within discharge limits. Saktaywin *et al.* (2005, 2006) proposed the recovery of P solubilised during ozonation for use as a resource, by a crystallisation process – where P is recovered as a crystalline product with Mg or Ca ions –, developed as an alternative technology rather than a chemical precipitation process. The advantages of the crystallisation process are that P can be recovered as an usable product and that sludge generation from coagulant in the WWTP is reduced (Saktaywin et al., 2005, 2006).

Table 13.2. Increase of effluent COD concentration in activated sludge systems integrated with ozonation.

O_3 dosage ($gO_3/gTSS_{produced}$)	O_3 dosage ($gO_3/gTSS_{treated}$)	Sludge reduction	Increase of effluent COD concentration	References
0.048	0.019	50%	Slight	Huysmans et al. (2001)
x = 0.02 ÷ 0.08 $gO_3/gCOD_{removed}$		y (%) = 1344 x	increase of 15–40 mgCOD/L	Paul and Debellefontaine (2007)
ozonation rate = 0.058 gO_3 $gTSS^{-1}$ d^{-1}		29%	increase of 13 mgCOD/L (from 39 to 52 mgCOD/L)	Egemen Richardson et al. (2009)
–	–	–	increase of 19 mgCOD/L (from 50 to 69 mgCOD/L)	Egemen et al. (1999)
		30%	increase of 25%	Böhler and Siegrist (2007)

In this configuration, tested at lab-scale and fed with synthetic substrate, a 60% reduction of excess sludge production (applying 0.03–0.04 $gO_3/gTSS$ and operating with a ratio of sludge flow rate to ozonation of 1.1% of wastewater inflow) and a potential P recovery of 70% was achieved (Saktaywin et al., 2006).

Although P recovery from sludge is technically feasible, its economic viability is still considered doubtful, due to the high costs of the techniques, compared to the cost of phosphate rock (Roeleveld et al., 2004). However, the recovery of P will become more important in the course of the century due to its forecast scarcity.

13.9.4 Influence on sludge pH

Several authors have reported that an increase of ozone dosages gradually decreased sludge pH, but the effect is only significant for high ozone dosages. Park et al. (2003) shows the relationship between pH and ozone dosage up to very high values of 1 gO_3/gTS (Figure 13.16). These authors observed pH values below 3.0 at ozone dosages greater than 0.5 gO_3/gTS, but these dosages are quite high for cost-effective applications. A low pH may cause the mobilisation of heavy metals from the sludge and a significant release of them. Park et al. (2008) observed that, with decreases in sludge pH, the level of zinc released increased.

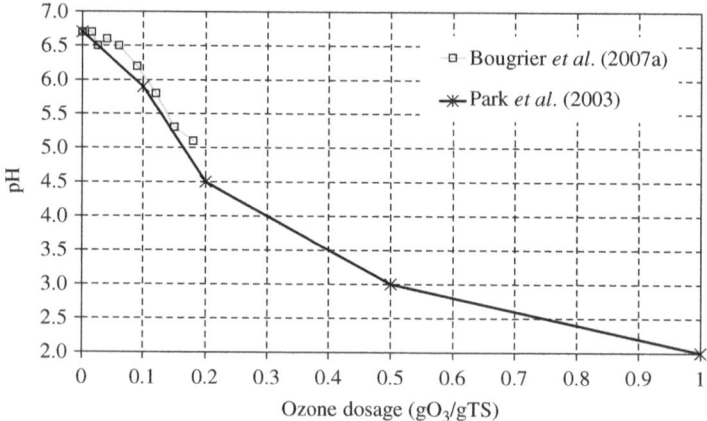

Figure 13.16. Effect of ozonation on sludge pH.

Sludge pH decreased from 6.7 to 5.1 as ozone dosages rose from 0.015 to 0.18 gO_3/gTS as shown in Figure 13.16 (Bougrier et al., 2007a). This was due to the oxidation of organic matter into more oxygenated molecules, such as carboxylic acids.

At an ozone dosage of 0.16 gO_3/gTSS, the pH value decreased from 8.0 to 7.2, which indicated the formation of acidic compounds (Chu et al., 2008). The results of HPLC analysis showed that lactic acid was detected and its concentration gradually increased with ozone dosages up to 0.16 gO_3/gTSS (Chu et al., 2008).

13.9.5 Influence on sludge flocs and microorganisms

Ozone penetrates into the microorganisms, increases the osmosis of cell membranes, damages the uniformity of the cell walls, and releases the intra-cellular components, protein and DNA, into water, causing permeabilisation of cells and loss of culturability (Komanapalli and Lau, 1998; Zhang et al., 2009). Soluble COD also increases as a consequence of the release of all these substances (Zhang et al., 2009).

An ozone dosage lower than 0.015 $gO_3/gVSS$ is not enough to promote cell rupture. Only the use of higher ozone concentrations improves cell rupture (Albuquerque et al., 2008).

PCR-DGGE fingerprinting was used to evaluate the evolution of the death of the bacteria and the destruction of the bacterial DNA contained in the sludge during the ozonation process (Yan et al., 2009). For ozone dosages below 0.02 $gO_3/gTSS$, no alteration in bacterial DNA was detected by DGGE, indicating that the DNA of the sludge was not attacked by ozone. At concentrations greater than 0.02 $gO_3/gTSS$, the bacteria began to break down and several bands in the DGGE fingerprint disappear. The destruction of bacteria occurs mainly at ozone dosages above 0.08 $gO_3/gTSS$ (Yan et al., 2009). Some bacteria can survive even at these relatively high ozone dosages due to the growth of the cocci in clusters, which could protect their DNA from the ozone attack (Yan et al., 2009).

In depth investigations of single bacteria dynamics during ozonation are difficult to carry out due to the aggregation of bacteria in flocs.

Respirometry – The evaluation of the whole microbial activity was also performed using respirometry (by measuring Oxygen Uptake Rate, OUR) (Chu et al., 2008; Zhang et al., 2009). The sludge OUR significantly decreased during ozonation, following first-order reaction (Zhang et al., 2009). Measurement of OUR showed that at an ozone dosage of around 0.02 $gO_3/gTSS$, nearly 80% of microbial respiration activity was lost (Chu et al., 2008).

In the study by Saktaywin et al. (2005) around 70% of sludge bacteria activity – measured by a single-OUR test – was lost at ozone dosage of 0.03–0.04 $gO_3/gTSS$.

Enzymatic activity – In some studies, protease of the sludge was chosen as the indicator of enzymatic microbial activity during sludge ozonation. The gradual decrease in total protease and catalase activity at ozone dosages from 0 to 0.05 $gO_3/gTSS$ indicated that enzymes began to lose their activity after the addition of any amount of ozone and were destroyed by ozone oxidation (Yan et al., 2009). When the ozone dosage was higher than 0.02 $gO_3/gTSS$, the ozone began to oxidize the proteins released from the treated sludge (Yan et al., 2009).

Filamentous microorganisms – A clear reduction of the filamentous organisms was often observed in ozonated lines. In the experiences of Kamiya and

Hirotsuji (1998) and Paul and Debellefontaine (2007), ozonation also caused the disappereance of networks of filamentous bacteria which were squeezed and bundled. As a consequence, typical problems of activated sludge systems such as bulking and foaming could be reduced and an improvement in sludge settling characteristics would be observed. Furthermore, the flocs were more compact in the activated sludge treated with ozone (Paul and Debellefontaine, 2007).

We can cite some early experiences, which were carried out in the 1988–1994 period aimed at the application of ozone to control bulking problems and improve settleability, using ozone at low levels (0.05–1.0 mgO$_3$ gVS^{-1} h^{-1}) (Collignon et al., 1993, 1994; van Leeuwen, 1988a and 1988b).

Pathogen reduction – For a significant effectiveness of ozone in the inactivation of fecal coliforms, an ozone dosage above 0.1 gO$_3$/gTS is needed, but the complete reduction of fecal coliforms, *Streptococcus* and *Salmonella* was observed only above 0.2–0.4 gO$_3$/gTS (Park et al., 2008). We must consider that these values are often uneconomic in practice.

Floc structure – Recently, some studies have observed that ozone does not violently disrupt floc structure and its effect is poor in dissolving the sludge matrix (Zhang et al., 2009). During ozonation the particle size distribution in sludge was found to be relatively stable and ozonation seems to reduce the numbers of small flocs (3 μm) and increase those for medium flocs (7.5–30 μm). Similar observations have been reported by Bougrier et al. (2006a), which indicated that ozonation did not seem to affect particle size.

The effect is evident only at higher dosages, over 0.1 gO$_3$/gTS, where the peak of the particle size distribution gradually moved to a smaller value. If the mean particle size was about 70 μm before sludge ozonation, at 0.5 g O$_3$/gTS the size was reduced to around 40 μm (Park et al., 2003).

At dosages up to 0.16 gO$_3$/gTSS, microscope observations showed that the flocs had been broken into fine, dispersed particles and the ESEM images showed a distinct difference in the appearance of the cells, which appeared deformed, indicating the cells had been permeabilised by ozone (Chu et al., 2008).

EPS content – Comparing activated sludge before and immediately after ozonation, an increase of soluble EPS was observed at ozone dosages from 0.022 to 0.070 gO$_3$/gTSS of initial sludge (Dytczak et al., 2006). Ozone destroys part of the flocs, thus a fraction of the bound polymeric material is released and solubilized, increasing soluble EPS. Only at higher dosages (0.088 gO$_3$/gTSS), does ozonation cause a decrease of the amount of EPS.

With regard to prolonged, everyday ozonation, both the total and bound EPS levels were higher in the ozonated reactors in comparison to the initial (control) reactor and stronger flocs were created. Ozonation integrated in a biological treatment with long SRT system favours not only EPS production, but also the degree of binding into the floc structure (Dytczak *et al.*, 2006).

A synthesis of the effect of ozonation on microrganisms, floc size, enzymatic activity and respiration, as a function of ozone dosage, is shown in Figure 13.17.

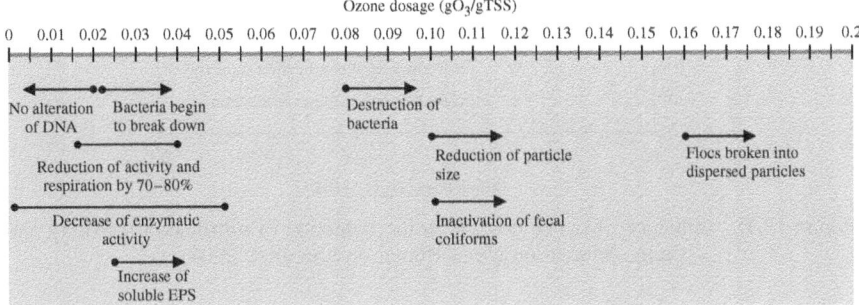

Figure 13.17. Effect of ozonation on microorganisms and sludge constituents as a function of ozone dosage.

13.9.6 Influence on nitrification

The influence on nitrification is one of the major critical point of ozonation integrated in activated sludge processes. It depends on various factors such as the ozone dosage, the treatment configuration, the effective SRT and the nitrifying biomass.

The reduction of the nitrification capacity after partial ozonation was investigated by Böhler and Siegrist (2004), who demonstrated that the entity of nitrification reduction was similar to the entity of sludge reduction, as indicated in Figure 13.18. For example, by applying an ozone dosage of 0.042 $gO_3/gTSS$ a decrease of nitrification capacity of 25% was observed compared to a sludge reduction of 21%.

The decrease of nitrification rates proportional to increasing ozone dosages was also observed by Dytczak *et al.* (2007), who calculated the *Ammonia Utilisation Rate* and *Nitrite + Nitrate Production Rate* parameters monitoring the changes in ammonium or nitrite + nitrate concentrations. The authors observed that in an activated sludge reactor operating with alternating anoxic/ aerobic conditions, nitrification decreased with ozonation (up to 8%) but was affected far less than in a continuously aerated reactor (25%).

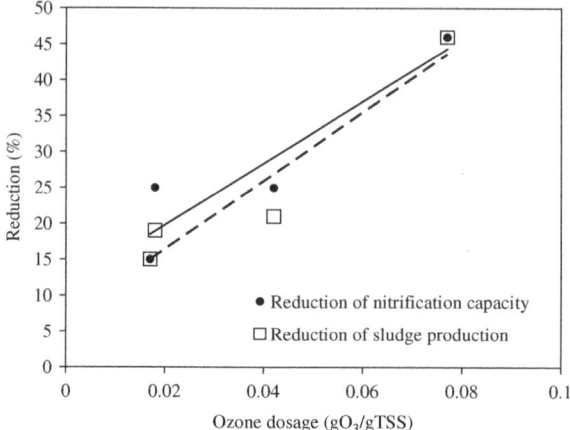

Figure 13.18. Influence of ozone dosage on the reduction of nitrification capacity and sludge production (*from* Böhler and Siegrist, 2004).

The deterioration of the nitrification rate caused by ozonation is due to the direct influence of ozone on nitrifiers and floc structure, but it may also be influenced by the higher soluble COD released during ozonation. In fact, since both nitrification and heterotrophic COD removal occur simultaneously, in the presence of solubilised organic matter, nitrifiers can easily be outcompeted by heterotrophs for oxygen and ammonia. The consequence is that nitrifiers could be partially suppressed by faster growing heterotrophs.

Awareness of the effective SRT of the nitrifiers is important to ensure full nitrification as indicated theoretically in § 7.8. On the basis of the expression presented in § 7.8 (pag. 103), the reduction of the activity of the nitrifiers due to ozonation is therefore just about compensated by the increase in apparent SRT due to the lower excess sludge production (Böhler and Siegrist, 2004).

Due to the increase in apparent SRT as a consequence of sludge reduction, the fraction of nitrifying biomass may increase from 5% (conventional value) to more than 10%, at ozone dosages of 0.04 $gO_3/gCOD_{removed}$ and at excess sludge reduction of 60% (Paul and Debellefontaine, 2007). Figure 13.19 shows the nitrification efficiency in the plant yield by ozone dosage, always between 90–98%, in spite of the high excess sludge reduction.

Déléris *et al.* (2002) also indicated that the percentage of $N-NH_4$ removed in an ozonated line and in a control line (both operating at pilot-scale, with activated sludge processes, SRT of 10 days, configuration of nitrification/denitrification, ozone dosages around 0.062 $gO_3/gTSS_{treated}$) was similar. The nitrification process in this case was not altered by the ozonation treatment.

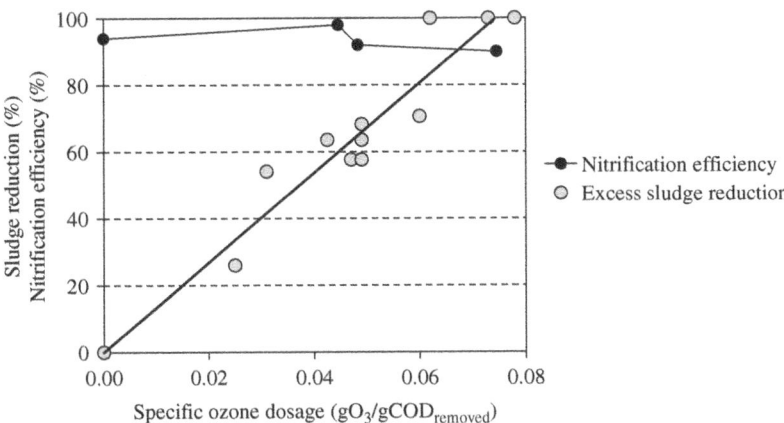

Figure 13.19. Nitrification efficiency compared to the entity of sludge reduction (*from* Paul and Debellefontaine, 2007).

Chiavola *et al.* (2007), applying ozonation at 0.05–0.07 $gO_3/gTSS_{treated}$ in a lab-scale SBR system, observed that nitrification and carbon removal efficiencies in the SBR were not appreciably modified.

13.9.7 Influence on denitrification

One of the main factors limiting denitrification is the availability of a carbon source in the influent wastewater. If the ozonated sludge is recycled to a pre-denitrification stage, the soluble biodegradable COD generated by ozonation can be used to improve denitrification capacity. Conversely, ozonation may reduce the population of denitrifiers and could decrease the rate of denitrification.

Dytczak *et al.* (2007) indicated that denitrification was enhanced rather than inhibited by ozonation. In fact, an increase in denitrification rate (measured as *Nitrite + Nitrate Utilisation Rate*) was observed for increasing ozone dosages (Dytczak *et al.*, 2007).

Böhler and Siegrist (2007) reported that for a sludge production of 100 mgTSS/$L_{wastewater}$ and 30% sludge reduction after ozonation, denitrification increased by approximately 7 mgN/$L_{wastewater}$. This benefit was partly reduced due to the additional ammonium (2 mgN/$L_{wastewater}$) released during sludge ozonation.

The denitrification rates measured by using ozonated sludge as a carbon source are indicated in Table 13.3 and compared with the denitrification rate obtained in the presence of municipal wastewater.

Table 13.3. Denitrification rate of ozonated sludge compared to municipal wastewater.

Carbon source	Denitrification rate [gNO$_3$–N gVSS^{-1} d^{-1}]		Reference
	range	typical value	
municipal wastewater		0.07	
solubilised organics by ozonation	0.011–0.081	0.05	Ahn et al. (2002a)
ozonated sludge	Readily biodegradable fraction of lysate = 0.088±0.013		Park et al. (2004a)
	Slowly biodegradable fraction of lysate = 0.036±0.004		

The values indicated in Table 13.3 show that the ozonated sludge could be an additional carbon source giving a specific denitrification rate (0.05 gNO$_3$-N gVSS^{-1} · d^{-1} on average) slightly lower than raw organic matter in municipal wastewater.

13.9.8 Influence on sludge settleability

Sludge disintegration by ozonation creates smaller flocs, fine and non-settleable particles as a consequence of deflocculation. In general, ozonation tends to increase sludge settleability: in fact, sludge settles much more easily, resulting in a much more compact sludge blanket compared to the untreated sludge (Paul and Debellefontaine, 2007; Weemaes et al., 2000). This can be explained by the changes in the water content of sludge, which decreases with the increase of ozone dosage (Zhao et al., 2007). Conversely, the resulting supernatant could become more turbid due to the presence of some non-settleable particles (Weemaes et al., 2000).

After ozonation, the recirculation of ozonated sludge in the biological reactors leads to re-flocculation of small particles and tends to enhance sludge settleability (Battimelli et al., 2003; Weemaes et al., 2000), improving SVI (Böhler and Siegrist, 2004; Egemen Richardson et al., 2009).

Some experiences show that ozonation has a direct influence on the SVI (Déléris et al., 2002; Liu et al., 2001) even in the case of bulking. The improvements in sludge settling characteristics can be related to two phenomena: (1) the effect of ozone on filamentous bacteria growth, because a significant reduction in filamentous bacteria is observed; (2) the effects of ozone on sludge water content, because sludge disintegration changes water distribution in biological flocs (Déléris et al., 2002).

Similar findings were found by Chiavola *et al.* (2007), who observed that ozonation at 0.05–0.07 $gO_3/gTSS_{treated}$, was able to improve sludge settleability in a lab-scale SBR system, by reducing abundance of filamentous bacteria within microbial population.

Ried *et al.* (2002) observed, in full-scale studies, a significant SVI decrease at ozone dosage of 0.08 $gO_3/gTSS_{produced}$, while the SVI descrease was less significant at very low ozone dosage around 0.003 $gO_3/gTSS$.

On the other hand, an excessive disintegration of flocs at high ozone dosages causes a large amount of non-settleable micro-particles (up to 13.8% of TSS at a dosage of 0.1 $gO_3/gTSS$), which remains in the supernatant and may have also an adverse effect on the filterability and dewaterability of sludge (Ahn *et al.*, 2002a; Liu *et al.*, 2001).

The maximum value of micro-particles and turbidity in supernatant was observed at a dosage of 0.2 gO_3/gTS, but they sharpy decreased at higher ozone dosages, due to subsequent mineralisation of disintegrated solids to carbon dioxide (Park *et al.*, 2003; Ahn *et al.*, 2002a).

13.10 INTEGRATION OF OZONATION IN THE SLUDGE HANDLING UNITS

Scientific interest in ozonation integrated into the sludge handling units has increased in recent years, mostly in anaerobic digestors but to a certain extent also in aerobic digestors (Figure 13.20).

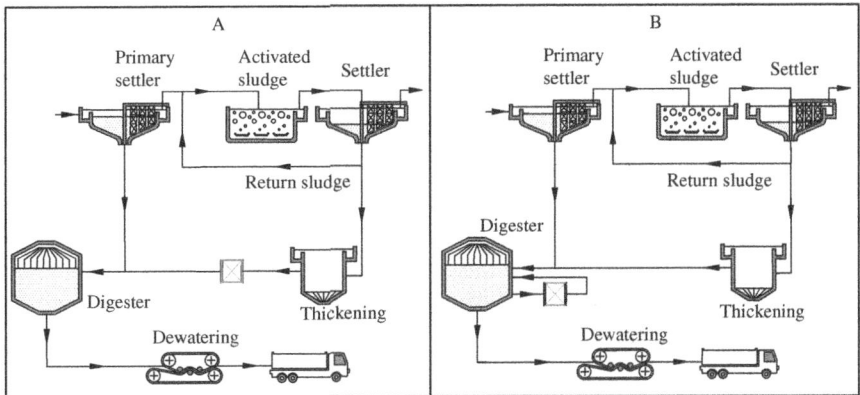

Figure 13.20. Alternatives for the integration of ozonation in the sludge handling units. The symbol ⊠ indicates the ozonation equipment.

The enhancement of sludge biodegradability by ozonation has been estimated in anaerobic and aerobic experiments by several authors (*inter alia* Yeom et al., 2002). Measuring oxygen consumption by ozonated sludge, Yeom *et al.* (2002) estimated a biodegradation after 15 d of 36%, 45.4%, 63%, 77.1% at ozone dosages of 0, 0.02, 0.05, and 0.1 $gO_3/gTSS$ respectively. Most of the biodegradation of the ozonated sludge occurred within 5 d, indicating the conversion of sludge into readily degradable substrate by ozone treatment. The biodegradation rate levelled off at relatively high ozone dosages (0.1 $gO_3/gTSS$) as observed by Yeom *et al.* (2002) both in aerobic and anaerobic experiments. The reason is because ozone is used to oxidise the solubilised compounds rather than to solubilise the remaining particulate solids.

The process concept for the integration of ozonation in the sludge handling units is similar to that employed for the reduction of excess sludge production in activated sludge systems described in § 13.9.

The main configurations for coupling anaerobic digestion and ozonation are indicated in Figure 13.20:

- *pre-ozonation*: the sludge undergoes ozonation before entering the digester (Figure 13.20A); this configuration is aimed at enhancing the biodegradability of raw sludge;
- *post-ozonation*: part of the digested sludge undergoes ozonation and is then recirculated into the anaerobic digester together with new sludge (Figure 13.20B);
- *combination of pre- and post-ozonation*: combination of the above.

13.10.1 Ozonation + anaerobic digestion

With regard to the combination of ozonation with anaerobic mesophilic digestion, ozone treatment was found to be effective in partially solubilizing the sludge solids and leading to subsequent improvements in anaerobic degradability and biogas production (*inter alia* Battimelli *et al.*, 2003; Goel *et al.*, 2003a; Goel *et al.*, 2004; Yasui *et al.*, 2005, Weemaes *et al.*, 2000). Since in anaerobic digestion long retention times, of the order of 20–30 days, are generally required to reach only moderate solid reduction (around 30–50%), ozonation of sludge can be used advantageously.

Anaerobic digestion with ozonation has been studied using lab batch tests, lab continuous experiments with different process configurations fed with either synthetic or real wastewater, pilot plants and at full-scale. As expected, the extent of solubilisation and digestion efficiency depend on the ozone dosages applied.

With regard to lab batch tests, Weemaes et al. (2000) reported an increase of methane production per unit of COD fed by a factor of 1.5 to 1.8 at ozone dosages of 0.04 and 0.08 gO_3/gTS respectively, compared to the untreated sludge. The rates of methane production were enhanced by a factor 1.7 at a dosage of 0.04 gO_3/gTS and 2.2 at 0.08 gO_3/gTS. Thus, the biodegradability of the sludge was significantly enhanced by ozonation. At higher ozone dosages a lag phase in methane production was observed, caused by the high redox potential produced by the oxidative process (Weemaes et al., 2000). The removal of organic matter was 36% for the untreated sludge, 58% for the sludge pre-treated at an ozone dosage of 0.04 gO_3/gTS, 68% at 0.08 gO_3/gTS and 62% at 0.16 gO_3/gTS (Weemaes et al., 2000).

In lab batch experiments simulating anaerobic digestion (seed:treated = 1:1), Yeom et al. (2002) observed that methane production increased by about two times with ozone dosages up to 0.2 gO_3/gTSS, while a further increase to 0.5 gO_3/gTSS did not enhance the biodegradation.

In lab batch tests, Bougrier et al. (2007a) recently reported that ozone pre-treatment led to:

– an acceleration of biogas production: at an ozone dosage of 0.15 gO_3/gTS the same quantity of biogas is produced in only 2 d, instead of the 18 d required by the untreated sludge;
– an increase in the quantity of biogas produced: for the lower ozone dosages (0.015–0.04 gO_3/gTS) a 23% increase in biogas production was observed. For higher ozone dosages (0.06 gO_3/gTS–0.15 gO_3/gTS), the biogas production increases in line with the ozone dosage, up to a factor of 2.4 in comparison with untreated sludge.

A comparison of some of these studies monitoring the increase of methane or biogas production in anaerobic digestion is shown in Figure 13.21.

Pre-ozonation was investigated by Goel et al. (2003a) in lab continuous experiments. A solubilisation of 19% and 37% of solids at ozone dosages of 0.015 and 0.05 gO_3/gTS respectively was obtained on synthetic sludge, composed of fructose and yeast extracts (Goel et al., 2003a). At 0.05 gO_3/gTS, the solid reduction efficiency in anaerobic digestion increased to about 59% as compared to 31% for the control run. At a lower ozone dosage of 0.015 gO_3/gTS, the solid reduction efficiency showed only slight improvements with respect to control. No significant inhibitory effects were observed in the pre-ozonation configuration.

Investigations at pilot-scale on municipal sewage sludge showed that an accumulation of a high fraction of inorganic solids can occur when treating real sludge. In fact, even though the organic removal efficiencies were similar both for synthetic and real sludge (Goel et al., 2003b; Goel et al., 2004), the concentration of accumulated solids was different, because synthetic wastewater usually contains low levels of inerts (Saktaywin et al., 2006).

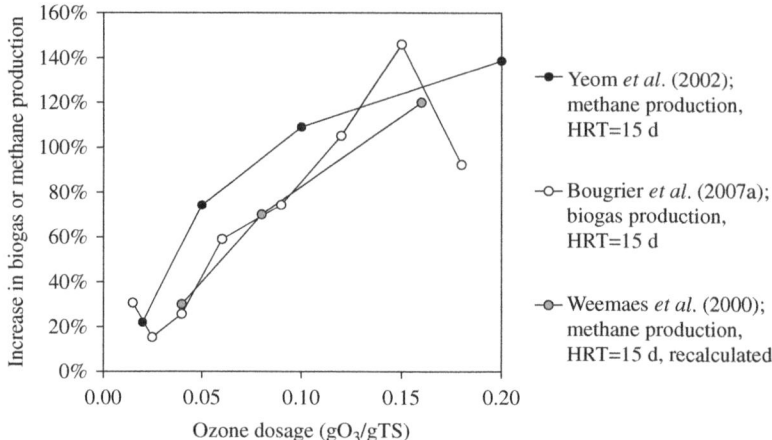

Figure 13.21. Increase of methane or biogas production in ozonation + anaerobic digestion with respect to a control without ozonation (HRT is the retention time in anaerobic digestion).

Post-ozonation was investigated at pilot scale treating municipal sewage sludge by Goel et al. (2004). At dosages of 0.055 gO$_3$/gTS, a recycled flow rate of 3.5 times the influent sludge and a very long SRT (275 d for the ozonated sludge compared to 58–144 d of the control reactors), the process resulted in TS removal efficiency of 61% (with respect to 39–43% of the controls).

Full-scale pre-ozonation + anaerobic digestion has been carried out at Schermbeck WWTP (17,000 PE), in parallel with a control line without ozonation (Sievers et al., 2004; Winter et al., 2002). The ozonation plant used consists of an ozone generator with an ozone capacity of 1 kgO$_3$/h, an ozone contactor of 6 m^3 with a feed sludge flow rate of 25 m^3/d. At ozone dosages of about 0.05 kg O$_3$/kgTSS$_{treated}$, the VSS reduction passed from 46.6% to 55.5% at SRT of 24 days.

The main results of ozonation + anaerobic digestion, described above, are summarised in Table 13.4.

As the sludge reduction process by ozonation needs additional energy input, the economic benefits of this technology depend mainly on the balance between the cost of this additional energy input and the resulting saving in biogas exploitation and in sludge disposal costs.

Table 13.4. Performance of ozonation + anaerobic digestion for sludge reduction.

Scale	Type of wastewater*	Type of test	O_3 dosage (gO_3/gTS)	Increase in methane production	Sludge reduction (TS)	References
lab	sw	batch test	0.04–0.08	+50–80%	58–68% (36% for the untreated sludge)	Weemaes et al. (2000)
lab	rw	batch test (feed:sludge = 1:1)	0.05	+60%	–	Yeom et al. (2002)
lab	rw	batch test (feed:sludge = 1:1)	0.2	+120%	–	Yeom et al. (2002)
lab	sw	continuous experiment (pre-ozonation)	0.015	–	slight	Goel et al., 2003a
lab	rw	continuous experiment (post-ozonation); r = 0.25**	0.16	–	+35%	Battimelli et al., 2003
pilot	rw	post-ozonation; r = 3.5**	0.055	–	61% (39–43% for the untreated sludge)	Goel et al. (2004)
full	rw	pre-ozonation	0.05	–	55.5% (46.6% for the untreated sludge)	Sievers et al. (2004)
lab	rw	batch test	0.015–0.04	+23% (biogas)	–	Bougrier et al., 2007a
lab	rw	batch test	0.15	+144% (biogas)	–	Bougrier et al., 2007a
			0.05	–	59% (31% for the untreated sludge)	

*sw = synthetic wastewater; rw = real wastewater.
**r = recirculated flow rate/inlet flow rate.

Goel *et al.* (2004), considering a post-ozonation process integrated with anaerobic digestion, estimated the theoretical minimum value of an economic index, defined as the ratio of sludge disposal cost (U_s, €/kg wet sludge) to energy cost (U_e, €/kWh). The obtained U_s/U_e was 0.44 kWh/kg wet sludge, which can be used to make a quick identification of the areas where the ozonation process could be economically applied. The authors concluded that the process could be applicable to situations where the actual value of this ratio is higher than the calculated minimum. In other words, for situations where the real cost of sludge disposal is higher than the energy cost.

Goel *et al.* (2004) reported an example of the economy index for Japan, which was approximately 1.5 kWh/kg wet sludge (unit sludge disposal cost for landfill of 0.15 €/kg wet sludge and unit energy cost of 0.10 €/kWh). These values do not differ significantly from the typical costs reported for some European countries in recent years.

Greater energy recovery with power generation is expected if ozonation + anaerobic digestion is applied in a larger plant with low-loaded anaerobic digestion tanks, since, in general, a bigger power production unit and ozone generators have better specific energy consumption efficiency (Yasui *et al.*, 2005).

The combination of ozonation + anaerobic digestion was also investigated in the removal of polycyclic aromatic hydrocarbons (PAH), which may accumulate in sludge during wastewater treatment because of their low biodegradability and their hydrophobic characteristics (Bernal-Martinez *et al.*, 2007). Ozonation pre-treatment of sludge allowed the biodegradability or bioavailability of PAH to be increased and an enhancement of PAH removal during anaerobic digestion was observed.

13.10.2 Ozonation + aerobic digestion

The integration of ozonation in aerobic digestion was investigated by Caffaz *et al.* (2005) at full-scale (12,000 PE). The configuration is similar to option B of Figure 13.20, in which a part of the digested sludge is fed to the ozonation contactor (V = 3.7 m^3) and then the ozonated sludge returns to the aerobic digester (V = 400 m^3). After the introduction of ozonation, the TSS reduction increased up to 89% by applying an ozone dosage of around 0.010 gO$_3$/gTSS$_{treated}$, compared to a reduction of 30% without ozonation. Other effects of ozonation integrated in aerobic digester were (Caffaz *et al.*, 2005):

- improvement of sludge settleability, which allowed to guarantee settlement even at very high solid concentration in the aerobic digester (15–20 g/L);

- reduction of filamentous microorganisms in digestion, although their presence remained in the activated sludge stage;
- improvement of denitrification process in the digester, resulting in a lower nitrogen load recirculated;
- increase of P in the surnatant.

13.10.3 Influence on sludge dewaterability

The COD solubilised during ozonation has negative effects on sludge filtration due to the surface charge and the consequent need for additional cations to destabilise the sludge flocs. Additionally, the fraction of smaller solid particles, which increases with rising ozone dosages, leads to a more compact filtration layer with reduced permeation of liquids (Sievers and Schaefer, 2007).

Several researchers report that sludge filterability, quantified by capillary suction time (CST), is deteriorated by ozonation, and, in particular an increase in CST value results at increasing ozone dosages (Liu et al., 2001; Weemaes et al., 2000). The CST increased strongly from 33–44 to 596–716 for sludge ozonated at 0.04 gO_3/gTS and to 688–862 at 0.08 gO_3/gTS dosage. At very high ozone dosages of 0.16 gO_3/gTS the value of CST decreased to 309–373 (Weemaes et al., 2000).

With regards to the specific resistance to filtration (SRF), in general filtration is better when the SRF is lower; on the contrary, filterability worsens when SRF increases. Ahn et al. (2000b) measured SRF to quantify sludge filterability before and after ozonation, indicating that the SRF value rapidly increased for ozone dosages above 0.05 gO_3/gTS and decreased only at very high dosages over 0.5 gO_3/gTS (Park et al., 2003). This last result was due to the decrease in the micro-particle fraction and turbidity at these high ozone dosages (see 13.9.8).

The effect of ozone dosage on SRF was also evaluated by Battimelli et al. (2003) for post-ozonation. For this type of sludge the SRF values improved below 0.09 $gO_3/gTSS$, while filterability worsened at higher ozone doses, up to 0.18 $gO_3/gTSS$. Battimelli et al. (2003) proposed an optimum dosage of 0.09 $gO_3/gTSS$ to minimise the SRF value in digested sludge.

A significant increase in the polymer demand for dewatering is observed after ozonation of sludge at very high dosages up to 0.5 $gO_3/gTSS$ (Scheminski et al., 2000).

The drawback regarding the worsening in filtration can be overcome by coupling ozonation with a biological treatment, in which biodegradation of solubilised COD occurs and the negative effects on settlement and filtration are contained (Sievers and Schaefer, 2007). In fact, in the full scale experiments of

Ried et al. (2002) ozonated sludge was mixed with the return sludge, which led to an improvement of CST values. Weemaes et al. (2000) reported that dewaterability was significantly enhanced after anaerobic digestion, because the CST of ozonated sludge decreased from 716 to 86, from 862 to 115.2 and from 372 to 74, respectively, at dosages of 0.04, 0.08 and 0.16 gO_3/gTS. The dewaterability after ozonation + digestion was comparable with the dewaterability of untreated sludge (Weemaes et al., 2000).

When ozonation is coupled with anaerobic digestion, significantly lower production of dewatered sludge cake with lower water content was consistently obtained (Yasui et al., 2005). The amount of dewatered sludge cake was less than 30% of that produced from the untreated sludge and the water content of the cake was correspondingly reduced from 80% to 68%. This decrease in water content contributed to a significant reduction in volume of the dewatered sludge cake prior to disposal with corresponding cost reduction (Yasui et al., 2005).

Experimental observations show that sludge ozonation reduces sludge viscosity, probably due to the breakdown of floc structure and reduced interfloc resistance (Goel et al., 2004). But in the case of anaerobic digestion combined with post-ozonation, where an accumulation of inorganic solids in the digester is observed, an increase of TS concentration in the reactor and thus greater viscosity of the digested sludge is expected (Goel et al., 2004).

14
Comparison of performance of sludge reduction techniques

> *"The most important aspects to be considered in operation and maintenance of sludge management system are [...] – in one word, the sustainability of the entire system"*
> (Spinosa, 2007b)

14.1 INTRODUCTION

The performances of the various techniques for sludge reduction, described in the previous chapters, are compared in the following sections. When comparing performance, the following factors are considered in the literature to a greater or lesser extent:

(1) COD (or TSS, VSS) solubilisation (as defined in § 7.2);
(2) degree of disintegration (as defined in § 7.3);
(3) biodegradability (described in § 7.4, § 7.5, § 7.6);
(4) bacteria inactivation (described in § 7.7);

© 2010 IWA Publishing. *Sludge Reduction Technologies in Wastewater Treatment Plants.* By Paola Foladori, Gianni Andreottola and Giuliano Ziglio. ISBN: 9781789065305. Published by IWA Publishing, London, UK.

(5) reduction of sludge production (usually in terms of TSS or TS) or reduction of observed sludge yield, which can be evaluated when a control line (without the sludge reduction technique) is used for comparison (as defined in § 7.9 and § 7.10).

Not all these factors can be used for all the sludge reduction techniques, as schematically shown in Table 14.1.

Table 14.1. Parameters suitable for evaluating the performance of sludge reduction techniques.

See section	Technique for sludge reduction	COD (or TSS, VSS) solubilisation (see § 7.2)	Degree of disintegration, DD_{COD} (see § 7.3)	Biodegradability (see § 7.4 § 7.5, § 7.6)	Bacteria inactivation (see § 7.7)	Reduction of sludge production or of Y_{obs} (see § 7.9, § 7.10)
§ 5.1, 6.1	Enzymatic hydrolysis with added enzymes					☑
§ 5.2	Enzymatic hydrolysis by thermophilic bacteria	☑		☑	☑	☑
§ 5.3, 6.2	Mechanical disintegration	☑	☑	☑	☑	☑
§ 5.4, 6.3	Ultrasonic disintegration	☑	☑	☑	☑	☑
§ 5.5, 6.4	Thermal treatment	☑	☑	☑	☑	☑
§ 6.5	Microwave treatment	☑	☑	☑	☑	☑
§5.6, 6.6	Chemical and thermo-chemical hydrolysis	☑		☑	☑	☑
§ 5.7, 6.7	Ozonation	☑	☑	☑	☑	☑

(*continued*)

Table 14.1. *Continued*

See section	Technique for sludge reduction	COD (or TSS, VSS) solubilisation (see § 7.2)	Degree of disintegration, DD_{COD} (see § 7.3)	Biodegradability (see § 7.4 § 7.5, § 7.6)	Bacteria inactivation (see § 7.7)	Reduction of sludge production or of Y_{obs} (see § 7.9, § 7.10)
§ 5.8, 6.8	Oxidation with strong oxidants	☑	☑	☑	☑	☑
§ 5.9, 6.9	Electrical treatment	☑	☑	☑	☑	☑
§ 5.10	Addition of chemical metabolic uncouplers				☑	☑
§ 5.11	Side-stream anaerobic reactor				☑	☑
§ 5.12	Extended aeration process					☑
§ 5.13	Membrane bioreactors					☑
§ 5.14	Granular sludge					☑
§ 5.15, 6.16	Microbial predation					☑
§ 6.10	Aerobic digestion					☑
§ 6.11	Digestion with alternating aerobic/anoxic/ anaerobic conditions					☑
§ 6.12	Dual digestion	☑		☑	☑	☑
§ 6.13	Autothermal thermophilic aerobic digestion	☑		☑	☑	☑

(*continued*)

Table 14.1. *Continued*

See section	Technique for sludge reduction	COD (or TSS, VSS) solubilisation (see § 7.2)	Degree of disintegration, DD_{COD} (see § 7.3)	Biodegradability (see § 7.4, § 7.5, § 7.6)	Bacteria inactivation (see § 7.7)	Reduction of sludge production or of Y_{obs} (see § 7.9, § 7.10)
§ 6.14	Anaerobic digestion			☑		☑
§ 6.15	Thermophilic anaerobic digestion			☑	☑	☑
§ 6.17	Wet air oxidation	☑		☑		☑
§ 6.18	Supercritical water oxidation	☑		☑		☑

The COD solubilisation indicates the release of soluble compounds after the treatment and is suitable (and widely used) to evaluate the performance of almost all the treatments which are based on the mechanism of cell lysis-cryptic growth (see § 4.2): mechanical or ultrasonic disintegration, thermal treatment (including microwave), chemical or thermo-chemical treatment, oxidation with ozone or other strong oxidants or electrical treatment.

Instead of TSS (or VSS) solubilisation, the parameter TSS disintegration is often used, which takes into account both solid solubilisation + mineralisation, which may potentially occur during the treatment. However, only a few treatments are able to achieve an appreciable mineralisation – usually when a high stress level is applied – and ozonation is one of the treatments more responsible for appreciable TSS reduction due to mineralisation when high ozone dosages are applied.

The degree of disintegration (DD_{COD}) is another widely used parameter for evaluating mechanical treatment, ultrasonic disintegration, thermal treatment or ozonation, while it is not used for chemical or thermo-chemical treatments, because a strong chemical treatment (alkali total fusion) is used as reference in the calculation of the DD_{COD} value.

For biological treatments – which can be based on enzymatic hydrolysis, uncoupled metabolism, endogenous metabolism or microbial predation – the

COD solubilisation or DD_{COD} are not always suitable parameters, because biological treatments are not all able to generate an appreciable disintegration effect. However, in the case of enzymatic hydrolysis by thermophilic aerobic bacteria, VSS solubilisation can be considered, especially for evaluating the performance of the thermophilic aerobic reactor under microaerobic conditions.

In the following sections, COD solubilisation (§ 14.2) and DD_{COD} (§ 14.3) are compared for the various techniques, using specific energy (E_s) as a reference parameter.

Not much is referred to in the literature on enhanced sludge biodegradability after the application of sludge reduction techniques. The solubilisation of the inert fraction of sludge does not necessarily release biodegradable compounds, on the contrary, solubilised compounds may maintain their original non biodegradability. Biodegradability can be evaluated under aerobic conditions (respirometry, OUR measurements), anoxic conditions (denitrification rate, NUR measurements) or anaerobic conditions (biogas or methane production). Most applications in the literature are related to sludge biodegradability evaluated through anaerobic batch tests to quantify biogas (or methane) production.

Bacteria inactivation is a suitable factor for investigating the performance of sludge reduction techniques, but it is not so frequently used and thus will not be further considered in this chapter.

The reduction of sludge production (measured in terms of TS or TSS), or the reduction of the observed sludge yield (Y_{obs}), can only be correctly evaluated by comparing a technique integrated in the wastewater handling units or in sludge handing units with a control line. In fact, only by adopting two lines in parallel – one of which is used as control and the other is identical but integrated with the specific treatment for sludge reduction – can the exact entity of sludge mass reduction be quantified, and, obviously, this approach is suitable for all techniques (Table 14.1). This evaluation is more feasible at lab- or pilot-scale, while investigations in full-scale plants are less common.

Another important aspect for evaluating or choosing a sludge reduction technique is the side-effects generated after introduction of the technique, which may have potentially negative impacts in the wastewater handling units or sludge handling units. In fact, to obtain sludge reduction, phenomena such as floc disintegration, cell lysis, hydrolysis, solubilisation, etc... are induced, resulting in a potential worsening of settleability or dewaterability of sludge, or increases in N and P loads recirculated in the activated sludge stages. These aspects are presented and discussed in section § 14.5.

Finally some installation/operational aspects are compared in § 14.6. The various techniques differ largely in operational requirements, passing from some simple biological treatments which do not need significant additional equipment/ work, to other more complex techniques, with high investment and/or operational costs, or potentially affected by problems of wear, corrosion, erosion, or foaming and/or odours.

14.2 COMPARISON OF COD SOLUBILISATION

COD solubilisation is the immediate result of a disintegration treatment. In Figure 14.1, the COD solubilisation induced by the application of some techniques for sludge reduction is compared, using the specific energy applied per mass of solids treated (E_s, expressed as kJ/kgTS).

The data of the mechanical treatments shown in Figure 14.1, are the same as indicated in Figure 9.12, but converted from DD_{COD} into S_{COD}, taking into account the approximate conversion $S_{COD} \approx 0.5 \cdot DD_{COD}$ (see § 7.3.1).

For ozonation, a conversion factor of 20 kWh (20×3600 kJ) for the production of 1 kg of ozone was assumed (see § 13.4, pag. 260) to convert the ozone dosage in Figure 13.6 into E_s values.

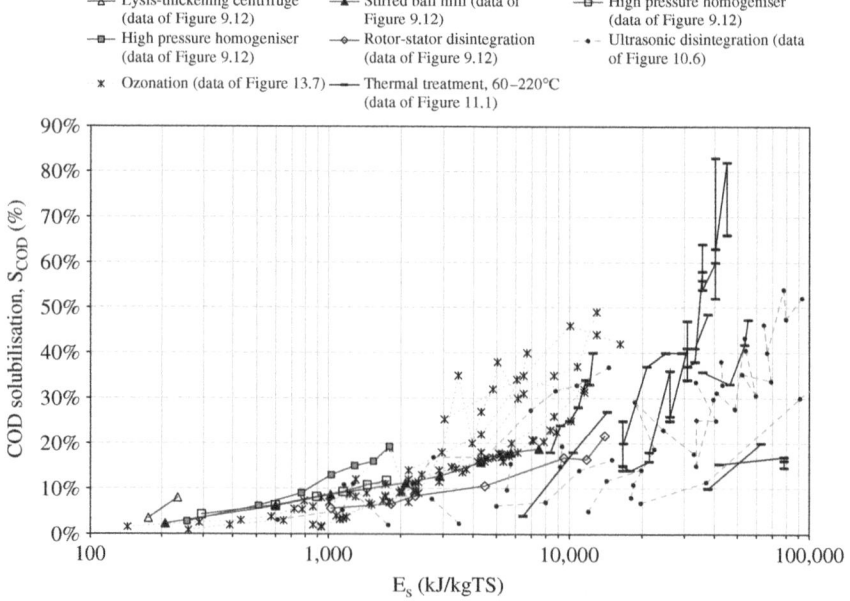

Figure 14.1. COD solubilisation as a function of E_s with the techniques applied. Conversions of data into E_s are explained in the text.

The E_s levels applied in thermal treatment were calculated from data reported in Figure 11.1. Heating energy expressed as E_s was calculated considering the specific heat capacity of liquid (4.18 kJ kg water^{-1} °C^{-1}), divided by the TS concentration of sludge (generally thickened) and multiplied by the increase of temperature from 20°C (environmental temperature) to the desired final temperature. The heat losses in the contact reactor during the treatment were not taken into account, due to the short contact time ranging from 1 min to 1.5 h.

Within the 20–30% band of COD solubilisation, a 1.5 log variation is observed on the E_s scale, considering all the techniques. The data are also scattered due to the influence of the solid content of sludge, which can differ in the various experiences in the literature (in the range 5–60 gTS/L). Solid content particularly influences mechanical, thermal and thermochemical treatments and the E_s required to obtain the same COD solubilisation generally increases as the solid content decreases. In the case of ozonation, E_s is similar for both concentrated and non concentrated sludge because ozone reacts only with solids.

The highest COD solubilisation is obtained in thermal treatment, especially when applied at high temperatures (>170°C).

Ozonation and ultrasonic disintegration generate high COD solubilisation, reaching values of around 50%, but ozonation requires less E_s to reach the same solubilisation level.

Thermal treatment requires high levels of E_s compared to other treatments to achieve the same solubilisation, but if thermal energy is available at the WWTP – for example, released during drying or from other thermal sources – the costs for the energy required can be reduced significantly with respect to other techniques using external electrical energy supplies.

The COD solubilisation level achieved by ozonation is not as high as thermal treatment, because the ozone transferred in the bulk liquid continues to react with the solubilised compounds, leading to their mineralisation (especially at high ozone dosages). In the case of ultrasonic disintegration, it is not economically viable to reach very high solubilisation levels, because this treatment is too energy hungry.

Among mechanical treatments, lysis-thickening centrifuge and high pressure homogenisation require the lowest E_s, but considerably lower COD solubilisation is reached, below 20%.

The application of biological treatments does not generally lead to a significant solubilisation yield and thus this category is not included in the comparison of COD solubilisation.

14.3 COMPARISON OF DEGREE OF DISINTEGRATION

The degree of disintegration (DD_{COD}) is applied especially for sludge reduction techniques which cause disintegration effects. Figure 14.2 shows data of DD_{COD} for some techniques available in literature as a function of E_s. Data are very scattered due also to the influence of the solid content of sludge (20–50 gTS/L).

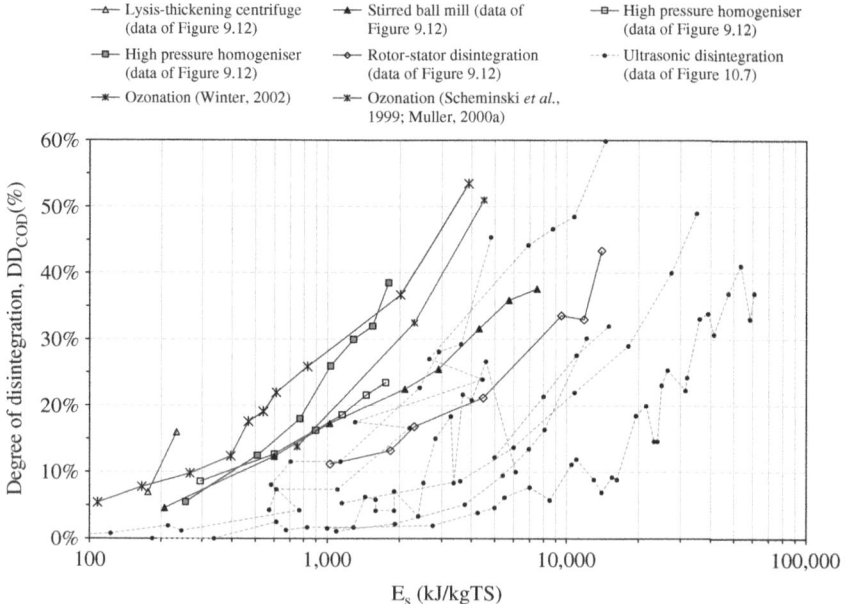

Figure 14.2. Disintegration degree (DD_{COD}) as a function of E_s with the techniques applied.

The lysis-thickening centrifuge achieves a DD_{COD} of only 15% (Müller, 2000a) but it is offset by the lowest energy consumption.

The other mechanical treatments (excluding the rotor-stator disintegration system) require levels of E_s similar to ozonation to obtain DD_{COD} of 15–20%. To reach higher values of DD_{COD} (20–40%), high pressure homogenisers and ozonation require the lowest levels of E_s. The rotor-stator disintegration system reaches the highest value of DD_{COD} among mechanical treatments, but it consumes much more E_s. Comparing all the treatments, only ozonation and ultrasonic disintegration obtain DD_{COD} above 50%. Data obtained from ultrasonic treatment differ among the various experiences, but the estimated average energy requirements are higher than for other treatments.

14.4 COMPARISON OF REDUCTION OF SLUDGE PRODUCTION

Table 14.2 summarises the entity of sludge reduction which can be obtained with the various techniques and the main parameters affecting the treatment. The data indicated present wide ranges, due to the variable operational conditions which can be applied: dosages of reagents, temperatures, contact times, stress frequency, etc ...

Furthermore, the entity of sludge reduction can also depend on:

- whether the wastewater which enters the activated sludge process is settled or not, due to the different level of inert particulate material in excess sludge and the presence or absence of primary sludge. This aspect, recently highlighted by Paul et al. (2006b), is not always taken into account in the various experiences in the literature;
- the solid content of treated sludge. Using the same level of E_s, the performance of ultrasonic disintegration, mechanical or thermal treatments depends on the solid concentration in the sludge.

In Table 14.2, the advisability of operating with a high solid content for the economic viability of the treatment is indicated for each technique. When a high solid content is advisable, a thickening unit should be inserted before the additional treatment, even when the sludge reduction technique is integrated in the wastewater handling units. However, ozonation is not affected by the solid content because ozone acts only against solids.

The data indicated in Table 14.2 should be considered carefully, because the ranges indicated originates from various experiences reported in the literature which are not always aimed to economically viability, but often aimed to evaluate the highest performance of the a technique. Furthermore results expressed as TSS, TS, VSS, VS, COD can not be correlated because originated from different experimental studies.

Among physico-mechanical treatments, sonication demonstrates the highest sludge reduction, but causes an excessive energy consumption.

Oxidative and thermal (or thermo-chemical) treatments demonstrate high performance in sludge reduction, and, in particular, ozonation can theoretically reach 100% of sludge reduction, which, however, is not achievable in practice for economic reasons.

Among biological treatments integrated in wastewater handling units (extended aeration process, MBR, granular sludge, microbial predation), MBR and granular sludge systems permit to reach 80–100% of sludge reduction but these systems can not be easily integrated in an existing WWTP, since their introduction implies the renovation of most of the biological units.

Table 14.2. Comparison of the entity of sludge reduction among the various techniques.

Technique for sludge reduction	See section	Main parameters affecting the technique	Sludge reduction induced by the technique (compared to a control)**	Advisability of high solid content***
Enzymatic hydrolysis with added enzymes	§ 5.1, 6.1	Dosage of products, type of enzymes, temperature, contact time	<16.5% (TSS)	No
Enzymatic hydrolysis by thermophilic bacteria	§ 5.2	Temperature, retention time, DO, presence of thermophilic bacteria	<75% (TS)	Yes
Mechanical disintegration	§ 5.3, 6.2	(all mechanical techniques) E_s		
Lysis-thickening centrifuge		Thickening efficiency, inlet sludge quality, wear level of rotating parts	3.5–21% (VS)	Yes
Stirred ball mills		Grinding time, rotation speed, sphere properties (size, material, filling ratio)	12–14% (VS)	No
High pressure homogenisers		Pressure, number of passages, type of valve, sludge temperature	20–24% (TSS)	Yes
High pressure jet and collision		Pressure, number of passages	5–13% (VS)	–

(continued)

Table 14.2. *Continued*

Technique for sludge reduction	See section	Main parameters affecting the technique	Sludge reduction induced by the technique (compared to a control)**	Advisability of high solid content***
Rotor-stator disintegration	§ 5.4, 6.3	Rotation speed, distance stator/rotor	–	Yes
Ultrasonic disintegration	§ 5.4, 6.3	E_s, ultrasound frequency, reactor configuration, transducer type	*(wastewater handling units)* 25–90% (TSS) *(sludge handling units)* 4.5–12.2% (VS)	Yes
Thermal treatment	§ 5.5, 6.4	Temperature, contact time, SF*	*(wastewater handling units)* <55% (TSS) *(sludge handling units)* 4–27% (TS), 20–45% (VS)	Yes
Microwave treatment	§ 6.5	Temperature, contact time	–	Yes
Chemical and thermo-chemical hydrolysis	§ 5.6, 6.6	Temperature, pH, contact time	*(wastewater handling units)* <37% (TSS)	Yes
Ozonation	§ 5.7, 6.7	Ozone dosage, SF	*(lab-scale)* <100% (TSS) *(full-scale, wastewater handling units)* 30–100% (TSS) *(full-scale, sludge handling units)* 8.9–32%, 59% (TSS)	No

(*continued*)

Table 14.2. Continued

Technique for sludge reduction	See section	Main parameters affecting the technique	Sludge reduction induced by the technique (compared to a control)**	Advisability of high solid content***
Oxidation with strong oxidants	§ 5.8, 6.8			
Peroxidation		H_2O_2 dosage, contact time, temperature, SF*	(wastewater handling units) <50% (TSS) (sludge handling units) 2.7–26.9% (VSS)	–
Chlorination		Chlorine dosage, contact time, SF*	(synthetic wastewater) <65% (TSS)	–
Electrical treatment	§ 5.9, 6.9	Voltage, treatment time, temperature, chamber and electrodes characteristics	27–45%	Yes
Addition of chemical metabolic uncouplers	§ 5.10	Type of uncouplers, dosage	16–86.9% (TSS)	No
Side-stream anaerobic reactor	§ 5.11	Volume, ORP, pH, HRT, SRT, interchange rate	(synthetic wastewater) <63% (TSS)	–

(continued)

Table 14.2. Continued

Technique for sludge reduction	See section	Main parameters affecting the technique	Sludge reduction induced by the technique (compared to a control)**	Advisability of high solid content***
Extended aeration process	§ 5.12	F/M ratio, SRT	<30%(TS)	No
Membrane bioreactors	§ 5.13	F/M ratio, SRT	<100% (TSS)	No
Granular sludge	§ 5.14	Organic load, biomass concentration	44-80% (TSS)	Yes
Microbial predation	§ 5.15, 5.16	Type of predators, area or volume of predation-reactors, DO, SRT	(industrial wastewater) <80% (TSS) (wastewater handling units) <10% (TSS) (sludge handling units) 16-46.4% (TSS)	No
Aerobic digestion	§ 6.10	HRT	<30% (VSS)	No
Digestion with alternating aerobic/anoxic/anaerobic conditions	§ 6.11	HRT, length of phases, cycle time	<10%	No

(continued)

Table 14.2. Continued

Technique for sludge reduction	See section	Main parameters affecting the technique	Sludge reduction induced by the technique (compared to a control)**	Advisability of high solid content***
Dual digestion	§ 6.12	HRT, temperature, DO	(in the 1st aerobic reactor) <40% (VSS)	Yes
Autothermal thermophilic aerobic digestion	§ 6.13	HRT, temperature, DO, ORP	50% (VS)	Yes
Anaerobic digestion	§ 6.14	HRT, temperature	30–50% (VSS)	Yes
Thermophilic anaerobic digestion	§ 6.15	HRT, temperature	40–45% (VSS), 30% (TS)	Yes
Wet air oxidation	§ 6.17	Temperature, pressure, oxygen, retention time, solid content	80% (COD), 70% (TS)	–
Supercritical water oxidation	§ 6.18	Temperature, pressure, oxygen, retention time, solid content	99% (COD), 80% (TS)	–

*SF is the treatment frequency described in § 7.11; E_s is specific energy usually expressed as kJ/kgTSS.
**Results expressed as TSS, TS, VSS, VS, COD can not be correlated because originated from different experimental studies.
***High solid content advisable for economic viability.

The side-stream anaerobic reactor integrated in wastewater handling units reach a high sludge reduction, but data available in the literature refers to lab-plants fed with synthetic wastewater with negligible inert solids.

Biological treatments integrated in the sludge handling units (aerobic and anaerobic digestion, dual digestion, autothermal digestion) do not reach VSS reduction higher than 50%. The limit of sludge reduction in biological treatments is mainly due to the difficulty in solubilising sludge and increasing the hydrolysis rate. However, biological treatments involve relatively low maintenance costs when compared to mechanical, thermal or thermo-chemical treatments, which are faster and more efficient but involve additional costs for mechanical energy, heating and/or chemical reagents (see Table 14.5).

Further improvement in sludge reduction to very high levels will only be possible, considering the sludge fractionation described in § 2.2.1, by transforming the inert (refractory) particulate organic matter of sludge into biodegradable compounds and only a high stress action can transform the original structure of inert particulate matter.

The highest sludge reduction is obtained, as expected, with the hydrothermal oxidation (wet air oxidation and supercritical water oxidation), which are considered as final disposal routes and thus not described in the book.

14.5 COMPARISON OF IMPACTS

The impacts originated by the introduction of a sludge reduction technique in a WWTP differ according to whether the technique is integrated in the wastewater handling units or in the sludge handling units.

When a technique is integrated in the wastewater handling units it generally causes an immediate increase in oxygen requirements, a modification of sludge settleability and in rare cases – especially when a high stress action is applied – even causing changes in the effluent quality regarding COD, N and P.

When a technique is integrated in the sludge handling units, the organic and nutrient loads recirculated in the wastewater handling units may increase, and sludge dewaterability may be positively or negatively affected.

The main impacts are summarised and compared in the following sections.

14.5.1 Impacts of techniques integrated in the wastewater handling units

When a sludge reduction technique is integrated in the wastewater handling units, certain impacts may occur, due to the existence of compounds originally contained in the reduced sludge and due to modifications of floc structure and changes in microbial populations (Table 14.3).

Some sludge reduction techniques cause an increase of soluble organic matter (COD, TOC) in the effluents (column 3 of Table 14.3); however, even in these cases the overall performance of the WWTPs and the quality of effluents discharged are not always compromised and total COD concentration is often still far below the limit for discharge. For example, ozonation is one of the treatments which causes a slight increase in effluent concentration of soluble COD and TOC, due to the release of inert soluble and colloidal COD, but this increase only becomes problematic if very high ozone dosages (uneconomic) are applied.

Because N and P are required for organic matter removal in biological reactors and are incorporated in the new biomass, during the application of a sludge reduction technique they are released again into the bulk liquid (column 4 of Table 14.3). So an unavoidable consequence of some techniques is thus an increase of P concentration in the effluent, or a higher N load to be nitrified/denitrified in the biological reactors to meet the required discharge limits. For example, when using ozonation at very high ozone dosages the effluent P may become a problem, and a further treatment for P removal may be required, also aimed at P recovery, which important considering the forecast P scarcity in the course of this century.

An increased oxygen requirement in the activated sludge process is expected for several sludge reduction techniques (column 5 of Table 14.3), due to the lower assimilation of organic carbon in the biomass and the conversion of an important part of it to carbon dioxide. Also less N is assimilated into the biomass, causing higher oxygen consumption for nitrification. In synthesis, the increase of oxygen demand (causing an increase in the aeration costs) is due to:

– oxidation of the COD of reduced sludge;
– nitrification of the TKN released by the reduced sludge.

Settleability is strongly affected by: (i) microbial population characteristics (presence of floc-forming or filamentous microorganisms), (ii) floc surface physico-chemical properties, (iii) exocellular polymers produced by microorganisms, (iv) cation concentration, involved in the formation of polymeric networks of sludge flocs. The application of a sludge reduction technique may thus affect settleability, acting on one or more of the above properties (column 6 of Table 14.3). Because activated sludge processes need good settleability in order to achieve the desired effluent quality, the introduction of stresses to an activated sludge system requires care to ensure that the quality of the effluent is not compromised. However, many sludge reduction techniques, especially some of those based on cell lysis-cryptic growth, are able to improve sludge settleability.

Table 14.3. Comparison of potential impacts originated from the integration of sludge reduction techniques in the wastewater handling units.

Technique for sludge reduction	See section	Increase of effluent COD, TOC	Increase of N, P loads	Significant increase of oxygen requirements	Worsened sludge settleability	Reduction of filamentous bacteria (bulking and foaming)
Enzymatic hydrolysis with added enzymes	§ 5.1	No	No	Potential	No (improved)	Potential
Enzymatic hydrolysis by thermophilic bacteria	§ 5.2	Potential	Yes	Yes	–	Yes
Mechanical disintegration	§ 5.3	Potential	Yes (nitrogen) No (phophorus)	Yes	Yes (at low E_s) Improved (for HPH*)	Yes
Ultrasonic disintegration	§ 5.4	Potential	Potential	Yes	No (improved)	Yes
Thermal treatment	§ 5.5	Yes	Yes	Yes	No (improved)	Yes
Chemical and thermo-chemical hydrolysis	§ 5.6	Yes	Yes	Yes	No (improved)	Yes
Ozonation	§ 5.7	Yes	Yes	Yes	No (improved)	Yes
Oxidation with strong oxidants	§ 5.8	Yes	Potential	–	Yes	Yes
Electrical treatment	§ 5.9	Potential	Potential	Potential	–	Potential
Addition of chemical metabolic uncouplers	§ 5.10	Yes	–	Yes	Yes	No
Side-stream anaerobic reactor	§ 5.11	No	Potential	Yes	Yes	–

(*continued*)

Table 14.3. Continued

Technique for sludge reduction	See section	Increase of effluent COD, TOC	Increase of N, P loads	Significant increase of oxygen requirements	Worsened sludge settleability	Reduction of filamentous bacteria (bulking and foaming)
Extended aeration process	§ 5.12	No	No	Yes	Yes (pin-point)	No
Membrane bioreactors	§ 5.13	No	No	Yes	–	No
Granular sludge	§ 5.14	No	No	–	–	–
Microbial predation	§ 5.15	No	Potential	Yes	No (potentially improved)	No

Note: Yes/No = confirmed in the literature
Potential = although not necessarily confirmed in the literature, reasonable theoretical probability. Also depending on the applied stress level.
*HPH = High pressure homogeniser

Furthermore, in the application of a sludge reduction technique a modification of the growth rates of the various microbial species and a competitive selection may occur, altering population composition. This may also lead to a positive effect in reducing the presence of filamentous bacteria which cause bulking and foaming (column 7 of Table 14.3).

14.5.2 Impacts of techniques integrated in the sludge handling units

The integration of a sludge reduction technique in the sludge handling units may affect (Table 14.4):

- sludge dewaterability, causing modifications in sludge filterability, solid content of filtered cake, increase/decrease of polymers needed for sludge conditioning before dewatering;
- organic and nutrient loads in the return flows recirculated into the wastewater handling units;
- pathogen inactivation, occurring under appropriate thermal conditions or using strong oxidants;
- odour formation.

It is well known that optimisation of sludge dewaterability depends on fundamental properties of sludge such as solid content, pH, size distribution of flocs, zeta-potential, EPS, etc. The various sludge reduction techniques do not have the same effect on sludge dewaterability, which depends on their action mechanism (column 3 of 14.4). For example, sonication, which decreases particle size, leads to a worsening of filterability, probably explained by the increase of the particle surface area and the bound water linked to the particles surface. Conversely, thermal and thermo-chemical treatments lead to the release of the initial bound water, with a consequent improvement in filterability and dewaterability.

Almost all the sludge reduction techniques, especially those based on thermal or chemical treatments or anaerobic digestion, may cause a significant release of COD, N and P in the water separated during dewatering and recirculated in the wastewater handling units (column 4 of 14.4), where it may constitute an additional overload to be treated.

With regard to pathogen inactivation, only sludge reduction techniques involving a thermal or a thermo-chemical treatment carried out at a suitable temperature, are able to contribute significantly (column 5 of 14.4).

Table 14.4. Comparison of potential impacts originated from the integration of sludge reduction techniques in the sludge handling units.

Technique for sludge reduction	See section	Worsened sludge dewaterability	Significant increase of COD, N, P loads in return flows	Pathogen inactivation	Odour formation
Enzymatic hydrolysis with added enzymes	§ 6.1	No (improved)	No	No	Potential
Mechanical disintegration	§ 6.2	No (improved) Yes (at low E_s)	Yes	No	No
Ultrasonic disintegration	§ 6.3	Yes (at high E_s)	Potential	No	No
Thermal treatment	§ 6.4	No (improved)	Yes	Yes	Yes
Microwave treatment	§ 6.5	No (potentially improved)	Potential	Yes	Potential
Chemical and thermo-chemical hydrolysis	§ 6.6	No (improved)	Yes	Yes	Yes
Ozonation	§ 6.7	Potential	Yes	No	No
Oxidation with strong oxidants	§ 6.8	No (improved)	Yes	Potential	No
Electrical treatment	§ 6.9	–	–	No	No
Aerobic digestion	§ 6.10	Yes	Yes	No	No

(*continued*)

Table 14.4. Continued

Technique for sludge reduction	See section	Worsened sludge dewaterability	Significant increase of COD, N, P loads in return flows	Pathogen inactivation	Odour formation
Digestion with alternating aerobic/anoxic/anaerobic conditions	§ 6.11	Potential	Potential	No	No
Dual digestion	§ 6.12	Potential	Yes	Yes	Potential
Autothermal thermophilic aerobic digestion	§ 6.13	Potential	Yes	Yes	Potential
Anaerobic digestion	§ 6.14	Yes	Yes	No	Yes
Thermophilic anaerobic digestion	§ 6.15	Potential	Yes	Yes	Yes
Microbial predation	§ 6.16	No (improved)	Yes	No	–
Wet air oxidation	§ 6.17	Yes	Yes	Yes	Potential
Supercritical water oxidation	§ 6.18	Yes	Yes	Yes	Potential

Note: Yes/No = confirmed in the literature
Potential = although not necessarily confirmed in the literature, reasonable theoretical probability. Also depending on the applied stress level.

Odour formation is related mainly to sludge reduction techniques based on thermal or thermo-chemical treatments or biological processes developed under anaerobic conditions for long retention times (column 6 of 14.4), while mechanical treatments or biological processes under aerobiosis at environmental temperatures do not generally cause severe odour problems.

14.6 COMPARISON OF INSTALLATION/OPERATIONAL ASPECTS

We can imagine that the perfect technique for sludge reduction should meet the following requirements:

- ensuring a high sludge mass reduction or permitting flexibility in selecting the desired level of sludge reduction;
- maintaining the desired treatment efficiency, guaranteeing the required effluent quality, especially in the case of nutrient removal;
- transferability to existing WWTPs and readily adaptable to existing equipment;
- requiring low/moderate investment costs;
- requiring low/moderate operational costs for reagents, energy, etc...;
- easy to manage, with limited use of highly skilled workers;
- simple, robust, durable, not subject to excessive or rapid deterioration of equipment;
- maintaining an adequate residual sludge quality, with the following features:
  - sludge quality in terms of settling and dewatering,
  - sanitary quality: very low pathogen and heavy metal content;
  - no accumulation of inert material in sludge, maintaining a high VSS/TSS ratio and good bacterial biomass content, which can guarantee good performance in subsequent biological reactors when treated sludge is recirculated;
- fully researched to ensure extensive knowledge and experience.

In Table 14.5 a synthesis of these installation/operational aspects is identified for each sludge reduction technique.

Comparison of performance of sludge reduction techniques

Table 14.5. Comparison of construction/operational aspects of sludge reduction techniques.

Technique for sludge reduction	See section	Flexibility to adapt to high/very high levels of sludge reduction (information from Table 14.2)	Maintain the desired treatment efficiency	Readily adaptable to existing WWTPs (compact system)	Low/moderate investment costs	Low/moderate operational costs	Ease of management	Deterioration of equipment	Process fully researched or full-scale applications
Enzymatic hydrolysis with added enzymes	§ 5.1, 6.1	No	Yes	Yes	Yes	No	Yes	No	No
Enzymatic hydrolysis by thermophilic bacteria	§ 5.2	Yes	Potential	No	No	No	Yes	No	Yes
Mechanical disintegration	§ 5.3, 6.2	No	Yes	Yes	No	Yes/No*	Potential	Yes	Yes
Ultrasonic disintegration	§ 5.4, 6.3	Yes	Yes	Yes	Yes	No	Yes	Yes	Yes
Thermal treatment	§ 5.5, 6.4	Yes	Yes	No	No	No	Potential	Yes	Yes
Microwave treatment	§ 6.5	Potential	–	No	No	No	–	–	No
Chemical and thermo-chemical hydrolysis	§ 5.6, 6.6	Potential	Yes	No	No	Yes/No**	No	Yes	Yes
Ozonation	§ 5.7, 6.7	Yes	Yes	Yes	No	No	Yes	No	Yes
Oxidation with strong oxidants	§ 5.8, 6.8	Yes	Yes (H_2O_2 + thermal) No (chlorination, THM)	Yes	Yes	No (H_2O_2 + thermal) Yes (chlorination)	Yes	Yes	No

(continued)

Table 14.5. Continued

Technique for sludge reduction	See section	Flexibility to adapt to high/very high levels of sludge reduction (information from Table 14.2)	Maintain the desired treatment efficiency	Readily adaptable to existing WWTPs (compact system)	Low/moderate investment costs	Low/moderate operational costs	Ease of management	Deterioration of equipment	Process fully researched or full-scale applications
Electrical treatment	§§ 5.9, 6.9	Potential	Potential	Yes	–	No	–	Yes	No
Addition of chemical metabolic uncouplers	§ 5.10	Yes	No	Yes	Yes	No	Yes	No	No
Side-stream anaerobic reactor	§ 5.11	Potential	Yes	No	No	Yes	Yes	No	Yes/No
Extended aeration process	§ 5.12	No	Yes	No	No	No	Yes	No	Yes
Membrane bioreactors	§ 5.13	Yes	Yes	No	No	No	Yes/No	No	Yes
Granular sludge	§ 5.14	Yes	Yes	No	No	Potential	No	No	Yes
Microbial predation	§§ 5.15, 6.16	Potential	Potential	No	No	Yes	Potential	No	No
Aerobic digestion	§ 6.10	No	Yes	Yes	Yes	No	Yes	No	Yes
Digestion with alternating aerobic/anoxic/anaerobic conditions	§ 6.11	No	Yes	No	Yes	Yes	Yes	No	Yes
Dual digestion	§ 6.12	No	Yes	No	No	Yes	No	No	Yes

(continued)

Table 14.5. Continued

Technique for sludge reduction	See section	Flexibility to adapt to high/very high levels of sludge reduction (information from Table 14.2)	Maintain the desired treatment efficiency	Readily adaptable to existing WWTPs (compact system)	Low/moderate investment costs	Low/moderate operational costs	Ease of management	Deterioration of equipment	Process fully researched or full-scale applications
Autothermal thermophilic aerobic digestion	§ 6.13	No	Yes	No	No	Yes	No	No	Yes
Anaerobic digestion	§ 6.14	No	Yes	No	No	Yes	No	No	Yes
Thermophilic anaerobic digestion	§ 6.15	No	Yes	No	No	Yes	No	No	Yes
Wet air oxidation	§ 6.17	Yes	Potential	No	No	No	No	Yes	Yes
Supercritical water oxidation	§ 6.18	Yes	Potential	No	No	No	No	Yes	Yes

Note: Yes/No = confirmed in the literature
Potential = although not necessarily confirmed in the literature, reasonable theoretical probability. Also depending on the applied stress level.
* depending on the mechanical disintegration device chosen: mills are relatively simple, while high pressure homogenisers are more complicated.
** chemical treatment (only alkaline treatment) without heating is relatively chep.

The flexible nature of a technique, capable of adaptation to high/very high levels of sludge reduction, allows the desired level of reduction to be selected by operators (for efficiency or economy) and affects the initial choices of equipment, installed power, foot print, etc... (column 3 of Table 14.5). This adaptability was evaluated on the basis of the results of sludge reduction indicated in Table 14.2. For example, ozonation allows theoretically high levels of sludge reduction between 0–100% to be achieved and this technique is very flexible and adaptable; the desired level of sludge reduction depends directly on the ozone dosage selected according to economic parameters.

Almost all the techniques guarantee the maintenance of the desired treatment efficiency in WWTPs in terms of the effluent quality to meet discharge limits (column 4 of Table 14.5). Some problems may occur in the maintenance of a stable nitrification process when the technique is integrated in the wastewater handling units. However, even the sludge reduction techniques based on the mechanisms of cell lysis-cryptic growth, which may act against bacterial cells (including nitrifiers) do not compromise the biological process at the stress level of treatments generally suitable for economic viability.

With regard to installation/operational aspects (columns 5–9 of Table 14.5), the proposed techniques are very different, passing from very easily installed treatments which are expensive to operate (enzyme addition) to expensive to install treatments which are cheaper to operate (side-stream anaerobic reactors). In fact, among biological processes, the use of a side-stream anaerobic reactor has very low operational costs, but the drawbacks are the initial investment costs (additional tank and some other equipment proposed in the processes commercialised).

Equipment for mechanical disintegration is often considered to have high investment costs, while only ultrasonic disintegration is very energy intensive.

In thermal and thermo-chemical treatments, the need to reach high temperatures and high pressures, leads to the need for equipment to raise temperatures and pressures and expensive construction materials to prevent corrosion problems. Operational costs are also high due to the need for sludge heating, when a thermal source is not available within the WWTP.

Both operational and investment costs are high in sludge ozonation due to the need for on-site ozone production (ozone generator and destroyer, requirement for liquid oxygen, energy for pumping, maintenance, etc.).

Some indications about the level of research/knowledge or the availability of full-scale applications, as far as we know, are indicated in column 10 of Table 14.5. At present, several techniques are already successfully applied in practice – such as ozonation, ultrasonic treatment, thermal and thermo-chemical treatments, some biological, etc... –, but other treatments, especially some biological processes based on microbial predation by worms (which seems to be promising and environmental friendly) are still at the lab or pilot scale stage.

It is evident that the perfect technique for sludge reduction – which satisfies all the desired requirements – does not exist yet, and in practice it is important to consider that the choice of the technique is case specific, requiring a detailed techno-economic evaluation of the candidate technologies for the specific WWTP. Furthermore, alongside the benefits of the technology chosen, some drawbacks may always remain.

Therefore a general recommendation for one sludge reduction technique or another cannot be given, and cost effectiveness is not achieved with every application and has to be carefully assessed.

Nowadays, the application of the best-performing technique is of great interest, especially for WWTPs located in critical areas where sludge disposal costs are and will become very high.

In conclusion, a perfect system, valid in all areas, does not yet exist, however ... *a perfect system does not stimulate new ideas.*

15
Nomenclature

ASM1	Activated Sludge Model No.1
ATAD	autothermal thermophilic aerobic digestion
ATP	adenosine triphosphate
b	decay rate
$b_{H,T}$	decay rate of heterotrophic bacteria at the operative temperature
$b_{N,T}$	decay rate of nitrifying bacteria at the operative temperature
BOD_5	Biochemical Oxygen Demand
cm	centimeter
cm^2	square centimeter
CNP	4-chloro-2-nitrophenol
C_{O3}	ozone concentration
COD	chemical oxygen demand
COD_{nbp}	particulate non-biodegradable COD

COD_b	biodegradable COD
CST	capillary suction time
°C	degree Celsius
ρ	density of water
d	day
DCP	2,4-dichlorophenol
DD_{COD}	degree of disintegration as COD solubilisation
DD_{O2}	degree of isintegration as oxygen consumption
DNP	2,4-dinitrophenol
DO	dissolved oxygen
D_{O3}	ozone dosage
DOC	dissolved organic carbon
DS	dry solids
E	enhancement factor
EDTA	ethylenediaminetetraacetic acid
EPS	extracellular polymeric substances
E_s	specific energy
f	endogenous fraction
F/M	Food/Microorganisms
FCM	flow cytometry
f_{cv}	conversion factor of VSS into COD
fr	resonance frequency
g	gram
GHz	gigahertz
gww	gram wet weight
h	hour
HRT	hydraulic retention time
J	joule
K	Kelvin
K_a	acid constant
kg	kilogram
kHz	kilohertz
kJ	kilojoule
k_{La}	mass transfer coefficient

Nomenclature

kPa	kilopascal
kV	kilovolt
kW	kilowatt
kWh	kilowatt hour
L	liter
μm	micrometer
m	meter
m^3	cubic meter
MBBR	moving bed biofilm reactor
MBR	membrane biological reactor
mCP	meta-chlorophenol
meq	milliequivalent
mg	milligram
MHz	megahertz
min	minute
mL	milliliter
mm	millimeter
mM	millimolar
mNP	m-nitrophenol
M_{O3}	mass of ozone
mol	mole
MPa	megapascal
mV	millivolt
Nm^3	normal cubic metres
NUR	Nitrate Utilisation Rate
OC	oxygen consumption
oCP	ortho-chlorophenol
ORP	oxidation-reduction (red-ox) potential
OSA	Oxic-Settling-Anaerobic
OUR	Oxygen Uptake Rate
%	percent
P_0	pressure exerted on the liquid
PAH	polycyclic aromatic hydrocarbon
PCOD	particulate COD
pCP	pentachlorophenol

PE	population equivalent
PEF	pulsed electric field
FP	focused pulsed
PHA	polyhydroxy-alkanoate
pNP	para-nitrophenol
Q	flow rate
γ	ratio of the specific heat of gases
ω_r	resonance angular frequency
R_{AE}	reduction under aerobic conditions
R_{AE+AN}	reduction under aerobic + anoxic conditions
r_{O3}	actual flux of ozone
rpm	revolutions per minute
R_r	resonant bubble radius
RSP	reduction of sludge production
RSRR	recycled sludge reduction reactor
rw	real wastewater
s	second
S	soluble part of COD
SBR	sequencing batch reactor
S_{COD}	COD solubilisation
S_{COD}	soluble COD
SCWO	supercritical water oxidation
SF	stress frequency (or treatment frequency)
S_I	soluble non-biodegradable COD
SRF	specific resistance to filtration
SRT	sludge retention time or sludge age
S_S	soluble biodegradable COD
STPP	tripoliphosphate sodium
SVI	sludge volume index
sw	synthetic wastewater
δ	thickness of the liquid film
δ_E	effective film thickness
T	temperature
t	time
T_{AE}	aerobic period
T_{AN}	anoxic period

Tc	cycle time
TCP	2,4,5-trichlorophenol
TCS	3,3',4',5-tetrachlorosalicylanilide
THM	trihalomethanes
TOC	total organic carbon
TS	total solids
TSS	total suspended solids
U	uncoupler concentration
V	volume
V_a	volume of air supplied
VFA	volatile fatty acids
v_r	volume per unit of volume of the reactor
VS	volatile solids
VSS	volatile suspended solids
W	watt
w/w	weight per weight
WAO	wet air oxidation
ww	wastewater
WWTP	wastewater treatment plant
x	TSS concentration
X_0	initial biomass concentration
X_{BA}	nitrifying or autotrophic biomass
X_{BH}	heterotrophic biomass
X_I	inert particulate COD
X_P	endogenous COD
X_S	biodegradable particulate COD
x_s	TSS concentration
Y_{ATP}	biomass growth per gram of ATP consumed
Y_H	maximum growth yield of heterotrophic bacteria
$Y_{H,AE}$	aerobic growth yield of heterotrophic bacteria
$Y_{H,AX}$	anoxic growth yield of heterotrophic bacteria
Y_N	maximum growth yield of nitrifying bacteria
Y_{obs}	observed sludge yield

16
References

Ahn K.-H., Yeom I.-T., Park K.-Y., Maeng S.-K., Lee Y., Song K.-G., and Hwang J.-H. (2002a) Reduction of sludge by ozone treatment and production of carbon source for denitrification. *Water Science and Technology*, **46**(11–12), 121–125.

Ahn K.-H., Park K.Y., Maeng S.K., Hwang J.H., Lee J.W., Song K.-G., and Choi S. (2002b) Ozonation of wastewater sludge for reduction and recycling. *Water Science and Technology*, **46**(10), 71–77.

Albuquerque J.S., Domingos J.C., Sant'Anna Jr. G. L., and Dezotti M. (2008) Application of ozonation to reduce biological sludge production in an industrial wastewater treatment plant. *Water Science and Technology*, **58**(10), 1971–1975.

Al-Ghusain I., Hamoda M.F., and El-Ghany M.A. (2002) Nitrogen transformations during aerobic/anoxic sludge digestion. *Bioresource Technology*, **85**, 147–154.

Al-Ghusain I.A. and Hao O.J. (1995) Using pH as a real-time control parameter for wastewater treatment and sludge digestion processes. *J. Environ. Eng. Div., Am. Soc. Civ. Eng.*, **121**, 225–235.

Akerboom R., Lutz P., and Berger H. (1994) Folic acid reduces the use of secondary treatment additives in treating wastewater from paper recycling. *TAPPI*

© 2010 IWA Publishing. *Sludge Reduction Technologies in Wastewater Treatment Plants*. By Paola Foladori, Gianni Andreottola and Giuliano Ziglio. ISBN: 9781789065305. Published by IWA Publishing, London, UK.

International Environmental Conference Proceedings, Boston, vol. 2. pp. 941–946 (*referred to by* Mahmood and Elliott, 2006).

Anderson N.J., Dixon D.R., Harbour P.J., and Scales P.J. (2002) Complete characterisation of thermally treated sludges. *Water Science and Technology*, **46**(10), 51–54.

Andreottola G. and Foladori P. (2007) Chapter 3.2: Treatability evaluation. In: *Wastewater Quality Monitoring and Treatment*. Quevauviller, P. Thomas, O. and van der Beken A., (eds.). © 2007, John Wiley & Sons, Ltd.

Andreottola G., Foladori P., and Vian M. (2006) An integrated approach for the evaluation of bacteria damage and biodegradability increase in sonication of excess sludge. *Proceedings of the International IWA specialized Conference – Sustainable Sludge Management: State of the Art, Challenges and Perspectives*. Moscow, 29–31 May 2006.

Andreottola G., Foladori P., and Ziglio G. (2007) Experimental comparison of physico-chemical treatments for excess sludge reduction in wastewater treatment plants. *Proceedings of the 12th International Gothenburg Symposium on Chemical Treatment of Water and Wastewater*. Ljubljana, Slovenia, 20–23 May 2007. (Series: Chemical Water and Wastewater Treatment, N. IX. IWA Publishing, London, UK).

Appels L., Baeyens J., Degrève J., and Dewil R. (2008) Principles and potential of the anaerobic digestion of waste-activated sludge. *Progress in Energy and Combustion Science*, **34**, 755–781.

Ayol A., Filibeli A., Sir D., and Kuzyaka E. (2008) Aerobic and anaerobic bioprocessing of activated sludge: Floc disintegration by enzymes. *Journal of Environmental Science and Health, Part A*, **43**(13), 1528–1535.

Baier U. and Schmidheiny P. (1997) Enhances anaerobic degradation of mechanically disintegrated sludge. *Water Science and Technology*, **36**(11), 137–143.

Banu J.R., Uan D.K., and Yeom I.-T. (2009) Nutrient removal in an A2O-MBR reactor with sludge reduction. *Bioresource Technology*, **100**, 3820–3824.

Barjenbruch M., Hoffmann H., Kopplow O., and Tränckner J. (2000) Minimizing of foaming in digesters by pre-treatment of the surplus-sludge. *Water Science and Technology*, **42**(9), 235–241.

Barlindhaug J. and Ødegaard H. (1996) Thermal hydrolysis for the production of carbon source for denitrification. *Water Science and Technology*, **34**(1–2), 371–378.

Barrios J.A. (2007) Latin America and the Caribbean. In: *Wastewater Sludge: A Global Overview of the Current Status and Future Prospects*. Spinosa, L. (ed.). IWA Publishing Ltd, London, UK. pp. 19–22, ISBN: 1843 391422.

Battimelli A., Millet C., Delgenès J.P., and Moletta R. (2003) Anaerobic digestion of waste activated sludge combined with ozone post-treatment and recycling. *Water Science and Technology*, **48**(4), 61–68.

Battistoni P., Pezzoli S., Bolzonella D., and Pavan P. (2002) The AF–BNR–SCP process as a way to reduce global sludge production: Comparison with classical approaches on a full scale basis. *Water Science and Technology*, **46**(10), 89–96.

Bauerfeld K., Dockhorn T., and Dichtl N. (2008) Sludge treatment and reuse considering different climates and varying other conditions – Export-oriented research for developing and threshold countries. *Journal of Environmental Science and Health, Part A*, **43**(13), 1556–1561.

Benabdallah El-Hadj T., Dosta J., Márquez-Serrano R., and Mata-Álvarez J. (2007) Effect of ultrasound pre-treatment in mesophilic and thermophilic anaerobic digestion with emphasis on naphthalene and pyrene removal. *Water Research*, **41**, 87–94.

Bernal-Martinez A., Carrère H., Patureau D., and Delgenès J.-P. (2007) Ozone pretreatment as improver of PAH removal durino anaerobic digestion of urban sludge. *Chemosphere*, **68**, 1013–1019.

Bhatta C.P., Matsuda A., Kawasaki K., and Omori D. (2004) Minimisation of sludge production and stable operational condition of a submerged membrane activated sludge process. *Water Science and Technology*, **50**(9), 121–128.

Bisogni J.J. and Lawrence A.W. (1971) Relationship between biological solids retention time and settling characteristics of activated sludge. *Water Res.*, **5**(9), 753–763.

Böhler M. and Siegrist H. (2004) Partial ozonation of activated sludge to reduce excess sludge, improve denitrification and control scumming and bulking. *Water Science and Technology*, **49**(10), 41–49.

Böhler M. and Siegrist H. (2006) Potential of activated sludge disintegration. *Water Science and Technology*, **53**(12), 207–216.

Böhler M. and Siegrist H. (2007) Potential of activated sludge ozonation. *Water Sci. Technol.*, **55**(12), 181–187.

Boero V.J., Eckenfelder W.W. Jr., and Bowers A.R. (1991) Soluble microbial product formation in biological systems. *Water Science and Technology*, **23**(4–6), 1067–1076.

Bougrier C., Carrère H., and Delgenès J.P. (2005) Solubilisation of waste-activated sludge by ultrasonic treatment. *Chem. Eng. J.*, **106**, 163–169.

Bougrier C., Albasi C., Delgenès J.P., and Carrère H. (2006a) Effect of ultrasonic, thermal and ozone pre-treatments on water activated sludge solubilization and anaerobic biodegradability. *Chemical Engineering and Processing*, **45**, 711–718.

Bougrier C., Delgenès J.P., and Carrère H. (2006b) Combination of thermal treatments and anaerobic digestion to reduce sewage sludge quantity and improve biogas yield. *Process Safety and Environmental Protection*, **84**(B4), 280–284.

Bougrier C., Battimelli A., Delgenès J.P., and Carrère H. (2007a) Combined ozone pretreatment and anaerobic digestion for the reduction of biological sludge production in wastewater treatment. *Ozone: Science & Engineering*, **29**(3), 201–206.

Bougrier C., Delgenès J.P., and Carrère H. (2007b) Impacts of thermal pre-treatments on the semi-continuous anaerobic digestion of waste activated sludge. *Biochemical Engineering Journal*, **34**, 20–27.

Bougrier C., Delgenès J.P., and Carrère H. (2008) Effects of thermal treatments on five different waste activated sludge samples solubilisation, physical properties and anaerobic digestion. *Chemical Engineering Journal*, **139**, 236–244.

Braguglia C.M., Mininni G., Tomei M.C., and Rolle E. (2006) Effect of feed/inoculum ratio on anaerobic digestion of sonicated sludge. *Water Science and Technology*, **54**(5), 77–84.

Bruss J.H., Christensen J.R., and Rasmussen H. (1993) Anaerobic storage of activated sludge: Effects on conditioning and dewatering performance. *Water Science and Technology*, **28**, 350–357.

Buys B.R., Klapwijk A., Elissen H., and Rulkens W.H. (2008) Development of a test method to assess the sludge reduction potential of aquatic organisms in activated sludge. *Bioresource Technology*, **99**, 8360–8366.

Cacho Rivero J.A., Madhavan N., Suidan M.T., Ginestet P., and Audic J.-M. (2005) Oxidative and Thermo-oxidative co-treatment with anaerobic digestion of excess municipal sludge. *Water Science and Technology*, **52**(1–2), 237–244.

Caffaz S., Santianni D., Cerchiara M., Lubello C., and Stecchi R. (2005) Reduction of excess biological sludge with ozone: Experimental investigation in a full-scale plant. In: IOA 17th World Ozone Congress, Strasbourg, France.

Camacho P., Geaugey V., Ginestet P., and Paul E. (2002a) Feasibility study of mechanically disintegrated sludge and recycle in the activated-sludge process. *Water Science and Technology*, **46**(10), 97–104.

Camacho P., Déléris S., Geaugey V., Ginestet P., and Paul E. (2002b) A comparative study between mechanical, thermal and oxidative disintegration techniques of waste activated sludge. *Water Science and Technology*, **46**(10), 79–87.

Camacho P., Ginestet P., and Audic J.M. (2005) Understanding the mechanism of thermal disintegrating treatment in the reduction of sludge production. *Water Science and Technology*, **52**(10–11), 235–245.

Canales A., Pareilleux A., Rolls J.L., Goma G., and Huyard A. (1994) Decreased sludge production strategy for domestic wastewater treatment. *Water Science and Technology*, **30**, 96–106.

Carballa M., Omil F., Alder A.C., and Lema J.M. (2006) Comparison between the conventional anaerobic digestion of sewage sludge and its combination with a chemical or thermal pre-treatment concerning the removal of pharmaceuticals and personal care products. *Water Science and Technology*, **53**(8), 109–117.

Carrère H., Bougrier C., Castets D., and Delgenès J.P. (2008) Impact of initial biodegradability on sludge anaerobic digestion enhancement by thermal pre-treatment. *Journal of Environmental Science and Health Part A*, **43**, 1551–1555.

Cassini S.T., Andrade M.C.E., Abreu T.A., Keller R., and Gonçalves R.F. (2006) Alkaline and acid hydrolytic processes in aerobic and anaerobic sludges: Effect on total EPS and fractions. *Water Science and Technology*, **53**(8), 51–58.

Cesbron D., Déléris S., Debellefontaine H., Roustan M., and Paul E. (2003) Study of competition for ozone between soluble and particulate matter during activated sludge ozonation. *Trans IChemE, Part A*, **81**, 1165–1170.

Cetin S. and Erdincler A. (2004) The role of carbohydrate and protein parts of extracellular polymeric substances on the dewaterability of biological sludges. *Water Science and Technology*, **50**(9), 49–56.

Chang C.N., Ma Y.S., and Lo C.W. (2002) Application of oxidation – reduction potential as a controlling parameter in waste activated sludge hydrolysis. *Chemical Engineering Journal*, **90**, 273–281.

Chase H.A., Dennis J.S., Woodgate J. (2007) Scaling up of uncoupling route. In: *Comparative Evaluation of Sludge Reduction Routes*. Ginestet, P. (ed.). IWA Publishing, London, UK.

Chauzy J., Graja S., Gerardin F., Crétenot D., Patria L., and Fernandes P. (2005) Minimisation of excess sludge production in a WWTP by coupling thermal hydrolysis and rapid anaerobic digestion. *Water Science and Technology*, **52**(10–11), 255–263.

Chauzy J., Cretenot D., Bausseron A., and Déléris S. (2008) Anaerobic digestion enhanced by thermal hydrolysis: First reference BIOTHELYS® at Saumur, France. *Water Practice & Technology*, doi:10.2166/wpt.2008.004.

Chen G.H., Mo H.K., Saby S., Yip W.K., and Liu Y. (2000) Minimization of activated sludge production by chemically stimulated energy spilling. *Water Science and Technology*, **42**(12), 189–200.

Chen G.H., Yip W.K., Mo H.K., and Liu Y. (2001a) Effect of sludge fasting/feasting on growth of activated sludge cultures. *Water Research*, **35**(4), 1029–1037.

Chen G.H., Saby S., Djaer M., and Mo H.K. (2001b) New approaches to minimize excess sludge in activated sludge systems. *Water Science and Technology*, **44**(10), 203–208.

Chen G.H., Mo H.K., and Liu Y. (2002) Utilization of a metabolic uncoupler, 3,3′,4′,5-tetrachlorosalicylanilide (TCS) to reduce sludge growth in activated sludge culture. *Water Research*, **36**(8), 2077–83.

Chen G.H., An K.J., Saby S., Brois E., and Djafer M. (2003) Possible cause of excess reduction in an oxic-settling-anaerobic activated sludge process (OSA process). *Water Research*, **37**(16), 3855–3866.

Chen Y., Jiang S., Yuan H., Zhou Q., and Gu G. (2007) Hydrolysis and acidification of waste activated sludge at different pHs. *Water Research*, **41**, 683–689.

Chiavola A., Naso M., Rolle E., and Trombetta D. (2007) Effect of ozonation on sludge reduction in a SBR plant. *Water Science and Technology*, **56**(9), 157–165.

Chiu Y.C., Chang C.N., Lin J.G., and Huang S.J. (1997) Alkaline and ultrasonic pretreatment of sludge before anaerobic digestion. *Water Science and Technology*, **36**(11), 155–162.

Choi H.B., Hwang K.Y., and Shin E.B. (1997) Effects on anaerobic digestion of waste activated sludge pre-treatment. *Water Science and Technology*, **35**, 207–11.

Chu L.-B., Xing X.-H., Yu A.-F., Zhou Y.-N., Sun X.-L., and Jurcik B. (2007) Enhanced ozonation of simulated dyestuff wastewater by microbubbles. *Chemosphere*, **68**, 1854–1860.

Chu L.-B., Yan S.-T., Xing X.-H., Yu A.-F., Sun X.-L., and Jurcik B. (2008) Enhanced sludge solubilization by microbubble ozonation. *Chemosphere*, **72**, 205–212.

Chu L., Yan S., Xing X.-H., Sun X., and Jurcik B. (2009) Progress and perspectives of sludge ozonation as a powerful pretreatment method for minimization of excess sludge production. *Water Research*, **43**, 1811–1822.

Chudoba P., Chevalier J.J., Chang J., and Capdeville B. (1991) Effect of anaerobic stabilization of activated sludge on its production under batch conditions at various S_0/X_0 ratios. *Water Science and Technology*, **23**(4–6), 917–926.

Chudoba P., Chudoba J., and Capdeville B. (1992a) The aspect of energetic uncoupling of microbial growth in the activated sludge process: OSA system. *Water Science and Technology*, **26**(9–11), 2477–2480.

Chudoba B., Morel A., and Capdeville B. (1992b) The case of both energetic uncoupling and metabolic selection of microorganisms in the OSA activated sludge system. *Environ. Technol.*, **13**, 761–770.

Chudoba J. (1983) Quantitative estimation in COD units of refractory organic compounds produced by activated sludge microorganisms. *Water Reserach*, **19**, 37–43.

Cicek N., Macomber J., Davel J., Suidan M.T., Audic J., and Genestet P. (2001) Effect of solids retention time on the performance and biological characteristics of a membrane bioreactor. *Water Science and Technology*, **43**(11), 43–50.

Clark P.B. and Nujjoo I. (2000) Ultrasonic sludge pre-treatment for enhanced sludge digestion. *Water Environ. Man.*, **14**(1), 66–71.

Collignon A., Martin G., Martin N., and Laplanche A. (1993) Treatment of bulking by ozonation; mechanisms of ozone on microorganisms. *Trib. Eau.*, **46**(562), 46–57.

Collignon A., Martin G., Martin N., and Laplanche A. (1994) Bulking reduced with the use of Ozone. *Ozone Sci. Eng.*, **16**(5), 385–402.

Copp J.B. and Dold P.L. (1998) Comparing sludge production under aerobic and anoxic conditions. *Water Science and Technology*, **38**(1), 285–294.

Csikor Z.S., Miháltz P., Hanifa A., Kovács R., and Dahab M.F. (2002) Identification of factors contributing to degradation in autothermal thermophilic sludge digestion. *Water Science and Technology*, **46**(10), 131–138.

Cui R. and Jahng D. (2006) Enhanced methane production from anaerobic digestion of disintegrated and deproteinized excess sludge. *Biotechnology Letters*, **28**, 531–538.

Datta T., Liu Y., and Goel R. (2009) Evaluation of simultaneous nutrient removal and sludge reduction using laboratory scale sequencing batch reactors. *Chemosphere*, **76**, 697–705.

Déléris S., Geaugey V., Camacho P., Debellefontaine H., and Paul E. (2002) Minimization of sludge production in biological processes: An alternative solution for the problem of sludge disposal. *Water Science and Technology*, **46**(10), 63–70.

Déléris S., Paul E., Audic J.M., Roustan M., and Debellefontaine H. (2000) Effect of ozonation on activated sludge solubilization and mineralization. *Ozone Sci. Eng.*, **22**(5), 473–487.

Delgenès J.P., Penaud V., Torrijos M., and Moletta R. (2000) Investigations on the changes in anaerobic biodegradability and biotoxicity of an industrial microbial biomass induced by a thermochemical pre-treatment. *Water Science and Technology*, **41**(3), 137–144.

Dentel S.K. (2004) Contaminants in sludge: Implications for management policies and land application. *Water Science and Technology*, **49**(10), 21–29.

Dentel S. (2007) North America (USA and Canada). In: *Wastewater Sludge: A Global Overview of the Current Status and Future Prospects*. Spinosa, L. (ed.). IWA Publishing Ltd, London, UK, pp. 15–18. ISBN: 1843 391422.

Dey E.S., Szewczyk E., Wawrzynczyk J., and Norrlöw O. (2006) A novel approach for characterization of exopolymeric material in sewage sludge. *J. Residuals Sci. Technol.*, **3**(2), 97–103.

Dimock R. and Morgenroth E. (2006) The influence of particle size on microbial hydrolysis of protein particles in activated sludge. *Water Research*, **40**(10), 2064–2074.

Dixon D. and Anderson T. (2007) Australia. In: *Wastewater Sludge: A Global Overview of the Current Status and Future Prospects*. Spinosa, L. (ed.). IWA Publishing Ltd, London, UK, pp. 35–38. ISBN: 1843 391422.

Dogruel S., Sievers M., and Germirli-Babuna F. (2007) Effect of ozonation on biodegradability characteristics of surplus activated sludge. *Ozone: Science & Engineering*, **29**(3), 191–199.

Dohányos M., Zábranská J., and Jeníček P. (1997) Enhancement of sludge anaerobic digestion by using of a special thickening centrifuge. *Water Science and Technology*, **36**(11), 145–153.

Dohányos M., Zábranská J., Jeníček P., Štěpová J., Kutil V., and Horejš J. (2000) The intensification of sludge digestion by the disintegration of activated sludge and the thermal conditioning of digested sludge. *Water Science and Technology*, **42**(9), 57–64.

Dohányos M., Zábranská J., Kutil J., and Jeníček P. (2004) Improvement of anaerobic digestion of sludge. *Water Science and Technology*, **49**(10), 89–96.

Dytczak M.A., Londry K., Siegrist H., and Oleszkiewicz J.A. (2006) Extracellular polymers in partly ozonated return activated sludge: Impact on flocculation and dewaterability. *Water Science and Technology*, **54**(9), 155–164.

Dytczak M.A., Londry K.L., Siegrist H., and Oleszkiewicz J.A. (2007) Ozonation reduces sludge production and improves denitrification. *Water Research*, **41**, 543–550.

Egemen E., Corpening J. and Nirmalakhandan N. (2001) Evaluation of an ozonation system for reduced waste sludge generation. *Water Science and Technology*, **44**(2–3), 445–452.

Egemen E., Corpening J., Padilla J., Brennan R., and Nirmalakhandan (1999) Evaluation of ozonation and cryptic growth for biosolids management in wastewater treatment. *Water Science and Technology*, **39**(10–11), 155–158.

Egemen Richardson E., Edwards F., and Hernandez J. (2009) Ozonation in sequencing batch reactors for reduction of waste solids. *Water Environment Research*, **81**(5), 506–513.

Elissen H.J.H., Hendrickx T.L.G., Temmink H., and Buisman C.J.N. (2006) A new reactor concept for sludge reduction using aquatic worms. *Water Research*, **40**, 3713–3718.

Elissen H.J.H., Peeters E.T.H.M., Buys B.R., Klapwijk A., and Rulkens W. (2008) Population dynamics of free-swimming Annelida in four Dutch wastewater treatment plants in relation to process characteristics. *Hydrobiologia*, **605**, 131–142.

El-Din M.G. and Smith D.W. (2001) Designing ozone bubble columns: A spreadsheet approach to axial dispersion model. *Ozone: Science & Engineering*, **23**(5), 369–384.

Engelhart M., Krüger M., Kopp J., and Dichtl N. (2000) Effects of disintegration on anaerobic degradation of sewage excess sludge in downflow stationary fixed film digesters. *Water Science and Technology*, **41**(3), 171–179.

Eskicioglu C., Kennedy K.J., and Droste R.L. (2006) Characterization of soluble organic matter of waste activated sludge before and after thermal pre-treatment. *Water Research*, **40**, 3725–3736.

Eskicioglu C., Terzian N., Kennedy K.J., Droste R.L., and Hamoda M. (2007a) Athermal microwave effects for enhancing digestibility of waste activated sludge. *Water Research*, **41**, 2457–2466.

Eskicioglu C., Droste R.L., and Kennedy K.J. (2007b) Performance of anaerobic waste activated sludge digesters after microwave pre-treatment. *Water Environment Research*, **79**(11), 2265–2273.

Eskicioglu C., Kennedy K.J., and Droste R.L. (2007c) Enhancement of batch waste activated sludge digestion by microwave pre-treatment. *Water Environment Research*, **79**(11), 2304–2317.

Eskicioglu C., Kennedy K.J., and Droste R.L. (2008) Initial examination of microwave pretreatment on primary, secondary and mixed sludges before and after anaerobic digestion. *Water Sci. Technol.*, **57**(3), 311–317.

Feijoo G., Soto M., Mendez R., and Lema J.M. (1995) Sodium inhibition in the anaerobic digestion process. *Enzym. Microbiol. Technol.*, **17**, 180–188.

Foladori P., Bruni L., Andreottola G., and Ziglio G. (2007) Effects of sonication on bacteria viability in wastewater treatment plants evaluated by flow cytometry – Fecal indicators, wastewater and activated sludge. *Water Research*, **41**, 235–243.

Foladori P., Bruni L., and Ziglio G. (2004) Potenzialità della citometria a flusso nell'analisi di matrici ambientali. Quaderni del Dipartimento SAN 2, Department of Civil and Environmental Engineering, University of Trento. ISBN 88-8443-078-X. *In Italian.*

Frost and Sullivan (2003) Strategic analysis of the European Sludge Disposal Market, Report #3815-15.

Gaudy A.F., Yang P.Y., and Obayashi A.W. (1971) Studies on the total oxidation of activated sludge with and without hydrolytic pretreatment. *J. Water Pollut. Control Fed.*, **43**(1), 40-54.

Gavala H.N., Yenal U., Skiadas I.V., Westermann P., and Ahring B.K. (2003) Mesophilic, thermophilic anaerobic digestion of primary and secondary sludge. *Effect of pre-treatment at elevated temperature. Water Res.*, **37**, 4561-4572.

Geciova J., Bury D., and Jelen P. (2002) Methods for disruption of microbial cells for potential use in the dairy industry. *International Dairy Journal*, **12**, 541-553.

Ghyoot W. and Verstraete W. (1999) Reduced sludge production in a two-stage membrane-assisted bioreactor. *Water Research*, **34**, 205-215.

Gibson J.H., Hon H., Farnood R., Droppo I.G., and Seto P. (2009) Effects of ultrasound on suspended particles in municipal wastewater. *Water Research*, **43**, 2251-2259.

Ginestet P. (2007a) Comparative evaluation of sludge reduction routes. IWA Publishing Ltd, London, UK. ISBN: 1843391236.

Ginestet P. (2007b) Chapter: Economical evaluation (basic hypothesis). In: *Comparative Evaluation of Sludge Reduction Routes*. IWA Publishing Ltd, London, UK, pp. 20-31. ISBN: 1843391236.

Ginestet P. (2007c) Chapter: Economical comparison of the routes. In: *Comparative Evaluation of Sludge Reduction Routes*. IWA Publishing Ltd, London, UK, pp. 62-81. ISBN: 1843391236.

Ginestet P. and Camacho P. (2007) Technical evaluation of sludge production and reduction. In: *Comparative Evaluation of Sludge Reduction Routes*. IWA Publishing Ltd, London, UK, pp. 1-15. ISBN: 1843391236.

Goel R., Komatsu K., Yasui H., and Harada H. (2004) Process performance and change in sludge characteristics during anaerobic digestion of sewage sludge with ozonation. *Water Science and Technology*, **49**(10), 105-113.

Goel R., Tokutomi T., and Yasui H. (2003a) Anaerobic digestion of excess activated sludge with ozone pre-treatment. *Water Science and Technology*, **47**(12), 207-214.

Goel R., Yasui H., and Noike T. (2003b) Closed Loop anaerobic digestion using pre/post sludge ozonation and effect of low temperature on process performance. *Water Science and Technology*, **47**(12), 261-267.

Goel R., Tokutomi T., Yasui H., and Noike T. (2003c) Optimal process configuration for anaerobic digestion with ozonation. *Water Science and Technology*, **48**(4), 85-96.

Goel R.K. and Noguera D.R. (2006) Evaluation of sludge yield and phosphorus removal in a Cannibal solids reduction process. *Journal of Environmental Engineering*, **132**(10), 1331-1337.

Gómez J., De Gracia M., Ayesa E., and García-Heras J.L. (2007) Mathematical modelling of autothermal thermophilic aerobic digester. *Water Research*, **41**, 959-968.

Gonze E., Pillot S., Valette E., Gonthier Y., and Bernis A. (2003) Ultrasonic treatment of an aerobic activated sludge in a batch reactor. *Chem. Eng. Process.*, **42**, 965-975.

Graja S., Chauzy J., Fernandes P., Patria L., and Cretenot D. (2005) Reduction of sludge production from WWTP using thermal pretreatment and enhanced anaerobic methanisation. *Water Science and Technology*, **52**(1–2), 267–273.

Grönroos A., Kyllönen H., and Korpijärvi K. (2005) Ultrasound assisted method to increase soluble chemical oxygen demand (SCOD) of sewage sludge for digestion. *Ultrason. Sonochem.*, **12**, 115–120.

Guo X.-S., Liu J.-X., Wei Y.-S., and Li L. (2007) Sludge reduction with Tubificidae and the impact on the performance of the wastewater treatment process. *Journal of Environmental Sciences*, **19**, 257–263.

Hamer G. (1985) Lysis and cryptic growth in wastewater and sludge treatment processes. *Acta Biotech.*, **2**, 117–127.

Hao O.J. and Kim M.H. (1990) Continuous pre-anoxic and aerobic digestion of waste-activated sludge. *J. Environ. Eng. Div., Am. Soc. Civ. Eng.*, **116**(5), 863–879.

Hao O.J., Kim M.H., and Al-Ghusain I. (1991) Alternating aerobic and anoxic digestion of waste-activated sludge. *J. Chem. Technol. Biotechnol.*, **52**(4), 457–472.

Harrison D.E.F. and Loveless J.E. (1971) The effect of growth conditions on respiratory activity and growth efficiency in facultative anaerobes grown in chemostat culture. *J. Gen. Microbiol.*, **68**, 35–43.

Harrison D.E.F. and Maitra P.K. (1969) Control of respiration and metabolism in growing *Klebsiella aerogens*: the role of adenine nucleotides. *Biochem. J.*, **112**, 647–656.

Hasegawa S., Shiota N., Katsura K., and Akashi A. (2000) Solubilisation of organic sludge by thermophilic aerobic bacteria as a pretreatment for anaerobic digestion. *Water Science and Technology*, **41**(3), 163–169.

Hashimoto S., Fujita M., and Teral K. (1982) Stabilization of waste activated sludge through the anoxic–aerobic digestion process. *Biotechnol. Bioeng.*, **24**(8), 1335–1344.

Haug R.T. (1977) Sludge processing to optimize digestibility and energy production. *J. Water Pollut. Control Fed.*, **49**, 1713–1721.

Haug R.T., Stuckey D.C., Gossett J.M., and McCarty P.L. (1978) Effects of thermal pretreatment on digestibility an dewaterability of organic sludges. *J. Water Pollut. Control Fed.*, **50**, 73–85.

Haug R.T., LeBrun T.J., and Tortorici L.D. (1983) Thermal pre-treatment of sludges, a field demonstration. *J. Water Pollut. Control Fed.*, **55**, 23–34.

Hayakawa K., Ueno Y., Kawamura S., Kato T., and Hayashi R. (1998) Microorganisms inactivation using high-pressure generation in sealed vessels under sub-zero temperature. *Appl. Microbiol. Biotechnol.*, **50**, 415–418.

He S.-B., Xue G., and Wang B.-Z. (2006), Activated sludge ozonation to reduce sludge production in a membrane bioreactor (MBR). *J. Haz. Mat.*, B **135**, 406–411.

Heinz (2007) Scaling up of electrical route. In: *Comparative Evaluation of Sludge Reduction Routes*. Ginestet, P. (ed.). IWA Publishing, London, UK.

Hendrickx T.L.G., Temmink H., Elissen H.J.H., and Buisman C.J.N. (2009a) The effect of operating conditions on aquatic worms eating waste sludge. *Water Research*, **43**, 943–950.

Hendrickx T.L.G., Temmink H., Elissen H.J.H., and Buisman C.J.N. (2009b) Aquatic worms eating waste sludge in a continuous system. *Bioresource Technology*, **100**, 4642–4648.

Henze M. (1992) Characterization of wastewater for modeling of activated sludge processes. *Water Science and Technology*, **25**(6), 1–15.

Heo N.H., Park S.C., Lee J.S., and Kang H. (2003) Solubilization of waste activated sludge by alkaline pretreatment and biochemical methane potential (BMP) tests for anaerobic co-digestion of municipal organic waste. *Water Science and Technology*, **48**(8), 211–219.

Hogan F., Mormede S., Clark P., and Crane M. (2004) Ultrasonic sludge treatment for enhanced anaerobic digestion. *Water Science and Technology*, **50**(9), 25–32.

Huang X., Liang P., and Qian Y. (2007) Excess sludge reduction induced by Tubifex tubifex in a recycled sludge reactor. *Journal of Biotechnology*, **127**, 443–451.

Huysmans A., Weemaes M., Fonseca P.A., and Verstraete W. (2001) Short communication Ozonation of activated sludge in the recycle stream. *J. Chem. Technol Biotechnol.*, **76**, 321–324.

Hwang K.Y., Shin E.B., and Choi H.B. (1997) A mechanical pretreatment of waste activated sludge for improvement of anaerobic digestion system. *Water Science and Technology*, **36**(12), 111–116.

Jenicek P. (2007) Eastern Europe. In: *Wastewater Sludge: A Global Overview of the Current Status and Future Prospects*. Spinosa, L. (ed.). IWA Publishing Ltd, London, UK, pp. 9–13. ISBN: 1843 391422.

Jenkins C.J. and Mavinic D.S. (1989a) Anoxic/aerobic digestion of waste activated sludge: part I – solids reduction and digested sludge characteristics. *Environ. Technol. Lett.*, **10**, 355–370.

Jenkins C.J. and Mavinic D.S. (1989b) Anoxic/aerobic digestion of waste activated sludge: part II—supernatant characteristics, ORP monitoring results and overall rating system. *Environ. Technol. Lett.*, **10**, 355–370.

Jones R., Parker W., Zhu H., Houweling D., and Murthy S. (2009) Predicting the degradability of waste activated sludge. *Water Environment Research*, **81**(8), 765–771.

Johnson B.R., Daigger G.T., and Novak J.T. (2008) The use of ASM based models for the simulation of biological sludge reduction processes. *Water Practice & Technology*, doi:10.2166/wpt.2008.074.

Jung J., Xing X.U., and Matsumoto K. (2001) Kinetic analysis of disruption of excess activated sludge by Dyno Mill and characteristics of protein release for recovery of useful materials. *Biochemical Engineering Journal*, **8**, 1–7.

Jung S.J., Miyanaga K., Tanji Y., and Unno H. (2006) Effect of intermittent aeration on the decrease of biological sludge amount. *Biochemical Engineering Journal*, **27**(3), 246–251.

Kamiya T. and Hirotsuki J. (1998) New combined system of biological process and intermittent ozonation for advanced wastewater treatment. *Water Science and Technology*, **38**(8–9), 145–153.

Kampas P., Parsons S.A., Pearce P., Ledoux S., Vale P., Churchley J., and Cartmell E. (2007) Mechanical sludge disintegration for the production of carbon source for biological nutrient removal. *Water Research*, **41**, 1734–1742.

Karam J. and Nicell, J.A. (1997) Potential applications of enzymes in waste treatment. *J. Chem. Technol. Biotechnol.*, **69**(2), 141–153.

Kepp U., Machenbach I., Weisz N., and Solheim O.E. (2000) Enhanced stabilisation of sewage sludge through thermal hydrolysis – 3 years of experience with full scale plants. *Water Science and Technology*, **42**(9), 89–96.

Kim T.-H., Nam Y.-K., Park C., and Lee M. (2009) Carbon source recovery from waste activated sludge by alkaline hydrolysis and gamma-ray irradiation for biological denitrification. *Bioresource Technology*, **100**(23), 5694–5699.

Kim Y.-K., Bae J.-H., Oh B.-K., Lee W.H., and Choi J.W. (2002) Enhancement of proteolytic enzyme activity excreted from *Bacillus stearothermophilus* for a thermophilic aerobic digestion process. *Bioresource Technology*, **82**, 157–164.

Kristensen G.H., Jørgensen P.E., and Henze M. (1992) Characterization of functional microrganism groups and substrate in activated sludge and wastewater by AUR, NUR e OUR. *Water Science and Technology*, **25**(6), 43–57.

Komanapalli I.R. and Lau B.H.S. (1998) Inactivation of bacteriophage λ, Escherichia coli, and Candida albicans by ozone. *Appl. Microbiol. Biotechnol.*, **49**, 766–769.

Kopp J., Müller J., Dichtl N., and Schwedes J. (1997) Anaerobic digestion and dewatering characteristics of mechanically disintegrated excess sludge. *Water Science and Technology*, **14**(11), 129–136.

Kovács R., Miháltz P., and Csikor Z. (2007) Kinetics of autothermal thermophilic aerobic digestion – application and extension of Activated Sludge Model No. 1 at thermophilic temperatures. *Water Science and Technology*, **56**(9), 137–145.

Kujawa K. and Klapwijk B.A. (1999) Method to estimate denitrification potential for predenitrification systems using NUR batch test. *Water Research*, **33**(10), 2291.

Kuroda A., Takiguchi N., Gotanda T., Nomura K., Kato J., Ikeda T., and Ohtake H. (2002) A simple method to release polyphosphate from activated sludge for phosphorus reuse and recycling. *Biotechnol. Bioeng.*, **78**, 333–338.

la Cour Jansen J., Davidsson Å., Dey E.S., and Norrlöw (2004) Enzyme assisted sludge minimisation. In: *Chemical Water and Wastewater Treatment VIII*. Hahn H.H., Hoffmann E., and Ødegaard H. (eds.). IWA Publishing, London, UK, pp. 345–353. ISBN 1 84339 068 X.

Laera G., Pollice A., Saturno D., Giordano C., and Lopez A. (2005) Zero net growth in a membrane bioreactor with complete sludge retention. *Water Research*, **39**, 5241–5249.

Lagerkvist A. and Chen H. (1993) Control of two step anaerobic degradation of municipal solid waste (MSW) by enzyme addition. *Water Science and Technology*, **27**(2), 47–56.

Lang N.L. and Smith S.R. (2008) Time and temperature inactivation kinetics of enteric bacteria relevant to sewage sludge treatment processes for agricultural use. *Water Research*, **42**, 2229–2241.

Lapinski J., and Tunnacliffe A. (2003) Reduction of suspended biomass in municipal wastewater using bdelloid rotifers. *Water Research*, **37**, 2027–2034.

Lee D.J. (2007) South Asia and China. In: *Wastewater Sludge: A Global Overview of the Current Status and Future Prospects*. Spinosa L. (ed.). IWA Publishing Ltd, London, UK, pp. 27–29. ISBN: 1843 391422.

Lee N.M. and Welander T. (1996a) Reducing sludge production in aerobic wastewater treatment through manipulation of the ecosystem. *Water Research*, **30**(8), 1781–1790.

Lee N.M. and Welander T. (1996b) Use of protozoa and metazoa for decreasing sludge production in aerobic wastewater treatment. *Biotechnol. Lett.*, **18**(4), 429–434.

Lehne G., Müller A., and Schwedes J. (2001) Mechanical disintegration of sewage sludge. *Water Science and Technology*, **43**(1), 19–26.

Li Y-.Y. and Noike T. (1992) Upgrading of anaerobic digestion of waste activated sludge by thermal pretreatment. *Water Science and Technology*, **26**, 857–866.

Li H., Jin Y., Mahar RB., Wang Z., and Nie Y. (2008) Effects and model of alkaline waste activated sludge treatment. *Bioresource Technology*, **99**, 5140–5144.

Liang P., Huang X., and Qian Y. (2006a) Excess sludge reduction in activated sludge process through predation of Aeolosoma hemprichi. *Biochemical Engineering Journal*, **28**, 117–122.

Liang P., Huang X., Qian Y., Wei Y., and Ding G. (2006b) Determination and comparison of sludge reduction rates caused by microfaunas' predation. *Bioresource Technology*, **97**, 854–861.

Liu Y. (2000) Effect of chemical uncoupler on the observed growth yield in batch culture of activated sludge. *Water Research*, **34**, 2025–2030.

Liu Y. and Tay J.H. (2001) Strategy for minimization of excess sludge production from the activated sludge process. *Biotechnol. Adv.*, **19**(2), 97–107.

Liu J.C., Lee C.H., Lai J.Y., Wang K.C., Hsu Y.C., and Chang B.V. (2001) Extracellular polymers of ozonized waste activated sludge. *Water Science and Technology*, **44**(10), 137–142.

Low E.W. and Chase H.A. (1998) The use of chemical uncouplers for reducing biomass production during biodegradation. *Water Science and Technology*, **37**(4–5), 399–402.

Low E.W. and Chase H.A. (1999a) Reducing production of excess biomass during wastewater treatment. *Water Research*, **33**(5), 1119–1132.

Low E.W. and Chase H.A. (1999b) The effect of maintenance energy requirements on biomass production during wastewater treatment. *Water Research*, **33**(3), 847–853.

Low E.W., Chase H.A., Milner M.G., and Curtis T.P. (2000) Uncoupling of metabolism to reduce biomass production in the activated sludge process. *Water Research*, **34**(12), 3204–3212.

Lundin M., Olofsson M., Pettersson G.J., and Zetterlund H. (2004) Environmental and economic assessment of sewage sludge handling options. *Resources, Conservation and Recycling*, **41**, 255–278.

Luxmy, B.S., Kubo, T., and Yamamoto, K. (2001), Sludge reduction potential of metazoan in membrane bioreactors. *Water Science and Technology*, **44**(10), 197–202.

Mahmood T. and Elliott A. (2006) A review of secondary sludge reduction technologies for the pulp and paper industry. *Water Research*, **40**, 2093–2112.

Manterola G., Uriarte I., and Sancho L. (2008) The effect of operational parameters of the process of sludge ozonation on the solubilisation of organic and nitrogenous compounds. *Water Research*, **42**, 3191–3197.

Mao T., Hong S.-Y., Show K.-Y., Tay J.-H., and Lee D.-J. (2004) A comparison of ultrasound treatment on primary and secondary sludges. *Water Science and Technology*, **50**(9), 91–97.

Mason C.A. and Hamer G. (1987) Cryptic growth in Klebsiella pneumoniae. *Appl. Microbiol. Biotechnol.*, **25**, 577–584.

Mason T.J., Lorimer J.P., and Bates D.M. (1992) Quantifying sonochemistry: Casting some light on a 'black art'. *Ultrasonics*, **30**(1), 40–42.

Matsua A., Ide T., and Fuji S. (1988) Behavior of nitrogen and phosphorus during batch aerobic digestion of waste-activated sludge-continuous aerobic and intermittent aeration by control of DO. *Water Research*, **22**(12), 1495–1501.

Mayhew M. and Stephenson T. (1997) Low biomass yield activated sludge: A review. *Environmental Technology*, **18**, 883–892.

Mayhew M. and Stephenson T. (1998) Biomass yield reduction: Is biochemical manipulation possible without affecting activated sludge process efficiency? *Water Science and Technology*, **38**(8–9), 137–144.

Middelberg A.P.J. (1995) Process-scale disruption of microorganisms. *Biotechnology Advances*, **13**(3), 491–551.

Mitani M.M., Keller A.A., Sandall O.C., and Rinker R.G. (2005) Mass Transfer of Ozone Using a Microporous Diffuser Reactor System. *Ozone: Science & Engineering*, **27**(1), 45–51.

Muller A., Wentzel M.C., Loewenthal R.E., and Ekama G.A. (2003) Heterotroph anoxic yield in anoxic aerobic activated sludge systems treating municipal wastewater. *Water Research*, **37**, 2435–2441.

Müller J. (2000a) Pre-treatment processes for the recycling and reuse of sewage sludge. *Water Science and Technology*, **42**(9), 167–174.

Müller J. (2000b) Disintegration as a key-step in sewage sludge treatment. *Water Science and Technology*, **41**(8), 123–130.

Müller J.A. (2001) Prospects and problems of sludge pretreatment processes. *Water Science and Technology*, **44**(10), 121–128.

Müller J.A. (2007) Western Europe. In: *Wastewater Sludge: A Global Overview of the Current Status and Future Prospects*. Spinosa L. (ed.). IWA Publishing Ltd, London, UK, pp. 5–8. ISBN: 1843 391422.

Müller J.A. and Strünckmann G. (2007) Scaling up of mechanical route. In: *Comparative Evaluation of Sludge Reduction Routes*. Ginestet P. (ed.). IWA Publishing, London, UK.

Müller J., Lehne G., Schwedes J., Battenberg S., Näveke R., Kopp J., Dichtl N., Scheminski A., Krull R., and Hempel D.C. (1998) Disintegration of sewage sludges and influence on anaerobic digestion. *Water Science and Technology*, **38**(8–9), 425–433.

Müller J.A., Winter A., and Strünkmann G. (2004) Investigation and assessment of sludge pre-treatment processes. *Water Science and Technology*, **49**(10), 97–104.

Nagare H., Tsuno H., Saktaywin W., and Soyama T. (2008) Sludge ozonation and its application to a new advanced wastewater treatment process with sludge disintegration. *Ozone: Science & Engineering*, **30**(2), 136–144.

Nah I.W., Kang Y.W., Hwang K.Y., and Song W.K. (2000) Mechanical pre-treatment of waste activated sludge for anaerobic digestion process. *Water Research*, **34**, 2362–2368.

Nebe-von-Caron G., Stephens P.J., Hewitt C.J., Powell J.R., and Badley R.A. (2000) Analysis of bacterial function by multi-colour fluorescence flow cytometry and single cell sorting. *Journal of Microbiological Methods*, **42**, 97–114.

Neis U., Nickel U.K., and Tiehm A. (2000) Enhancement of anaerobic digestion by ultrasonic disintegration. *Water Science and Technology*, **42**(9), 73–80.

Neis U., Nickel K., and Lundén A. (2008) Improving anaerobic and aerobic degradation by ultrasonic disintegration of biomass. *Journal of Environmental Science and Health, Part A*, **43**(13), 1541–1545.

Neyens E. and Baeyens J. (2003) A review of thermal sludge pre-treatment processes to improve dewaterability. *Journal of Hazardous Materials – Part B*, **98**(1–3), 51–67.

Neyens E., Baeyens J., and Creemers C. (2003a) Alkaline thermal sludge hydrolysis. *Journal of Hazardous Materials – Part B*, **97**, 295–314.
Neyens E., Baeyens J., Weemaes M., and De Heyder B. (2003b) Hot acid hydrolysis as a potential treatment of thickened sewage sludge. *Journal of Hazardous Materials – part B*, **98**, 275–293.
Neyens E., Baeyens J., Dewil R., and De Heyder B. (2004) Advanced sludge treatment affects extracellular polymeric substances to improve activated sludge dewatering. *Journal of Hazardous Materials – part B*, **106**, 83–92.
Nickel K. and Neis U. (2007) Ultrasonic disintegration of biosolids for improved biodegradation. *Ultrasonics Sonochemistry*, **14**, 450–455.
Nielsen B. and Petersen G. (2000) Thermophilic anaerobic digestion and pasteurisation. Practical experience from Danish wastewater treatment plants. *Water Science and Technology*, **42**(9), 65–72.
Nielsen P.H. and Keiding K. (1998) Disintegration of activated sludge flocs in presence of sulphide. *Water Research*, **32**, 313–320
Nishijima W., Fahmi Mukaidani T., and Okada M. (2003) DOC removal by multi-stage ozonation-biological treatment. *Water Research*, **37**, 150–154.
Novak O. (2000) Expenditure on the operation of municipal wastewater treatment plants for nutrient removal. *Water Science and Technology*, **41**(9), 281–289.
Novak J.T., Muller C.D., and Murthy S.N. (2001) Floc structure and the role of cations. *Water Science and Technology*, **44**(10), 209–213.
Novak J.T., Sadler M.E., and Murthy S.N. (2003) Mechanisms of floc destruction during anaerobic and aerobic digestion and the effect on conditioning and dewatering of biosolids. *Water Research*, **37**, 3136–3144.
Novak J.T., Dong H.C., Curtis B.A., and Doyle M. (2006) Reduction of sludge generation using the Cannibal® process: mechanisms and performance. In: *Residuals and Biosolids Management Conference*, Cincinnati, Ohio, 2006. WEF: Water Environmental Federation, USA.
Novak J.T., Chon D.H., Curtis B.-A., and Doyle M. (2007) Biological solids reduction using the Cannibal process. *Water Environment Research*, **79**(12), 2380–2386.
Ødegaard H. (2004) Sludge minimization technologies – an overview. *Water Science and Technology*, **49**(10), 31–40.
Ødegaard H., Paulsrud B., and Karlson I. (2002) Wastewater sludge as a resource – sludge disposal strategies and corresponding treatment technologies aimed at sustainable handling of wastewater sludge. *Water Science and Technology*, **46**(10), 295–303.
Oh Y.-K., Lee K.-R., Ko K.-B., and Yeom I.-T. (2007) Effects of chemical sludge disintegration on the performances of wastewater treatment by membrane bioreactor. *Water Research*, **41**, 2665–2671.
Okey R.W. and Stensel D.H. (1993) Uncouplers and activated sludge – the impact on synthesis and respiration. *Toxicol. Environ. Chem.*, **40**, 235–254.
Okuno N. (2007) East Asia (Japan and Korea). In: *Wastewater Sludge: A Global Overview of the Current Status and Future Prospects*. Spinosa L. (ed.). IWA Publishing Ltd, London, UK, pp. 23–26. ISBN: 1843 391422.
Onyeche T.I. (2004) Sludge as source of energy and revenue. *Water Science and Technology*, **50**(9), 197–204.

Orhon D., Artan N., and Cimsit Y. (1989) The concept of soluble residual product formation in the modelling of activated sludge. *Water Science and Technology*, **21**(4–5), 339–350.

Pagan R. and Mackey B. (2000) Relationship between membrane damage and cell death in pressare treated Escherichia coli cells: differences between exponential – and stationary-phase cells and variation among strains. *Appl. Environ. Microbiol.*, **66**(7), 2829–2834.

Palatsi J., Gimenez-Lorang A., Ferrer I., and Flotats X. (2009) Start-up strategies of thermophilic anaerobic digestion of sewage sludge. *Water Science and Technology*, **59**(9), 1777–1784.

Park K.Y., Ahn K.-H., Maeng S.K., Hwang J.H., and Kwon J.H. (2003) Feasibility of sludge ozonation for stabilization and conditioning. *Ozone: Science & Engineering*, **25**(1), 73–80.

Park K.Y., Lee J.W., Ahn K.H., Maeng S.K., Hwang J.H., and Song K.G. (2004a) Ozone disintegration of excess biomass and application to nitrogen removal. *Water Environ. Res.*, **76**(2), 162–167.

Park B., Ahn J.-H., Kim J., and Hwang S. (2004b) Use of microwave pretreatment for enhanced anaerobiosis of secondary sludge. *Water Science and Technology*, **50**(9), 17–23.

Park C. and Novak J.T. (2007) Characterization of activated sludge exocellular polymers using several cation-associated extraction methods. *Water Research*, **41**(8), 1679–1688.

Park K.Y., Maeng S.K., Song K.G., and Ahn K.H. (2008) Ozone treatment of wastewater sludge for reduction and stabilization. *Journal of Environmental Science and Health, Part A*, **43**(13), 1546–1550.

Park W.-J., Ahn J.-H., Hwang S., and Lee C.-K. (2009) Effect of output power, target temperature, and solid concentration on the solubilization of waste activated sludge using microwave irradiation. *Bioresource Technology*, doi:10.1016/j.biortech. 2009.02.062.

Parmar N., Singh A., and Ward O.P. (2001) Enzyme treatment to reduce solids and improve settling of sewage sludge. *J. Ind. Microbiol. Biotechnol.*, **26**(6), 383–386.

Paul E. and Debellefontaine H. (2007) Reduction of excess sludge produced by biological treatment processes: effect of ozonation on biomass and on sludge. *Ozone: Science & Engineering*, **29**(6), 415–427.

Paul E., Camacho P., Lefebvre D., and Ginestret P. (2006a) Organic matter release in low temperature thermal treatment of biological sludge for reduction of excess sludge production. *Water Science and Technology*, **54**(5), 59–68.

Paul E., Camacho P., Sperandio M., and Ginestet P. (2006b) Technical and economical evaluation of a thermal, and two oxidative techniques for the reduction of excess sludge production. *Process Safety and Environmental Protection*, **84**(B4), 247–252.

Peddie C. and Mavinic D. (1990) A pilot-scale evaluation of aerobic/anoxic sludge digestion. *Can. J. Civ. Eng.*, **17**(1), 68–78.

Penaud V., Delgenès J.P., and Moletta R. (1999) Thermo-chemical pretreatment of a microbial biomass: influence of sodium hydroxide addition on solubilization and anaerobic biodegradability. *Enzyme and Microbial Technology*, **25**, 258–263.

Peng D., Bernet N., Delgenes J.P., and Moletta R. (1999) Aerobic granular sludges: A case study. *Water Research*, **33**, 890–893.

Pérez-Elvira S.I., Nieto Diez P., and Fdz-Polanco F. (2006) Sludge minimisation technologies. *Reviews in Environmental Science and Bio/Technology*, **5**, 375–398.

Pickworth B., Adams J., Panter K., and Solheim O.E. (2006) Maximising biogas in anaerobic digestion by using engine waste heat for thermal hydrolysis pre-treatment of sludge. *Water Science and Technology*, **54**(5), 101–108.

Pinnekamp J. (1989) Effects of thermal pre-treatment of sewage sludge on anaerobic digestion. *Water Science and Technology*, **21**(4–5), 97–108.

Pino-Jelcic S.A., Hong S.M., and Park J.K. (2006) Enhanced anaerobic biodegradability and inactivation of fecal coliforms and Salmonella spp. in wastewater sludge by using microwaves. *Water Environmnetal Research*, **78**, 209–216.

Pirt S.J. (1965) The maintenance energy of bacteria in growing cultures. *Proc. Roy. Soc. London, Ser. B.*, **163**, 224–231

Pollice A., Laera G., and Blonda M. (2004) Biomass growth and activity in a membrane bioreactor with complete sludge retention, *Water Research*, **38**(7), 1799–1808.

Pollice A., Laera G., Saturno D., and Giordano C. (2008) Effects of sludge retention time on the performance of a membrane bioreactor treating municipal sewage. *Journal of Membrane Science*, **317**, 65–70.

Porter J., Deere D., Hardman M., Edwards C., and Pickup R. (1997) Go with the flow – use of flow cytometry in environmental microbiology. *FEMS Microbiology Ecology*, **24**, 93–101.

Qiu Y., Kuo C., and Appi M.E. (2001) Performance and simulation of ozone absorption and reactions in a stirred tank reactor. *Environ. Sci. Tech.*, **35**, 209–215.

Ramadori R., Di Iaconi C., Lopez A., and Passino R. (2006) An innovative technology based on aerobic granular biomass for treating municipal and/or industrial wastewater with low environmental impact. *Water Science and Technology*, **53**(12), 321–329.

Ramirez I., Mottet A., Carrère H., Déléris S., Vedrenne F., and Stayer J.P. (2009) Modified ADM1 disintegration/hydrolysis structures for modeling batch thermophilic anaerobic digestion of thermally pretreated waste activated sludge. *Water Research*, **43**, 3479–3492.

Rasmussen H., and Nielsen P.H. (1996) Iron reduction in activated sludge measured with different extraction techniques. *Water Research*, **30**, 551–558.

Ratsak C.H., Kooi B.W., and van Verseveld H.W. (1994) Biomass reduction and mineralisation increase due to the ciliate *Tetrahymena pyriformis* grazing on the bacterium *Pseudomonas fluorescent*. *Water Science and Technology*, **29**(7), 119–128.

Ratsak C.H., Maarsen K.A., and Kooijman S.A.L. (1996) Effects of protozoa on carbon mineralization in activated sludge. *Water Research*, **30**(1), 1–12.

Ratsak C.H., and Verkuijlen J. (2006) Sludge reduction by predatory activity of aquatic oligochaetes in wastewater treatment plants: science or fiction? A review. *Hydrobiologia*, **564**, 197–211.

Recktenwald M., Wawrzynczyk J., Dey E.S., and Norrlöw O. (2008) Enhanced efficiency of industrial-scale anaerobic digestion by the addition of glycosidic enzymes. *Journal of Environmental Science and Health, Part A*, **43**(13), 1536–1540.

Rensink J.H. and Rulkens W.H. (1997) Using metazoan to reduce sludge production. *Water Science and Technology*, **36**(11), 171–179.

Ried A., Mielcke J., Wieland A., Schaefer S., and Sievers M. (2007) An overview of the integration of ozone systems in biological treatment steps. *Water Science and Technology*, **55**(12), 253–258.
Ried A., Stapel H., Koll R., Schettlinger M., Wemhöner F., Hamann-Steinmeier A., Miethe M., and Brombach A. (2002) Optimierungsmóglichkeiten beim betrieb von biologischen kláranlagen durch deneinsatz von ozon. KA, **49**(5), 648–661.
Rittman B.E., Lee H.-S., Zhang H., Alder J., Banaszak J.E., and Lopez R. (2008) Full-scale application of focused-pulsed pre-treatment for improving biosolids digestion and conversion to methane. *Water Science and Technology*, **58**(10), 1895–1901.
Rocher M., Goma G., Pilas Begue A., Louvel L., and Rols J.L. (1999) Towards a reduction in excess sludge production in activated sludge processes: Biomass physicochemical treatment and biodegradation. *Applied Microbiology Biotechnology*, **51**, 883–890.
Rocher M., Roux G., Goma G., Pilas Begue A., Louvel L., and Rols J.L. (2001) Excess sludge reduction in activated sludge processes by integrating biomass alkaline heat treatment. *Water Science and Technology*, **44**(2–3), 437–444.
Roeleveld P., Loeffen P., Temmink H., and Klapwijk B. (2004) Dutch analysis for P-recovery from municipal wastewater. *Water Science and Technology*, **49**(10), 191–199.
Roš M. and Zupančič G.D. (2003) Two stage thermophilic anaerobic–aerobic mineralization-stabilization of excess activated sludge. *Journal of Environmental Science and Health Part A – Environmental Science and Engineering & Toxic and Hazardous Substance Control*, **A38**(10), 2381–2389.
Russel J.B. and Cook G.M. (1995) Energetics of bacterial growth: balance of anabolic and catabolic reactions. *Microbiol. Rev.*, **59**(1), 48–62.
Ryan F.J. (1959) Bacterial mutation in a stationary phase and the question of cell turnover. *J. Gen. Microbiol.*, **21**, 530–549.
Saby S., Djafer M., and Chen G.H. (2002) Feasibility of using a chlorination stepto reduce excess sludge in activated sludge process. *Water Research*, **36**(3), 656–66.
Saby S., Djafer M., and Chen G.H. (2003) Effect of low ORP in anoxic sludge zone on excess sludge production in oxic-settling-anoxic activated sludge process. *Water Research*, **37**(1), 11–20.
Sakai Y., Tani K., and Tkahashi F. (1992) Sewage treatment under conditions of balancing growth and cell decay with a high concentration of activated sludge supplemented with ferromagnetic powder. *J. Ferm. Bioeng.*, **76**(6), 413–415.
Sakai Y., Aoyagi T., Shiota N., Akashi A., and Hasegawa S. (2000) Complete decomposition of biological waste sludge by thermophilic aerobic bacteria. *Water Science and Technology*, **42**(9), 81–88.
Sakai Y., Fukase T., Yasui H., and Shibata M. (1997) An activated sludge process without sludge production. *Water Science and Technology*, **36**(11), 163–170.
Saktaywin W., Tsuno H., Nagare H., Soyama T., and Weerapakkaroon J. (2005) Advanced sewage treatment process with excess sludge reduction and phosphorus recovery. *Water Research*, **39**, 902–910.
Saktaywin W., Tsuno H., Nagare H., and Soyama T. (2006) Operation of a new sewage treatment process with technologies of excess sludge reduction and phosphorus recovery. *Water Science and Technology*, **53**(12), 217–227.

Salerno M.B., Lee H.S., Parameswaran P., and Rittmann B.E. (2009) Using a pulsed electric field as a pretreatment for improved biosolids digestion and methanogenesis. *Water Environment Research*, **81**(8), 831–839.

Salvado H. and Puigagut J. (2007) Scaling up of predation enhancement route. In: *Comparative Evaluation of Sludge Reduction Routes*. Ginestet P. (ed.). IWA Publishing, London, UK.

Scheminski A., Krull R., and Hempel D.C. (2000) Oxidative treatment of digested sewage sludge with ozone. *Water Science and Technology*, **42**(9), 151–158.

Shiota N., Akashi A., and Hasegawa S. (2002) A strategy in wastewater treatment process for significant reduction of excess sludge production. *Water Science and Technology*, **45**(12), 127–134.

Show K.Y., Mao T., and Lee D.J. (2007) Optimisation of sludge disruption by sonication. *Water Research*, **41**, 4741–4747.

Sievers M., Ried A., and Koll R. (2004) Sludge treatment by ozonation – evaluation of full-scale results. *Water Science and Technology*, **49**(4), 247–253.

Sievers M. and Schaefer, S. (2007) The impact of sequential ozonation-aerobic treatment on the enhancement of sludge dewaterability. *Water Science and Technology*, **55**(12), 201–205.

Siegrist H., Brunner I., Koch G., Linh Con Phan, and Van Chieu LE. (1999) Reduction of biomass decay rate under anoxic and anaerobic conditions. *Water Science and Technology*, **39**(1), 129–137.

Skiadas I.V., Gavala H.N., Lu J., and Ahring B.K. (2005) Thermal pre-treatment of primary and secondary sludge at 70°C prior to anaerobic digestion. *Water Science and Technology*, **52**(1–2), 161–166.

Smith G. and Göransson J. (1992) Generation of an effective internal carbon source for denitrification through termal hydrolysis of pre-precipitated sludge. *Water Science and Technology*, **25**(4–5), 211–218.

Snyman H.G. (2007) Africa. In: *Wastewater Sludge: A Global Overview of the Current Status and Future Prospects*. Spinosa L. (ed.). IWA Publishing Ltd, London, UK, pp. 31–34. ISBN: 1843 391422.

Song Y.-D. and Hu H.-Y. (2006) Isolation and characterization of thermophilic bacteria capable of lysing microbial cells in activated sludge. *Water Science and Technology*, **54**(9), 35–43.

Song K.-G., Choung Y.-K., Ahn K.-H., Cho J., and Yun H. (2003) Performance of membrane bioreactor system with sludge ozonation process for minimization of excess sludge production. *Desalination*, **157**, 353–359.

Spanjers H. and Vanrolleghem P. (1995) Respirometry as a tool for rapid characterisation of wastewater and activated sludge. *Water Science and Technology*, **31**(2), 105–114.

Spérandio M. and Paul E. (2000) Estimation of wastewater biodegradabile COD fractions by combining respirometric experiments in various So/Xo ratios. *Water Research*, **34**(4), 1233–1246.

Spinosa L. and Wichmann K. (2006) European developments in standardisation of sludge physical parameters. *Water Science and Technology*, **54**(5), 1–8.

Spinosa L. (2007a) *Wastewater Sludge: A Global Overview of the Current Status and Future Prospects*. IWA Publishing Ltd, London, UK. ISBN: 1843 391422.

Spinosa L. (2007b) Introduction and overview. In: *Wastewater Sludge: A Global Overview of the Current Status and Future Prospects*. Spinosa L. (ed.). IWA Publishing Ltd, London, UK, pp 1–3. ISBN: 1843 391422.

Steen H.B. (2000) Flow cytometry of bacteria: Glimpses from the past with a view to the future. *Journal of Microbiological Methods*, **42**, 65–74.
Strand S.E., Harem G.N., and Stensel H.D. (1999) Activated-sludge yield reduction using chemical uncouplers. *Water Environ Res*, **71**(4), 454–458.
Strünkmann G.W., Müller J.A., Albert F., and Schwedes J. (2006) Reduction of excess sludge production using mechanical disintegration devices. *Water Science and Technology*, **54**(5), 69–76.
Takiguchi N., Kishino M., Kuorda A., Kato J., and Ohtake H. (2004) A laboratory-scale test of anaerobic digestion and methane production after phosphorus recovery from waste activated sludge. *Journal of Bioscience and Bioengineering*, **97**(6), 365–368.
Tanaka S., Kobayashi T., Kamiyama K.I., and Bildan L.N.S. (1997) Effects of thermochemical pretreatment on the anaerobic digestion of waste activated sludge. *Water Science and Technology*, **35**(8), 209–215.
Tanaka S. and Kamiyama K. (2002) Thermochemical pretreatment in the anaerobic digestion of waste activated sludge. *Water Science and Technology*, **46**(10), 173–179.
Tchobanoglous G., Burton F.L., and Stensel H.D. (2003) Wastewater Engineering: Treatment and Reuse. Inc. Metcalf & Eddy, McGraw-Hill, 4th edition.
Tiehm A., Nickel K., Zellhorn M., and Neis U. (2001) Ultrasonic waste activated sludge disintegration for improving anaerobic stabilization. *Water Research*, **35**(8), 2003–2009.
Tiehm A., Nickel K., and Neis U. (1997) The use of ultrasound to accelerate the anaerobic digestion of sewage sludge. *Water Science and Technology*, **36**(11), 121–128.
Turovskiy I.S. and Mathai P.K. (2006) Wastewater sludge processing. Wiley-Interscience. John Wiley & Sons, Inc., Hoboken, New Jersey.
Van Leeuwen J. (1988a) Bulking control with ozonation in a nutrient removal activated sludge system. *Water SA*, **14**(3), 119–124.
Van Leeuwen J. (1988b) Improved sewage treatment with ozonated activated sludge. *J. Inst. Water Environ. Manage*, **2**(5), 493–499.
van Loosdrecht M.C.M. and Henze M. (1999) Maintenance, endogenous respiration, lysis, decay and predation. *Water Science and Technology*, **39**(1), 107–117.
Vanrolleghem, P.A., Spanjers, H., Petersen, B., Ginestet, P., and Takacs, I., 1999. Estimating (combinations of) Activated Sludge Model No.1 parameters and components by respirometry. *Water Science and Technology*, **39**(1), 195–215.
Vavilin V.A., Rytov S.V., and Lokshina L.Y. (1996) A description of hydrolysis kinetics in anaerobic degradation of particulate organic matter. *Biosour. Technol.*, **56**, 229–237.
Visvanathan C., Ben Aim R., and Parameshwaran K. (2000) Membrane separation bioreactors for wastewater treatment. *Critical Reviews in Environmental Science and Technology*, **30**(1), 1–48.
Vives-Rego J., Lebaron P., and Nebe-von Caron G. (2000) Current and future applications of flow cytometry in aquatic microbiology. *FEMS Microbiology Reviews*, **24**, 429–448.
Vlyssides A.G., and Karlis P.K. (2004) Thermal-alkaline solubilization of waste activated sludge as a pre-treatment stage for anaerobic digestion. *Bioresource Technol.*, **91**(2), 201–206.

Wagner J., and Rosenwinkel K.H. (2000) Sludge production in membrane bioreactors under different conditions. *Water Science and Technology*, **41**(10–11), 251–258.

Wagner M., Amann R., Lemmer H., and Schleifer K.H. (1993) Probing activated sludge with oligonucleotides specific for proteobacteria: inadequacy of culture-dependent methods for describing microbial community structure. *Appl. Environ. Microbiol.*, **59**(5), 1520–1525.

Wang Q., Kuninobu M., Kakimoto K., I.-Ogawa H., and Kato Y. (1999) Upgrading of anaerobic digestion of waste activated sludge by ultrasonic pretreatment. *Bioresource Technology*, **68**, 309–313.

Wang F., Wang Y., and Ji M. (2005) Mechanisms and kinetics models for ultrasonic waste activated sludge disintegration. *Journal of Hazardous Materials*, **B123**, 145–150.

Wang W., Jung Y.J., Kiso Y., Yamada T., and Min K.S. (2006a) Excess sludge reduction performance of an aerobic SBR process equipped with a submerged mesh filter unit. *Process Biochemistry*, **41**, 745–751.

Wang F., Lu S., and Ji M. (2006b) Components of released liquid from ultrasonic waste activated sludge disintegration. *Ultrasonics Sonochemistry*, **13**, 334–338.

Wang Y., Wei Y., and Liu J. (2009) Effect of H_2O_2 dosing strategy on sludge pretreatment by microwave-H_2O_2 advanced oxidation process. *Journal of Hazardous Materials*, **169**, 680–684.

Warner A., Ekama G., and Marais G. (1985) Comparison of aerobic and anoxic–aerobic digestion of waste-activated sludge. *Water Science and Technology*, **17**, 1475–1478.

Watson S.D., Akhurst T., Whiteley C.G., Rose P.D., and Pletschke B.I. (2004) Primary sludge floc degradation is accelerated under biosulphidogenic conditions: Enzymological aspects. *Enzyme Microb. Technol.*, **34**(6), 595–602.

Wawrzynczyk J., Recktenwald M., Norrlöw O., and Dey E.S. (2008) The function of cation-binding agents in the enzymatic treatment of municipal sludge. *Water Res. (2007)*, **42**(6–7), 1555–1562.

Wawrzynczyk J., Szewczyk E., Norrlöw O., and Dey, E.S. (2007) Application of enzymes, sodium tripolyphosphate and cation exchange resin for the release of extracellular polymeric substances from sewage sludge. *J. Biotechnol.*, **130**(3), 274–281.

Weemaes M. and Verstraete W.H. (1998) Evaluation of current wet sludge disintegration techniques. *J. Chem. Technol. Biotechnol.*, **73**, 83–92.

Weemaes M., Grootaerd H., Simoens F., and Verstraete W. (2000) Anaerobic digestion of ozonized biosolids. *Water Research*, **34**(8), 2330–2336.

Wei Y.S., Van Houten R.T., Borger A.R., Eikelboom D.H., and Fan Y.B. (2003a) Minimization of excess sludge production for biological wastewater treatment. *Water Research*, **37**(18), 4453–4467.

Wei Y.S., Van Houten R.T., Borger A.R., Eikelboom D.H., and Fan Y.B. (2003b) Comparison performances of membrane bioreactor and conventional activated sludge processes on sludge reduction induced by Oligochaete. *Environ. Sci. Technol.*, **37**, 3171–3180.

Wei Y. and Liu J. (2005) The discharged excess sludge treated by Oligochaeta. *Water Science and Technology*, **52**(10–11), 265–272.

Wei Y. and Liu J. (2006) Sludge reduction with a novel combined worm-reactor. *Hydrobiologia*, **564**, 213–222.

Wei Y., Wang Y., Guo X., and Liu J. (2009) Sludge reduction potential of the activated sludge process by integrating an oligochaete reactor. *Journal of Hazardous Materials*, **163**, 87–91.

Wilson C.A. and Novak J.T. (2009) Hydrolysis of macromolecular components of primary and secondary wastewater sludge by thermal hydrolytic pre-treatment. *Water Research*, **43**, 4489–4498.

Winter A. (2002) Minimisation of costs by using disintegration at a full-scale anaerobic digestion plant. *Water Science and Technology*, **46**(4–5), 405–412.

Woods N.C., Sock S.M, and Daigger G.T. (1999) Phosphorus recovery technology modelling and feasibility evaluation for municipal wastewater treatment plants. *Env. Tech.*, **20**, 663–679.

Xue T. and Huang X. (2007) Releasing characteristics of phosphorus and other substances during thermal treatment of excess sludge. *Journal of Environmental Sciences*, **19**, 1153–1158.

Yan S., Miyanaga K., Xing X.-H., and Tanji Y. (2008) Succession of bacterial community and enzymatic activities of activated sludge by heat-treatment for reduction of excess sludge. *Biochemical Engineering Journal*, **39**, 598–603.

Yan S.-T., Chu L.-B., Xing X.-H., Yua A.-F., Sunc X.-L., and Jurcik B. (2009) Analysis of the mechanism of sludge ozonation by a combination of biological and chemical approaches. *Water Research*, **43**, 195–203.

Yang X.-F., Xie M.-L., and Liu Y. (2003) Metabolic uncouplers reduce excess sludge production in an activated sludge process. *Process Biochemistry*, **38**, 1373–1377.

Yasui H. and Shibata M. (1994) An innovative approach to reduce excess sludge production in the activated sludge process. *Water Science and Technology*, **30**(9), 11–20.

Yasui H., Nakamura K., Sakuma S., Iwasaki M., and Sakai Y. (1996) A full-scale operation of a novel activated sludge process without excess sludge production. *Water Science and Technology*, **34**(3–4), 395–404.

Yasui, H., Komatsu, K., Goel, R., Li, Y.Y., and Noike, T. (2005) Full-scale application of anaerobic digestion process with partial ozonation of digested sludge. *Water Science and Technology*, **52**(1–2), 245–252.

Ye F.X. and Li Y. (2005) Reduction of excess sludge production by 3,3′,4′,5-tetrachlorosalicylanilide in an activated sludge process. *Appl. Microbiol. Biotechnol*, **67**, 269–274.

Yeom I.T., Lee K.R., Lee Y.H., Ahn K.H., and Lee S.H. (2002) Effects of ozone treatment on the biodegradability of sludge from municipal wastewater treatment plants. *Water Science and Technology*, **46**(4–5), 421–425.

Yoon S.-H., Kim H.-S., and Lee S. (2004a) Incorporation of ultrasonic cell disintegration into a membrane bioreactor for zero-sludge production. *Process Biochemistry*, **39**, 1923–1929.

Yoon S.-H., Kim H.-S., and Yeom I.-T. (2004b) The optimum operational condition of membrane bioreactor (MBR): cost estimation of aeration and sludge treatment. *Water Research*, **38**, 37–46.

Yoon S.-H., and Lee S. (2005) Critical operational parameters for zero sludge production in biological wastewater treatment processes combined with sludge disintegration. *Water Research*, **39**, 3738–3754.

Zábranská J., Štěpová J., Wachtl R., Jeníček P., and Dohányos M. (2000a) The activity of anaerobic biomass in thermophilic and mesophilic digesters at different loading rates. *Water Science and Technology*, **42**(9), 49–56.

Zábranská J., Dohányos M., Jenícek P., and Kutil J. (2000b) Thermophilic process and enhancement of excess activated sludge degradability – two ways of intensification of sludge treatment in the Prague central wastewater treatment plant. *Water Science and Technology*, **41**(9), 265–272.

Zábranská J., Dohányos M., Jeníček P., Růžičiková H., and Vránová A. (2003) Efficiency of autothermal thermophilic aerobic digestion and thermophilic anaerobic digestion of municipal wastewater sludge in removing Salmonella spp. and indicator bacteria. *Water Science and Technology*, **47**(3), 151–156.

Zábranská J., Dohányos M., Jeníček P., and Kutil J. (2006) Disintegration of excess activated sludge – evaluation and experience of full-scale applications. *Water Science and Technology*, **53**(12), 229–236

Zambrano J.A., Gil-Martinez M., Garcia-Sanz M., and Irizar I. (2009) Benchmarking of control strategies for ATAD technology: A first approach to the automatic control of sludge treatment systems. *Water Science and Technology*, **60**(2), 409–417.

Zhang G., Zhang P., Yang J., and Chen Y. (2007a) Ultrasonic reduction of excess sludge from the activated sludge system. *Journal of Hazardous Materials*, **145**, 515–519.

Zhang P., Zhang G., and Wang W. (2007b) Ultrasonic treatment of biological sludge: Floc disintegration, cell lysis and inactivation. *Bioresource Technology*, **98**, 207–210.

Zhang G., Zhang P., Gao J., and Chen Y. (2008) Using acoustic cavitation to improve the bio-activity of activated sludge. *Bioresource Technology*, **99**, 1497–1502.

Zhang G., Yang J., Liu H., and Zhang J. (2009) Sludge ozonation: Disintegration, supernatant changes and mechanisms. *Bioresource Technology*, **100**, 1505–1509.

Zhao Y.X., Yin J., Yu H.L., Han N., and Tian F.J. (2007) Observations on ozone treatment of excess sludge. *Water Science and Technology*, **56**(9), 167–175.

Zhou H.D. and Smith D.W. (2000) Ozone mass transfer in water and wastewater treatment: Experimental observations using a 2D laser particle dynamics analyzer. *Water Research*, **34**, 909–921.

Ziglio G., Andreottola G., Barbesti S., Boschetti G., Bruni L., Foladori P., and Villa R. (2002) Assessment of activated sludge viability with flow cytometry. *Water Research*, **36**(2), 460–468. Anche in: Modern Scientific Tools in Bioprocessing. Elsevier, UK. ISBN 0-444-51006-0.

Ziglio G., Andreottola G., Foladori P., and Ragazzi M. (2001) Experimental validation of a single-OUR method for wastewater RBCOD characterisation. *Water Science and Technology*, **43**(11), 119–126.

Zupančič G.D., and Roš M. (2008) Aerobic and two-stage anaerobic–aerobic sludge digestion with pure oxygen and air aeration. *Bioresource Technology*, **99**, 100–109.

Index

A
Aachen-Soers plant 174
acidic reagents 234–235
activated sludge model 1 (ASM1) 139
adenosine triphosphate (ATP) 13, 35–36, 57, 116, 121–122, 144
Aeolosoma hemprichi 151–152
Aeolosomatidae 151
aeration
  extended 161
  sludge reduction 59–60
aerobic conditions
  degradability 112–113
  heterotrophic maximum growth yield 115–121
aerobic digestion
  and ozonation 292–293
  sludge reduction 77–78

aerobic/anoxic digestion 120
  oxidation-reduction (red-ox) potential (ORP) 78
  sludge reduction 78–79
alkaline reagents 234–235
alkaline treatment
  hydrolysis 240
  and membrane biological reactors (MBR) 164
ammonia 216, 240, 267–268
anaerobic digestion
  biodegradability 99–100
  biogas production 99–100
  degradability 112–115
  hydraulic retention time (HRT) 223, 290, 308
  and ozonation 288–292
  sludge reduction 82–83

© 2010 IWA Publishing. *Sludge Reduction Technologies in Wastewater Treatment Plants*. By Paola Foladori, Gianni Andreottola and Giuliano Ziglio. ISBN: 9781789065305. Published by IWA Publishing, London, UK.

thermo-chemical hydrolysis 244–246
thermophilic 129–131
anaerobic reactors
   sludge reduction 57–59
   thermophilic 131–140
   oxidation-reduction (red-ox) potential (ORP) 121, 123
Anellida 150–151
ASM1 *see* activated sludge model 1
ATP *see* adenosine triphosphate
autothermal thermophilic aerobic digestion (ATAD)
   hydraulic retention time (HRT) 308
   oxidation-reduction (red-ox) potential (ORP) 308
   sludge handling units 81–82, 138–140
   sludge reduction 81–82
   sludge retention time (SRT) 138

B
Bacillus 131–133
Bacillus stearothermophilus 131, 133
   proteases 133
bacteria
   inactivation of 100–102
   thermophilic 131
ball mills 174–178, 185–188
*Bdelloi*d 150
biochemical oxygen demand (BOD) 16, 18–19
   biodegradability 215
   dewatering 248
   extended aeration 59, 161–162
   growth yield 116, 145, 148–149, 271
   wastewater handling units 136
biodegradability
   biochemical oxygen demand (BOD) 215
   biogas production 99–100
   inert solids 40–41
   oxygen uptake rate (OUR) 299
   ozonation 265
   respirometry 95–96
   thermal treatment 215–216

biogas production 99–100
biological processes
   and ozonation 268–269
   and thermal treatment 220–229
biological sludge production 13–18
biological treatments 109–112
   aerobic degradability 112–113
   anaerobic degradability 113–114
   chemical metabolic uncouplers 144–148
   enzymatic hydrolysis 140–144
   extended aeration 161
   floc disintegration 114–115
   granular sludge 164–166
   maximum growth yield 115–121
   membrane biological reactors (MBR) 162–164
   microbial predation 148–161
   side-stream anaerobic reactors 121–129
   sulphides 114–115
   thermophilic anaerobic digestion 129–131
   thermophilic anaerobic reactors 131–140
Biolysis 270
biomass
   oxygen uptake rate (OUR) 198
   sludge retention time (SRT) 103–104
   yield 59–63, 103–104
BioThelys process 227
   hydraulic retention time (HRT) 227
BOD *see* biochemical oxygen demand
Bodo caudatus 149
Bodo saltans 149
Branchnria sowerbyi 156
bubbles
   cavitation bubbles 49, 69, 193, 196–197, 199
   microbubbles 199, 254
   radius of 196

C
Cambi process 227
   hydraulic retention time (HRT) 227

Cannibal system 125–129
  oxidation-reduction (red-ox) potential
    (ORP) 57–59, 126
  sludge retention time (SRT) 126–129
capillary suction time (CST) 75, 106,
    219, 243, 293–294
cations 235
cavitation bubbles 49, 69, 193,
    196–197, 199
cell lysis 32–35
centrifuges 171–174, 185–188
Cercobodo 149
chemical hydrolysis 50–51, 73–74
chemical metabolic uncouplers
    144–148
  sludge reduction 57
chemical oxygen demand (COD) 10
  particulate COD (PCOD) 92
  sludge fractionation 10–12
chemical oxygen demand (COD)
    solubilisation 91–92
  chemical treatment 235–240
  contact time 238–239
  degree of disintegration (DD) 93
  effect of pH 237–238
  effect of temperature 236–237
  high temperature 214–215
  moderate temperature 212–214
  ozonation 261–266
  sludge reduction technique comparison
    300–301
  thermal treatment 211–215
  and total suspended solids (TSS)
    91–92, 261–266
  ultrasonic disintegration 197–200
chemical sludge 8
chemical treatment
  acidic reagents 234–235
  alkaline reagents 234–235
  chemical oxygen demand (COD)
    solubilisation 235–240
  contact time 238–239
  dewaterability 243
  and membrane biological reactors
    (MBR) 163–164
  nitrogen solubilisation 240–243

  phosphorus solubilisation 240–243
  sludge handling units 244–248
  wastewater handling units 243–244
4-chloro-2-nitrophenol (CNP) 144–145
COD *see* chemical oxygen demand
compressibility 107
costs
  disposal 26–27
  treatment 25–26
cryptic growth 32
CST *see* capillary suction time

D

DCP *see* 2,4-dichlorophenol
DD *see* degree of disintegration
degradability
  aerobic conditions 112–113
  anaerobic conditions 112–115
degree of disintegration (DD) 92–93
  chemical oxygen demand (COD)
    solubilisation 93
  and dissolved oxygen (DO) 94–96
  mechanical disintegration 177–178,
    185–187
  oxygen uptake rate (OUR) 94
  sludge reduction technique comparison
    302
  see also disintegration
denitrification
  and ozonation 285–286
  rate 97–99
dewaterability
  chemical treatment 243
  extracellular polymeric substances
    (EPS) 219, 243
  and ozonation 293–294
  specific resistance to filtration (SRF)
    107, 293
  thermal treatment 219–220
  ultrasonic disintegration 203
dewatering 168
  biochemical oxygen demand (BOD)
    248
  sludge 4
  thermo-chemical hydrolysis 246–248
2,6-dibromo-4-nitrophenol 144

2,4-dichlorophenol (DCP) 144
2,4-dinitrophenol (DNP) 144–147
disintegration
    rotor-stator systems 183–188
    ultrasonic *see* ultrasonic disintegration
    see also degree of disintegration
disposal costs 26–27
dissolved organic carbon (DOC) 133, 242
dissolved oxygen (DO)
    and degree of disintegration 94–96
    microbial predation 307–308
    oxygen uptake rate (OUR) 96
    predation-reactor 156, 159
    and sludge solubilisation 133–134
    thermophilic bacteria 133–134
divalent ions 112–113, 133
DNP *see* 2,4-dinitrophenol
dry mass reduction 4–5
dual digestion 137–138
    hydraulic retention time (HRT) 308
    sludge reduction 79–81

E
EDTA *see* ethylenediaminetetraacetic acid
efficiency estimate
    bacteria inactivation 100–102
    biodegradability 95–96, 99–100
    biomass yield 103–104
    capillary suction time (CST) 106
    chemical oxygen demand (COD) solubilisation 91–92
    compressibility 107
    degree of disintegration 92–93
    denitrification rate 97–99
    maximum growth yield 103–104
    physical properties 106–107
    settleability 107
    sludge reduction 105
    sludge retention time (SRT) 102–103
    sludge yield 103–104
    specific resistance to filtration (SRF) 107
    thickenability 107
    total suspended solids (TSS) solubilisation 91–92
    treatment frequency 105–106

electrical treatment 55–56, 76–77
endogenous metabolism
    membrane biological reactors (MBR) 39
    sludge reduction 37–39
    sludge retention time (SRT) 38–39, 59
energy
    recovery 4
    sludge disintegration 170–171
    total solids (TS) 170
    ultrasonic disintegration 193–195
enzymatic hydrolysis
    enzymes 45–46, 67–68, 140–144
    extracellular polymeric substances (EPS) 67
    rate limiting steps 133
    sludge reduction 45–47, 67–68
    thermophilic bacteria 46–47
    volatile suspended solids (VSS) 46
enzymes
    reactivity and ozonation 281
    total solids (TS) 142–143
equipment
    mechanical disintegration 168–170
    ultrasonic disintegration 191–193
ethylenediaminetetraacetic acid (EDTA) 142
Europe
    disposal costs 26–27
    sludge production 2
    treatment costs 25–26
extended aeration 59, 161–162
    sludge retention time (SRT) 161
extracellular polymeric substances (EPS)
    and cations 235
    dewaterability 219, 243
    enzymatic hydrolysis 67, 140–142
    flocs 112, 114–115, 198–199, 203
    microwave treatment 72, 229
    ozonation 277, 282–283
    and pH 235
    thermal treatment 50

F
farmland 3–4
FCM *see* flow cytometry
filamentous microorganisms 281–282

flocs
   extracellular polymeric substances (EPS) 112, 114–115, 198–199, 203
   and ozonation 282
   size reduction 168
flow cytometry (FCM) 101–102
foaming 168
focused pulsed (FP) treatment 76–77
full plants
   sludge composition 9–12
   sludge production 12–19
Fürstenfeldbruck plant 173–174

G

granular sludge 164–166
   sequencing batch reactor (SBR) 165
   sludge reduction 61–62
grinding chambers 174
growth
   biochemical oxygen demand (BOD) 116, 145, 148–149, 271
   cryptic growth 32
   maximum growth yield 103–104

H

Hexamita 149
high pressure jet and collision system 182–183
homogenisers
   high pressure 178–182, 185–188
   sludge volume index (SVI) 182
hydraulic retention time (HRT)
   aerobic digestion 134
   anaerobic digestion 75, 223, 290, 308
   autothermal thermophilic aerobic digestion (ATAD) 308
   BioThelys process 227
   Cambi process 227
   Cannibal system 126–127
   dual digestion 137
   microbial predation 63, 154
   pre-treatment 246
   rapid digestion 226
   S-TE process 135
   thermal treatment 225
   thermophilic reactors 46
hydrothermal oxidation 41–42

I

incineration 4
inert solids 40–41
iron 115

K

Krepro process 248

L

landfill sites 2–3
Liberec plant 172–173
Limnodrilns 156
Lumbriculus variegatus 152–153, 158–159
lysis-thickening centrifuge 171–174, 185–188

M

mass reduction 70, 76
maximum growth yield 103–104
MBR *see* membrane biological reactors
mechanical disintegration 167–168
   degree of disintegration (DD) 177–178, 185–187
   energy levels 170–171
   equipment 168–170
   high pressure homogenisers 178–182
   high pressure jet and collision treatment 182–183
   lysis-thickening centrifuges 171–174
   rotor-stator disintegration 183–185
   sludge reduction 47–48, 68–69
   stirred ball mill 174–178
   technique comparison 185–188
membrane biological reactors (MBR) 162–166
   and alkaline treatment 164
   and chemical treatments 163–164
   endogenous metabolism 39
   and ozonation 164, 269, 274–275
   and physical treatments 163–164
   sludge reduction 60–61
   sludge retention time (SRT) 162–163
   thermal treatment 221
   and ultrasonic disintegration 164
   wastewater handling units 44, 221, 303

mesophilic anaerobic digestion 111, 221–222
meta-chlorophenol (mCP) 144
meta-nitrophenol (mNP) 144
metozoa 148–161
microbial predation
   membrane biological reactors (MBR) 63
   sludge reduction 39–40, 62–63, 83–84
microbubbles 199, 254
microorganisms
   damage to 167–168
   filamentous 281–282
   oxygen uptake rate (OUR) 200, 281
   and ozonation 280–283
   thermal treatment 218–219
   ultrasonic disintegration 200–202
   wastewater treatment plants 110
microwave treatment
   extracellular polymeric substances (EPS) 72, 229
   sludge reduction 71–72
   thermal treatment 229–231
mineral fractionation 265–266

N
Naididae 151
Nematoda 150–151
nitrate
   concentration 268
   utilisation rate (NUR) test 97–98, 299
nitrification 117–118
   denitrification 117–118
   ozonation 283–285
nitrites 268
nitrogen solubilisation
   chemical treatment 240–243
   ozonation 267–268
   thermal treatment 216
North American sludge production 2
NUR *see* nitrate utilisation rate

O
o-nitro-p-chlorophenol 144, 146–147
observed biomass yield 103–104
observed sludge yield 103–104
   reduction of 146–148

OC *see* oxygen consumption
oCP *see* ortho-chlorophenol
Opercularia 149
ORP *see* oxidation-reduction (red-ox) potential
ortho-chlorophenol (oCP) 144
osmosis 280–281
OUR *see* oxygen uptake rate
oxic-settling-anaerobic (OSA) process 36, 121–125
   iron 115
   oxidation-reduction (red-ox) potential (ORP) 122, 124–125
   side-stream anaerobic reactors 57–59
oxidation
   hydrothermal 41–42
   oxidants 53–55, 74–75
   ozone 51–53, 74
   supercritical water oxidation (SCWO) 41, 86–87
oxidation-reduction (red-ox) potential (ORP)
   aerobic/anoxic digestion 78, 118, 120
   alkaline hydrolysis 240
   anaerobic digestion 121, 123
   autothermal thermophilic aerobic digestion (ATAD) 308
   Cannibal system 58–59, 126
   oxic-settling-anaerobic (OSA) process 122, 124–125
   sensors 140
oxygen consumption (OC) 94
   degree of disintegration 94
oxygen uptake rate (OUR)
   biodegradability 299
   biomass activity 198
   degree of disintegration (DD) 94
   dissolved oxygen (DO) 95–96
   microorganisms 200, 281
ozonation 249–251
   and aerobic digestion 292–293
   and anaerobic digestion 288–292
   biodegradability 265
   and biological processes 268–269
   chemical oxygen demand (COD) solubilisation 261–266
   and denitrification 285–286

and dewaterability 293–294
and enzyme reactivity 281
extracellular polymeric substances
 (EPS) 277, 282–283
and flocs 282
and membrane biological reactors
 (MBR) 164, 269, 274–275
and microorganisms 281–282
mineral fractionation 265–266
nitrification 283–285
nitrogen solubilisation 267–268
ozone dosage 257–260
parameters 251–252
and pathogens 282
phosphorus solubilisation 267–268
reactor configuration 252–257
and respirometry 281
sequencing batch reactor (SBR) 269,
 273–275, 278, 285, 287
and settleability 286–287
sludge handling units 74, 287–294
sludge reduction 51–53, 74, 274–278
sludge retention time (SRT) 273–275,
 283–284, 290
solids 260–261, 265
and solubilisation 266
total suspended solids (TSS)
 disintegration 261–266
wastewater handling units 51–53,
 269–287
wastewater treatment plants
 (WWTP) 278
ozone
 dosage 257–260, 271–272, 280
 factor 258
 and microorganisms 280–283
 sludge pH 280
 transfer 255–257

P

PAH *see* polycyclic aromatic
 hydrocarbon
para-nitrophenol (pNP) 144–147
particulate COD (PCOD) 10–12, 92
pathogens 282
PCOD *see* particulate COD
pCP *see* pentachlorophenol

PEF *see* pulsed electric field
pentachlorophenol (pCP) 144, 146–147
pH
 and thermo-chemical treatment COD
  solubilisation 237–238
 extracellular polymeric substances
  (EPS) 235
 and ozone dosage 280
PHA *see* polyhydroxy-alkanoate
phosphorus solubilisation
 chemical treatment 240–243
 ozonation 267–268
 thermal treatment 216–218
physical properties 106–107
physical treatments 163–164
pNP *see* para-nitrophenol
Podophrya 149
polycyclic aromatic hydrocarbon
 (PAH) 292
polyhydroxy-alkanoate (PHA) 37
power density 193
predation 148–161
predation-reactor
 dissolved oxygen (DO) 156, 159
 sludge handling units 157–159
 wastewater handling units 156–157
primary sludge 12–13
 wastewater treatment plants (WWTP)
  1, 7–8
proteases 281
 Bacillus stearothermophilus 133
protein hydrolysis 267
protozoa 148–161
pulsed electric field (PEF) 55–56, 76

R

rate limiting steps 133
raw sludge 8
reactor configuration
 ozonation 252–257
recycled sludge reduction reactor
 (RSRR) 156
reduction of sludge production (RSP)
 105, 118–119
respirometry
 biodegradability 95–96
 and ozonation 281

rotation speed 176
Rotifera 150–151
rotor-stator disintegration systems 183–188

S
SBR *see* sequencing batch reactor
SCWO *see* supercritical water oxidation
secondary sludge 1, 8
sensors 140
sequencing batch reactor (SBR)
  granular sludge 165
  ozonation 269, 273–275, 278, 285, 287
  side-stream anaerobic reactors 128
  sonication 200, 204
  surnatant extraction 126
settleability 107
  and ozonation 286–287
  sludge volume index (SVI) 107, 159, 203, 219, 286–287
  thermal treatment 219–220
  ultrasonic disintegration 203
settling 168
SF *see* stress frequency
side-stream anaerobic reactors
  ambient temperature 121–129
  oxic-settling-anaerobic (OSA) process 57–59, 121–125
  sequencing batch reactor (SBR) 128
  sludge reduction 57–59
sludge
  chemical oxygen demand (COD) 10–12
  composition 7–12
  dewatering 4
  disintegration 170–171
  disposal costs 26–27
  dry mass reduction 4–5
  energy levels 170–171
  energy recovery 4
  European production 2
  farmland 3–4
  flocs 114–115
  fractionation 10–12
  granular 164–166
  incineration 4

landfill sites 2–3
North American production 2
ozone transfer 255–257
physical properties 106–107
raw sludge 8
wastewater treatment plants (WWTP) 2
sludge handling units
  aerobic digestion 77–78
  aerobic/anaerobic digestion 78–79
  anaerobic digestion 82–83
  autothermal thermophilic aerobic digestion (ATAD) 81–82, 138–140
  chemical treatment 73–74, 244–248
  dual digestion 79–81
  electrical treatment 76–77
  enzymatic hydrolysis 67–68
  mechanical disintegration 68–69
  microbial predation 83–84
  microwave treatment 71–72
  oxidation 74–76
  ozonation 74, 287–294
  predation-reactor 157–159
  solubilisation by thermophilic enzyme (S-TE) process 137–138
  supercritical water oxidation 86–87
  thermal treatment 70–71, 221–223
  thermophilic anaerobic digestion 83
  ultrasonic disintegration 69–70, 205–207
  wet air oxidation 84–86
sludge production
  data 18–19
  full plants 12–18
  reduction of sludge production (RSP) 105, 118–119
  total solids (TS) 12
sludge reduction 4, 29–32
  aerobic digestion 77–78
  aerobic/anaerobic digestion 78–79
  anaerobic digestion 82–83
  anaerobic reactors 57–59
  autothermal thermophilic aerobic digestion (ATAD) 81–82
  cell lysis 32–35
  chemical hydrolysis 50–51, 73–74

chemical metabolic uncouplers 57
comparison 303–309
cryptic growth 32
dual digestion 79–81
electrical treatment 55–56, 76–77
endogenous metabolism 37–39
enzymatic hydrolysis 45–47, 67–68
enzymes 67–68
evaluation of 105
extended aeration 59–60
granular sludge 61–62
hydrothermal oxidation 41–42
inert solids 40–41
mechanical disintegration 47–48, 68–69
membrane biological reactors (MBR) 60–61
microbial predation 39–40, 62–63, 83–84
microwave treatment 71–72
oxidants 53–55, 74–75
oxidation 51–55, 74–75
ozonation 51–53, 74, 274–278
supercritical water oxidation 86–87
thermal treatment 50, 70–71
thermophilic anaerobic digestion 83
ultrasonic disintegration 48–49, 69–70
uncoupled metabolism 35–37
wet air oxidation 84–86
sludge reduction reactors 156
sludge reduction techniques 295–300
chemical oxygen demand (COD) solubilisation 300–301
degree of disintegration 302
impacts 309–316
installation 316–321
operation 316–321
sludge handling units 313–316
sludge reduction 303–309
wastewater handling units 309–313
sludge retention time (SRT) 8, 12, 14–19, 102–103
autothermal thermophilic aerobic digestion (ATAD) 138
biomass yield 59–63, 103–104
Cannibal system 126–129
endogenous metabolism 38–39, 59

extended aeration 161, 307
membrane biological reactors (MBR) 162–163, 307
ozonation 273–275, 283–284, 290
side-stream anaerobic reactors 306
sludge reduction 102–103, 273
thermal treatment 221, 225
two-stage reactors 154
ultrasonic disintegration 206
sludge solubilisation 168
and dissolved oxygen (DO) 133–134
temperature effects on 132
sludge treatment
costs 25–26
thermal 50, 70–71
total solids (TS) 25–27
sludge volume index (SVI) 50
homogenisation 182
settleability 107, 159, 203, 219, 286–287
sludge yield 103–104
reduction of 146–148
solids
concentration 265
mineralisation 260–261
solubilisation
by thermophilic enzyme (S-TE) process 134–135
hydraulic retention time (HRT) 135
and ozonation 266
total organic carbon (TOC) 54, 242
sonication 189, 207
specific energy 193–194, 170–171
specific resistance to filtration (SRF) 107
dewaterability 203, 219, 243, 293
spheres
filling ratio 175
material 175
size 175
SRT *see* sludge retention time
stirred ball mills 174–178, 185–188
STPP *see* tripolyphosphate sodium
stress frequency (SF) 105–106, 179, 305–306, 308
see also treatment frequency
sulphides 114–115

supercritical water oxidation (SCWO) 41, 86–87
surnatant extraction 126
SVI see sludge volume index

T

TCP see 2,4,5-trichlorophenol
TCS see 3,3',4',5-tetrachlorosalicylanilide
temperature
 ambient temperature 121–129
 and chemical oxygen demand (COD) solubilisation 212–215, 236–237
 side-stream anaerobic reactor 121–129
 and sludge solubilisation 132
3,3',4',5-tetrachlorosalicylanilide (TCS) 57, 144, 146–148
thermal hydrolysis
 mesophilic anaerobic digestion 221–222
 thermophilic anaerobic digestion 222–223
thermal treatment 209–211
 biodegradability 215–216
 biological processes 220–229
 chemical oxygen demand (COD) solubilisation 211–215
 dewaterability 219–220
 extracellular polymeric substances (EPS) 50
 hydraulic retention time (HRT) 225
 membrane biological reactors (MBR) 221
 microorganisms 218–219
 microwave treatment 229–231
 nitrogen solubilisation 216
 phosphorus solubilisation 216–218
 settleability 219–220
 sludge handling units 70–71, 221–223
 sludge reduction 50, 70–71
 sludge retention time (SRT) 221, 225
 wastewater handling units 220–221
thermo-chemical hydrolysis
 anaerobic digestion 244–246
 dewatering 246–248
 sludge reduction 50–51, 73–74

thermophilic anaerobic digestion 129–131
 hydraulic retention time (HRT) 308
 sludge reduction 83
 thermal hydrolysis 222–223
 total solids (TS) 129
thermophilic bacteria 131
 aerobic 133–134
 dissolved oxygen (DO) 133–134
 enzymatic hydrolysis 46–47
thermophilic enzymes 134
thermophilic reactors 46
 anaerobic 131–140
thickenability 107
THM see trihalomethanes
TOC see total organic carbon
Tokophrya 149
total organic carbon (TOC)
 solubilisation 54, 242
 treated effluent 278, 310–312
total solids (TS) 8–10
 energy levels 170
 enzymes 142–143
 sludge production 12
 sludge treatment 25–27
 thermophilic anaerobic digestion 129
 ultrasonic disintegration 69, 187, 197, 205
total suspended solids (TSS) 10
 and chemical oxygen demand (COD) 91–92, 261–266
 disintegration, ozonation 261–266
 sludge contact time 134
 solubilisation 91–92
 volatile suspended solids (VSS) ratio 265
treated effluent 278, 310–312
treatment
 costs 25–26
 frequency 105–106
 see also stress frequency (SF)
Trepomona 149
2,4,6-tribromophenol 144
2,4,5-trichlorophenol (TCP) 144–147
trihalomethanes (THM) 55, 317
tripolyphosphate sodium (STPP) 142
Tubifex tubifex 151–152, 156–157

Tubificidae 151–152, 156–157
two-stage reactors 154–155
   sludge retention time (SRT) 154

U

ultrasonic disintegration 185–191
   chemical oxygen demand (COD) solubilisation 197–200
   dewaterability 203
   energy evaluation 193–195
   equipment 191–193
   and membrane biological reactors (MBR) 164
   microorganisms 200–202
   settleability 203
   sludge handling units 69–70, 205–207
   sludge reduction 48–49, 69–70
   sludge retention time (SRT) 206
   total solids (TS) 69, 187, 197, 205
   ultrasound frequency 195–197
   wastewater handling units 203–205
ultrasound
   dose 193
   frequency 195–197
uncoupled metabolism, sludge reduction 35–37

V

VFA *see* volatile fatty acids
viscosity reduction 168
volatile fatty acids (VFA) 80, 131, 133–134, 137, 212, 230
volatile solids (VS) 9–10
   mass reduction 70, 76
   ultrasonic disintegration 206–207
volatile suspended solids (VSS) 8–10, 15
   enzymatic hydrolysis 46
   fractions 17–18
   total suspended solids (TSS) ratio 265
Vorticella 149

W

wastewater handling units
   anaerobic reactors 57–59
   biochemical oxygen demand (BOD) 136
   chemical hydrolysis 50–51
   chemical metabolic uncouplers 57
   chemical treatments 50–55
   electrical treatment 55–56
   enzymatic hydrolysis 45–47
   extended aeration 59–60
   granular sludge 61–62
   mechanical disintegration 47–48
   membrane biological reactors (MBR) 44, 60–61, 221
   microbial predation 62–63
   oxidation 51–55
   ozonation 51–53, 269–287
   predation-reactor 156–157
   solubilisation by thermophilic enzyme (S-TE) process 134–137
   thermal treatment 50, 220–221
   thermophilic bacteria 45–47
   ultrasonic disintegration 48–49, 203–205
wastewater treatment plants (WWTP)
   cell lysis 32–35
   chemical sludge 8
   cryptic growth 32
   effluent quality 278–279
   endogenous metabolism 37–39
   hydrothermal oxidation 41–42
   inert solids 40–41
   microbial predation 39–40
   microorganisms 110
   primary sludge 1, 7–8
   secondary sludge 1, 8
   sludge 2
   uncoupled metabolism 35–37
wet air oxidation 84–86

IWA Publishing's authorised EU representative for General Product Safety Regulations is Diane D'Arras, 15 rue Duret, 75116 Paris, France, e-mail: safety@iwap.co.uk.

Printed and bound by CPI Group (UK) Ltd, Croydon, CR0 4YY
22/04/2026
02094792-0001